Artificial Neural Networks

Artificial Neural Networks

Editor

Danyal Sadik

scitus
academics

Artificial Neural Networks
Edited by **Danyal Sadik**

ISBN: 978-1-68117-529-4
Library of Congress Control Number: 2016942929

SCITUS Academics LLC,
www.scitusacademics.com
Box No. 4766, 616 Corporate Way,
Suite 2, Valley Cottage,
NY 10989

Printed in United States of America on Acid Free Paper

Preface

An artificial neuron network (ANN) is a computer model based on the structure and functions of neural biological networks. Information moving through the system affects the structure of the ANN because a neural network changes-or maybe learns in a certain sense-according to the output and input. Artificial neural networks are raw electronic models based on the human brain's neural structure. The brain learns essentially from experience. It is clear that some problems beyond the scope of current computer systems can in fact be solved by small energy-efficient packages. It also promises a less complex way of developing machine solutions. This brand new computing method also offers a much more elegant degradation during overloading than its much more traditional counterparts. These biologically inspired computing methods are believed to be the next big advance in the computer industry. Even simple animal brains can perform functions that computers cannot currently perform. Computers rotate very well, such as keeping booklets or performing complex mathematical problems. However, computers have difficulty recognizing even simple patterns that generalize these patterns from the past into future actions. Artificial neural networks may be the best technology in the past two decades that has been popular in many applications.

Provides recent advances in architecture, methodology and artificial neural network applications; this book will therefore be a key source of recent developments and applications in artificial neural networks. This book is valuable for college students and practitioners.

Table of Contents

CHAPTER 1

SORN: A self-organizing recurrent neural network

Andreea Lazar1*, Gordon Pipa1, 2 and Jochen Triesch1

1 Frankfurt Institute of Advanced Studies,
Johann Wolfgang Goethe University, Frankfurt am Main, Germany

2 Department of Neurophysiology, Max Planck Institute for Brain Research, Frankfurt am Main, Germany

Understanding the dynamics of recurrent neural networks is crucial for explaining how the brain processes information. In the neocortex, a range of different plasticity mechanisms are shaping recurrent networks into effective information processing circuits that learn appropriate representations for time-varying sensory stimuli. However, it has been difficult to mimic these abilities in artificial neural network models. Here we introduce SORN, a self-organizing recurrent network. It combines three distinct forms of local plasticity to learn spatio-temporal patterns in its input while maintaining its dynamics in a healthy regime suitable for learning. The SORN learns to encode information in the form of trajectories through its high-dimensional state space reminiscent of recent biological findings on cortical coding. All three forms of plasticity are shown to be essential for the network's success.

1. INTRODUCTION

The mammalian neocortex is the seat of our highest cognitive functions. Despite much effort, a detailed characterization of its complex neural dynamics and an understanding of the relationship between these dynamics and cognitive processes remain elusive. Cortical networks present an astonishing ability to learn and adapt via a number of plasticity mechanisms which affect both their synaptic and neuronal properties. These mechanisms allow the recurrent networks in the cortex to learn representations of complex spatio-temporal stimuli. Interestingly, neuronal responses are highly dynamic in time (even when the stimulus is static) (Broome et al., 2006) and contain a rich amount of information about past events (Brosch and Schreiner, 2000 ; Bartlett and Wang, 2005 ; Broome et al., 2006 ; Nikolic et al., 2006).

But mimicking these features in artificial neural networks has proven to be very difficult. The first models that could address temporal tasks have incorporated in their structure an explicit representation of time (Elman and Zipser, 1988). Recurrent neural networks (RNNs) were the first models to represent time implicitly, through the effect that is has on processing (Hopfield, 1982 ; Elman, 1990). In the recently developed framework of 'reservoir' computing (Jaeger, 2001 ; Maass et al., 2002), a randomly structured RNN non-linearly transforms a time varying input signal into a spatial representation. At each time step, the network combines the incoming stimuli with a volley of recurrent signals containing a memory trace of recent inputs. For a network with N neurons, the resulting activation vector at a discrete time t, can be regarded as a point in a N-dimensional space. Over time, these points form a pathway through the state space also referred to as a *neural trajectory*. A separate read-out layer is trained, with supervised learning techniques, to map different parts of the state space to desired outputs. In real cortical networks, experimental evidence has shown that different stimuli elicit different trajectories while for a given stimuli the activity patterns evolve in time in a reproducible manner (Broome et al., 2006 ; Churchland et al., 2007). Furthermore, identical trials can present a high response variability, but the resulting trajectories are not dominated by noise (Mazor and Laurent, 2005 ; Broome et al., 2006 ; Churchland et al., 2007). Reservoir networks do not require classical attractor states and are compatible with the view that cortical computation is based on transient dynamics (Mazor and Laurent, 2005 ; Durstewitz and Deco, 2008 ;Rabinovich et al., 2008). It has been shown that neural systems may exhibit transients of long durations which carry more information about the stimulus then the steady states towards which the activity evolves (Mazor and Laurent, 2005).

Attempts to endow RNNs with unsupervised learning abilities by incorporating biologically plausible local plasticity mechanisms such as spike-timing-dependent plasticity (STDP) (Markram et al., 1997 ; Bi and Poo, 1998) have remained largely unsuccessful (and often unpublished). The problem is most difficult, because structural changes induced by plasticity will impact the network's dynamics giving rise to altered firing patterns between neurons. These altered firing patterns can further induce changes in connectivity through the plasticity mechanisms and so forth.

Understanding and controlling the ensuing self-organization of network structure and dynamics as a function of the network's inputs is a formidable challenge.

The key to the brain's solution to this problem may be the synergistic combination of multiple forms of neuronal plasticity. There has been extensive evidence that synaptic learning is accompanied by homeostatic mechanisms. Synaptic scaling regulates the total synaptic drive received by a neuron but maintains the relative strength of synapses established during learning (Turrigiano et al., 1998). At the same time, intrinsic plasticity (IP) was shown to directly regulate neuronal excitability (Desai et al., 1999 ;Zhang and Linden, 2003). In a RNN, IP induced robust homeostatic effects on the network dynamics (Steil, 2007 ; Schrauwen et al., 2008). But there is only little work combining several forms of plasticity in RNNs (Lazar et al., 2007).

In the following, we present a RNN of threshold units combining three different forms of plasticity that learns to efficiently represent and "understand" the spatio-temporal patterns in its input. The SORN model (self-organizing recurrent network) consists of a population of excitatory cells and a smaller population of inhibitory cells (Figure 1). The connectivity among excitatory units is sparse and subject to a simple STDP rule. Additionally, synaptic normalization (SN) keeps the sum of an excitatory neuron's afferent weights constant, while IP regulates a neuron's firing threshold to maintain a low average activity level. The network receives input sequences composed of different symbols and learns the structure embedded in these sequences in an unsupervised manner. The three types of plasticity mechanisms induce changes in network dynamics which we assess via hierarchical clustering and principal component analysis (PCA). In addition, we train a separate readout layer with supervised learning techniques and compare the performance of our network with that of fixed random networks constructed in the spirit of reservoir computing.

We show that only the combination of all three types of plasticity allows the network to (a) learn to effectively represent the spatio-temporal structure of its inputs, (b) maintain 'healthy' dynamics [1] that make efficient use of all the network's resources, and (c) perform much better on prediction tasks compared to random networks without plasticity. Furthermore, the network dynamics are consistent with a range of neurophysiological findings.

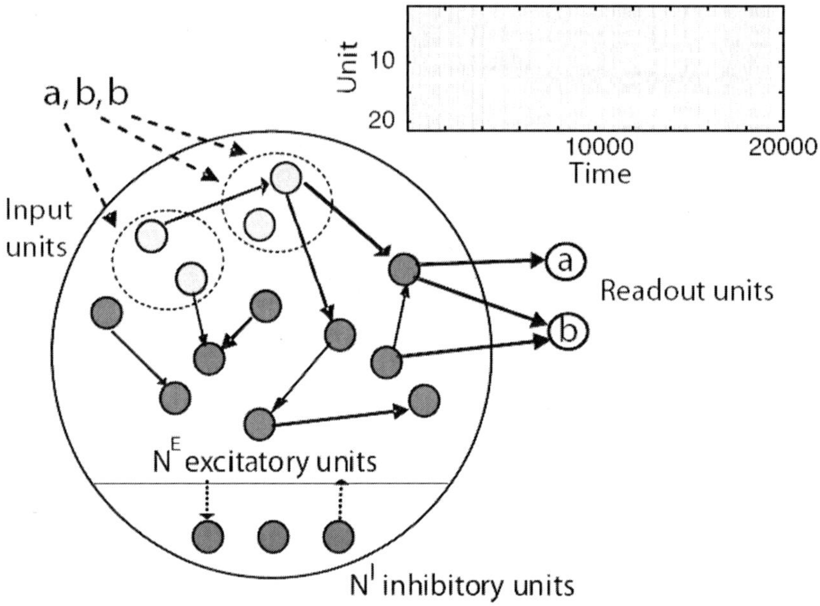

Figure 1. The self-organizing recurrent neural network (SORN) comprises populations of excitatory (blue) and inhibitory (red) cells. Directed connections with variable strength between neurons are indicated by black arrows. Some of the excitatory cells also receive external input (light blue). Three forms of plasticity interact to shape the dynamics of the network keeping them in a healthy regime and allowing the network to discover structure in its inputs. A population of readout units is trained with supervised learning methods.

2. MATERIALS AND METHODS

2.1 The Sorn Model

2.1.1 Network definition

We consider a network with N^E excitatory (E) and $N^I = 0.2 \times N^E$ inhibitory (I) threshold units. Neurons are coupled through weighted synaptic connections, where W_{ij} is the connection strength from unit j to unit i, with $i \neq j$. All possible connections between the excitatory and inhibitory neuron populations are present (W^{IE} and W^{EI}), while the excitatory–excitatory connections (W^{EE}) are sparse and random with a mean number λ^W of incoming and outgoing connections per neuron. Direct connections between inhibitory units are not present. The weight strengths are drawn from the interval [0, 1] and subsequently normalized such that the incoming connections to a neuron sum up to a constant value: $\sum_j W_{ij}^{IE} = 1$, $\sum_j W_{ij}^{EI} = 1$ and $\sum_j W_{ij}^{EE} = 1$. Inputs are time series $U(t)$ of different symbols (letters or digits). Each symbol is associated with a specific group of N^U input units which all receive a positive input drive ($v^U = 1$) when that particular symbol is active.

The network state, at a discrete time t, is given by the binary vectors $x(t) \in \{0,1\}^{N^E}$ and $y(t) \in \{0,1\}^{N^I}$ corresponding to the activity of the excitatory and inhibitory units, respectively. The evolution of the network state is described by:

$$x_i(t+1) = \Theta\left(\sum_{j=1}^{N^E} W_{ij}^{EE}(t) x_j(t) - \sum_{k=1}^{N^I} W_{ik}^{EI} y_k(t) + v_i^U(t) - T_i^E(t)\right) \quad (1)$$

$$y_i(t+1) = \Theta\left(\sum_{j=1}^{N^E} W_{ij}^{IE} x_j(t) - T_i^I\right). \quad (2)$$

The T^E and T^I are threshold values for the excitatory and inhibitory units. They are initially drawn from a uniform distribution in the interval $[0, T_{max}^E]$ and $[0, T_{max}^I]$, respectively. The heaviside step function $\Theta(.)$ constrains the activation of the network at time t to a binary representation: a neuron fires if the total drive it receives is greater then its threshold, otherwise it stays silent.

At each time step the activity of the network is determined both by the inputs $v_i^U(t)$ and the propagation of the previously emitted spikes through the network. This recurrent drive received by unit i is given by:

$$R_i(t+1) = \sum_{j=1}^{N^E} W_{ij}^{EE}(t) x_j(t) - \sum_{k=1}^{N^I} W_{ik}^{EI} y_k(t) - T_i^E(t). \quad (3)$$

Based on this, we define a "pseudo state" $x'(t)$ that only depends on the recurrent drive:

$$x_i'(t) = \Theta(R_i(t)); \quad (4)$$

This equation is identical to Eq. 1, but lacking the input drive $v_i^U(t)$. Most of our analysis focuses on the pseudo states $x'(t)$ as the network's internal representation of previous inputs, although it may contain less information than $R(t)$ due to the thresholding operation.

2.1.2 Plasticity mechanisms

The network relies on three forms of plasticity: STDP, synaptic scaling of the excitatory–excitatory connections, and IP regulating the thresholds of excitatory units.

Learning with STDP is constrained to the set of W^{EE} synapses. We use a simple model of STDP that strengthens the synaptic weight W_{ij}^{EE} by a fixed amount $\eta_{STDP} = 0.001$ whenever unit i is active in the time step following activation of unit j. When unit i is active in the time step preceding activation of unit j, W_{ij}^{EE} is weakened by the same amount:

$$\Delta W_{ij}^{EE}(t) = \eta_{STDP}\left(x_i(t)x_j(t-1) - x_i(t-1)x_j(t)\right); \quad (5)$$

STDP changes the synaptic strength in a temporally asymmetric "causal" fashion. The changes introduced by STDP can push the activity of the network to grow or shrink in an uncontrolled manner. To keep the activity balanced during learning we make use of additional homeostatic mechanisms that are sensitive to the total level of synaptic efficacy and the post-synaptic firing rate.

SN proportionally adjusts the values of incoming connections to a neuron so that they sum up to a constant value. Specifically, the W^{EE} connections are rescaled at every time step according to:

$$W_{ij}^{EE}(t) \leftarrow W_{ij}^{EE}(t) / \sum_j W_{ij}^{EE}(t). \quad (6)$$

This rule does not change the relative strengths of synapses established by STDP but regulates the total incoming drive a neuron receives.

An IP rule spreads the activity evenly across units, such that on average each excitatory neuron will fire with the same target rate H_{IP}. To this end, a unit that has just been active increases its threshold while an inactive unit lowers its threshold by a small amount:

$$T_i^E(t+1) = T_i^E(t) + \eta_{IP}\left(x_i(t) - H_{IP}\right), \quad (7)$$

where $\eta_{IP} = 0.001$ is a small learning rate. We set the target rate to $H_{IP} = 2 \times N^U/N^E$ in which the input spikes are approximately half of the total number of spikes. Other settings of H_{IP} do not necessarily lead to the desired improvements in prediction performance (see Appendix).

The implementation of the model described above and the simulations presented in Section "Results" were performed in Matlab.

3. RESULTS

3.1 SORNs Outperform Static Reservoirs

We demonstrate the SORN's ability to learn spatio-temporal structure in its inputs with a "counting" task, especially designed to test the memory property of the reservoir. To this end, we construct input sequences $U(t)$ as random alternations of two "words" 'abbb…bc' and 'eddd…df', composed of $n + 2$ "letters", with letters 'b' and 'd' repeating n times. In order to predict the next input letter correctly, the network has to learn to "count" how many repetitions of letters 'b' and 'd' it has already seen. Increasing n raises the difficulty of the task. We compare SORNs with all three forms of plasticity to static networks without plasticity. Networks of different sizes N^E have their initial parameters set to $N^U = 5\% \times N^E$, $T^E_{max} = 0.5$, $T^I_{max} = 1$ and $\lambda^W = 10$. For small static reservoirs, the parameters are tuned such that their dynamics is critical and the networks' firing rate is similar to the rate exhibited by SORNs structured by plasticity (see Supplementary Material and Section "Occluder Task"). It has been argued that a tuning of network dynamics to criticality should bring the performance of static reservoir networks close to the optimal performance (Bertschinger and Natschläger, 2004). To compute prediction performance, 5000 steps of network activity are simulated and a readout is trained in a supervised fashion to predict the next input $[U(t)]$, e.g., 'a', or 'c', or 5th repetition of 'b', etc., based on the network's internal state $[x'(t)]$ after presentation of the preceding letter $[U(t-1)]$. We use the Moore–Penrose pseudoinverse method that minimizes the squared difference between the output of the readout neurons and the target output value. The quality of the readout (the network performance) is assessed on a second sample of 5000 steps of activity using an independent input sequence.

The SORNs are exposed to the input sequences for 50,000 time steps. Then, all their weights and thresholds are frozen and a readout is trained in the same manner.

Since the input sequences are partly random – the order of letters within a word is fixed but the order of words is random – prediction performance is inherently limited. We define a normalized performance measure that obtains a score of 1 when the network always correctly predicts the next letter and its position within a word but is at chance level for guessing the first letter of the next word (either 'a' or 'e'). Figure 2 compares the performance of SORNs and static reservoir networks. For any given network size (N^E) and any given task difficulty (n), the plastic SORNs perform considerably better than their randomly structured counterparts (Figure 2 A). For the same task difficulty n, larger networks perform better then smaller networks. For a given network size the SORNs achieve a performance greater than 0.95 for much higher values of n compared to the static reservoirs (Figure 2 B). A more detailed analysis of the network performance as a function of various initial parameter settings is given in the Appendix.

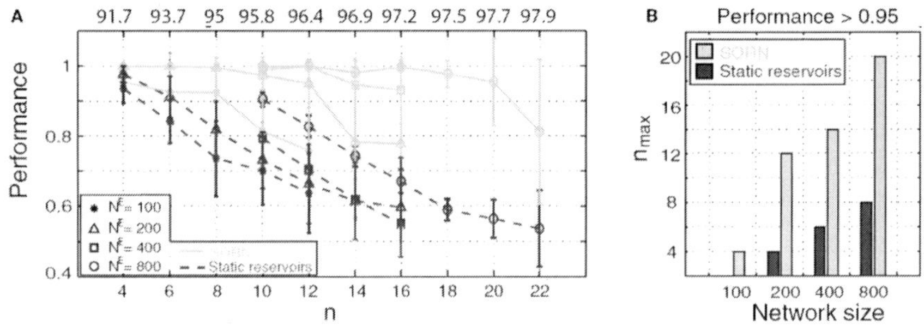

Figure 2. (A) Average normalized performance of 10 plastic SORNs and 10 static reservoir networks of size N^E, for different values of n. Numbers on top indicate optimal absolute performance achievable in the task. Error bars indicate standard deviation. **(B)** We show n_{max}, the highest value of n where normalized performance exceeds 95%, as a function of network size. Plastic networks succeed for substantially harder problems compared to random reservoirs.

3.2 SORNs Learn Effective Internal Representations

To better understand the reason underlying the performance advantage of SORNs over static reservoirs, we performed hierarchical clustering and PCA on the networks' internal representations.

We performed agglomerative hierarchical clustering of the networks' internal state representations (x'). Each pattern of activity $x'(t)$ is a point in a N^E-dimensional space. Agglomerative clustering starts by considering each of these points as centers of their own cluster. The distance between two clusters is computed as the Euclidean distance between their centers. Repeatedly, the two closest clusters are merged into a single cluster, until the entire data are collapsed.

In Figures 3 A,E we present a snapshot of the last 20 clusters of agglomerative clustering, for an example network with $N^E = 200$, $N^U = 10$, $T^E_{max} = 0.5$, $T^I_{max} = 0.8$, $\lambda = 10$ during a counting task with $n = 8$. In the case of randomly structured reservoir networks, the cluster structure of internal representations only weakly reflects the underlying input conditions

(Figure 3 A). Many of the emerging clusters combine network states resulting from distinct input conditions, i.e., the networks internal representation easily confuses, say, the 5th repetition of letter ‘b’ with its 6th repetition. In fact most clusters lump together as many as seven input conditions (Figure 3 B). In contrast after 50,000 steps of plasticity, the SORN learns an internal representation that tends to map different input conditions on to distinct network states falling into separate clusters (Figure 3 E). Here, each cluster will combine at most two different input conditions (Figure 3 F). For a parallel with the performance tests from the previous section, the analysis was performed on 5000 steps of activity with frozen weights and thresholds but the network presents similar clustering properties in the presence of plasticity.

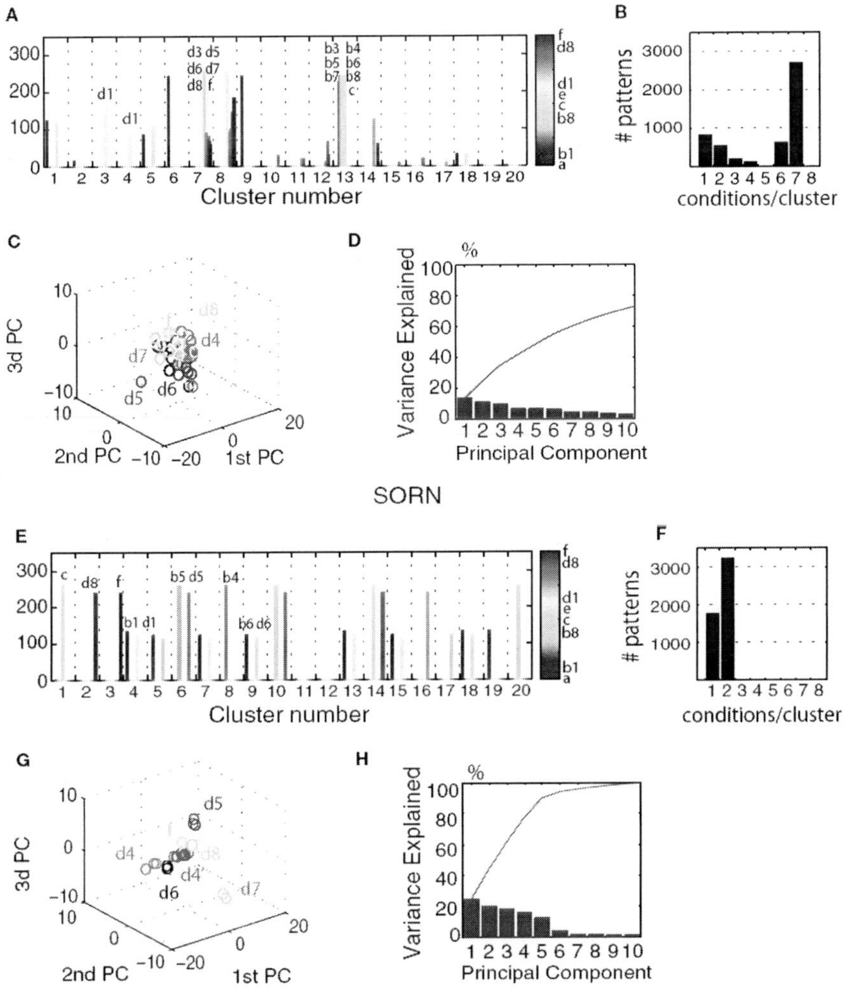

Figure 3. (A) Result of hierarchical clustering of the internal representation of a static random reservoir. Only a single stage with 20 clusters is shown. For each of the 20 clusters, a histogram depicts the different input conditions that contributed to the cluster. Clusters tend to mix many distinct input conditions, especially different repetitions of 'b' or 'd', instead of keeping them separate. **(B)** Histogram showing how many different input conditions contribute to each of the 20 clusters. **(C)** Result of PCA on the pseudo state x' corresponding to the last six letters of the input sequence 'edddddddf' which we refer to as 'd4', 'd5', 'd6', 'd7','d8' and 'f'. Identical input conditions are spread far apart and strongly overlap with other input conditions. **(D)** The amount of variance explained by the first principal components.**(E–H)** Same as **(A–D)** but for SORNs. **(E)** The cluster structure in SORNs reflects the different input conditions. **(F)** Representations of different inputs are comparatively distinct such that only one or two input conditions contribute to each cluster. **(G)** In PCA space, the different input conditions form compact clusters that are well separated for different input conditions. **(H)** Most of the variance is captured by only the first few principal components, suggesting more orderly dynamics in the SORNs.

3.3 Occluder Task

We demonstrate the ability of the SORN to learn effective representations on a second difficult task. Specifically, we consider an input sequence containing random alternations of the following four "words": '12345678', '87654321', '19999998', '89999991'. If we associate different spatial positions with the numbers 1–8, we can interpret these stimuli as left to right and right to left motion of an object along an axis. The symbol '9' can be interpreted as an occluder that obstructs the sight of the object at locations 2–7. This task is more difficult than the counting task in that several words share start and end letters and the repetitive symbol '9' is common in the last two sequences. The bidirectional quality of this stimuli might impose difficulties for the causal STDP rule. The interference of enforced synaptic pathways could decrease the prediction performance of SORNs. On the other hand, due to synaptic competition STDP might encourage one direction of motion and prune away the other. Our results suggest that both of these effects are avoided and SORNs present prediction advantages over random reservoirs.

We choose a network with $N^E = 200$, $N^U = 15$, $T^E_{max} = 0.75$, $T^I_{max} = 1.4$, and $\lambda^W = 10$. We run the SORN for 200,000 time steps and take snapshots of weights W and thresholds T at every 1000 steps of self-organization through plasticity. We evaluate each of these networks in terms of prediction performance for the one step prediction task. Similarly to the previous experiment, the performance drastically improves (Figure 4 A) and is close to the theoretical optimum for all the different time intervals of self-organization with plasticity. We also assess the criticality of the network dynamics by performing a perturbation analysis. For every state $x(t)$, we perturb the activation of a randomly chosen excitatory neuron (from active to inactive or from inactive to active) creating an altered state $\tilde{x}(t)$. The Hamming distance between $x(t)$ and its perturbed version $\tilde{x}(t)$ is 1 [$d(t) = 1$]. We calculate the successor states of $x(t)$ and $\tilde{x}(t)$ by applying Eq. 1 and obtain $x(t+1)$ and $\tilde{x}(t+1)$ with the Hamming distance $d(t+1)$. If the average distance $\bar{d}(t+1) > 1$ the network amplifies perturbations and is in a supercritical regime. If $\bar{d}(t+1) < 1$ the network has self-correcting properties and is in a subcritical dynamical regime. When $\bar{d}(t+1) \approx 1$ the dynamics is said to be critical. Performing perturbation analysis, we find that the network dynamics changes from a critical regime, in the case of static reservoirs, to a subcritical regime for SORNs (Figure 4 B). Interestingly, in the case of SORNs this corresponds to a higher network performance for prediction.

We also compare the tuning of the random reservoir network with the SORN after 50,000 steps of plasticity. For each of these two networks we consider 5000 time steps of network activity (in both cases without plasticity) and count the number of neuron responses corresponding to each of the 32 input conditions: left–right motion ('12345678'), left–right motion with occluder ('19999998'), right–left motion ('87654321') and right–left motion with occluder ('89999991'). For the random network we find that a number of neurons are silent and do not fire for any of the input conditions (Figure 4 C). Also the neurons responding to the occluder sequences are not very selective in terms of either location or direction. In contrast, for the SORN all neurons take part in the activity and their responses are input specific (Figure 4D). We calculated "tuning curves" of two example neurons to illustrate this point in more detail. To this end, we summed the neurons' responses for each of the eight locations of the visual space irrespective of motion direction or occluder presence. The neuron in (Figure 4 E) responded unselectively to all eight locations before any plasticity (static reservoir case, blue squares) and after learning it has developed a clear preference for location 4 (SORN case, green circles). The neuron in (Figure 4 F) was silent in the initial network setup (static reservoir case). Through plasticity, it developed selectivity for locations 3 and 7 (SORN case). Interestingly, this selectivity is also specific with regard to motion direction. The neuron fires when a stimulus is at location 3 moving to the right, or when the stimulus is at location 7 moving to the left (not shown).

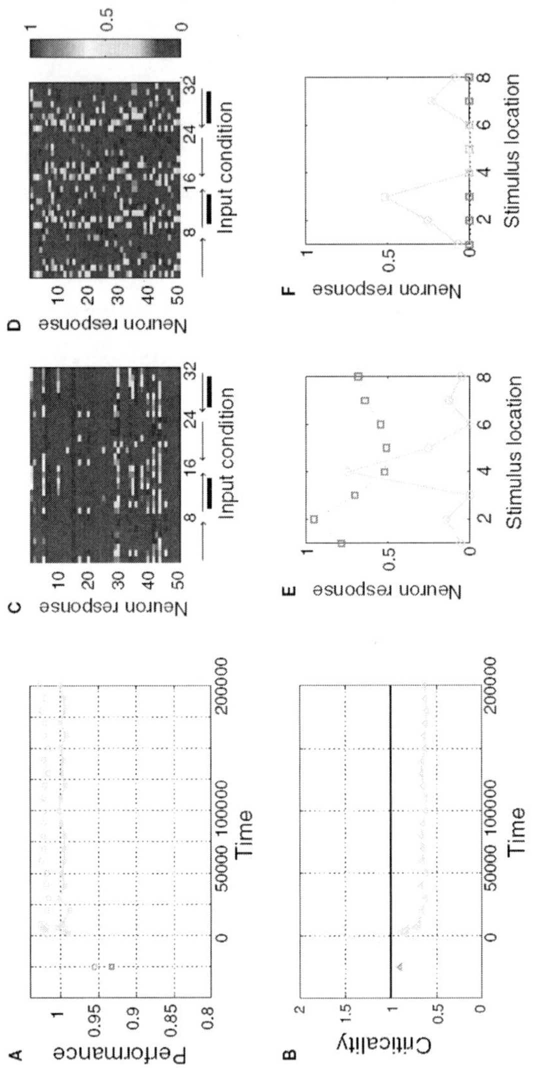

Figure 4. Performance of the SORN and a random reservoir on the "occluder" task. (A) Performance of a readout trained to predict the next input letter. Performance of a static reservoir network (blue symbols left of 0 on the time axis) is worse than that of the SORN. **(B)** As plasticity unfolds, the dynamics develop from a near-critical regime to a subcritical one. Probability of 50 randomly selected neurons firing for different input conditions in static reservoirs **(C)** and the SORN **(D)**. In the SORN firing is more specific to particular input conditions. **(E,F)** Tuning curves for two representative model neurons in a static network (blue squares) and the SORN (green circles). In the SORN, neurons are sparsely active and respond to specific input conditions. In the static networks, some neurons will respond rather unspecifically for all input conditions **(E)** or not at all **(F)**.

3.4 Homeostatic Plasticity Mechanisms are Critical for Maintaining Healthy Dynamics

To better understand the role of the homeostatic plasticity mechanisms accompanying STDP-learning in SORNs, we compare SORNs with plastic reservoirs in which either the synaptic scaling or the IP is switched off. We consider networks receiving unstructured inputs, here in the form of random alternations of six symbols. Thus, there is no specific spatio-temporal structure in the inputs that could be learned during these experiments. The networks ($N^E = 200$, $N^U = 10$, $T^E_{max} = 0.5$, $T^I_{max} = 1$ and $\lambda^W = 10$) are shaped in the presence of all three forms of plasticity for 50,000 steps.

The results are summarized in Figure 5 . When SN is missing, the network dynamics develop into a regime with seizure-like synchronous bursts of activity (Figure 5 A), even though the network is driven by random inputs. We compared the distribution of the total number of spikes per time step for 10 networks with and without SN (Figure 5 B). In networks with SN the distribution is unimodal and centered at a low activity level. In contrast, networks without SN will show a bimodal distribution such that most units are either active or inactive at the same time. This is also expressed in the average correlation coefficient between neurons. In networks with SN the average correlation coefficient remains close to 0 with an average value of 0.025. For networks that lack SN the average correlation coefficient increases as a function of time to values beyond 0.8 within 50,000 steps of simulation, indicating a high degree of synchronization (Figure 5C).

When IP is missing a number of neurons remain permanently silent, while others develop an unnaturally high activity (Figure 5 D). We calculated the distribution of average firing rates for 10 such networks. In networks with IP, all excitatory units develop average firing rates close to the desired target rate, which was 0.1 in these experiments. Without IP, the distribution is more spread out with some units staying completely silent and others being active in almost every time step (Figure 5 E). We quantified this effect by following the time evolution of the spike source entropy, which measures how much uncertainty there is about the origin of a spike in the network. It is defined as:

$$SSE = \left(\sum_{i=1..N^E} p_i \times \log_2 p_i)/\log_2 (1/N^E \right), \tag{8}$$

where p_i is the probability that a spike is generated by the unit i. SSE achieves its maximum value of 1 if all units fire at the same rate ($p_i \propto H_i$, where H_i is the firing rate of neuron i). SORNs show an abrupt increase in SSE to a value close to 1, which indicates identical rates across the neuronal population, compared to a smaller value of 0.94 for networks missing IP

(Figure 5 F). Due to IP, SORNs make efficient use of all the network's resources.

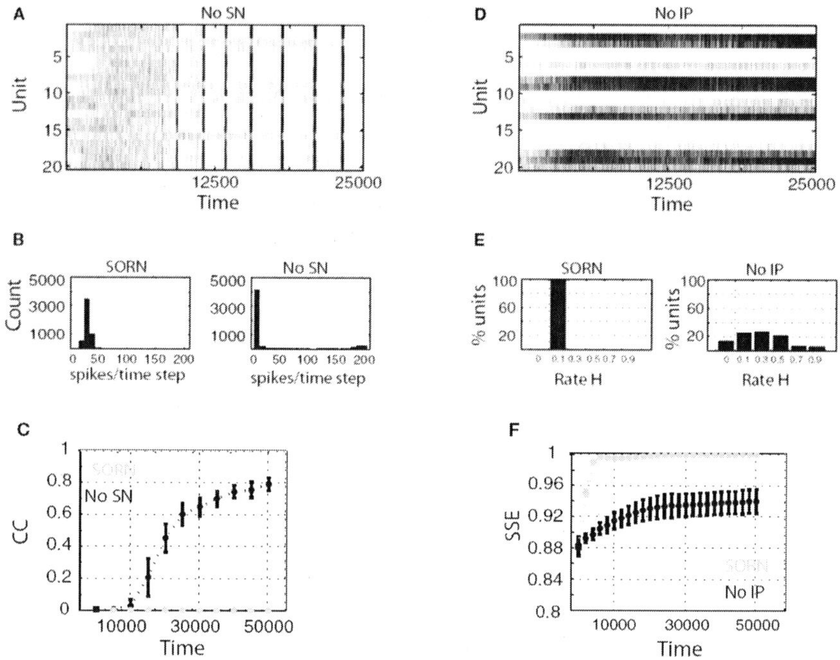

Figure 5. Activity of SORN networks compared to networks missing synaptic normalization (A–C) or intrinsic plasticity (D–F). (A,D) Snapshot of activity for 50 randomly selected reservoir units. **(B)** Distribution of population activity after 50,000 time steps of simulation. **(C)** Average correlation coefficient between neurons is dramatically increasing without SN. **(E)** Distribution of firing rates becomes very wide when IP is missing.**(F)** Spike source entropy increases to the maximum value for SORN networks, indicating a uniform division of labor across neurons.

4. DISCUSSION

Self-organizing recurrent networks are the substrate for neural information processing in the brain. Such networks are shaped by a wealth of plasticity mechanisms which affect synaptic as well as neuronal properties and operate over various time scales (from seconds to days and beyond). Somehow these mechanisms must work together to allow the brain to learn efficient representations for the various tasks it is facing. They shape the neural code and form the foundation on which our higher cognitive abilities are built.

While great progress has been made in characterizing these mechanisms individually, we only have a poor understanding of how they work together at the network level. In a non-linear system like the brain, any local change to, say, a synaptic efficacy potentially alters the pattern of activity at the level of the entire network and may induce further plastic changes to it. To investigate these processes, recent methods for observing the activities of large populations of neurons simultaneously need to be combined with careful measurements of the evolution of their synaptic and intrinsic properties – a formidable task for experimental neuroscience.

Computational modeling and theoretical analysis can contribute to this quest by providing simplified model systems that hopefully capture the essence of some of the brain's mechanisms and that can reveal underlying principles. In this article, we have introduced the SORN (self-organizing RNN). It combines three different kinds of plasticity and learns to represent and in a way "understand" the structure in its inputs. Maybe its most striking features is the ability to map identical inputs onto different internal representations based on temporal context. For example, it learns to distinguish the 5th repetition of an input from the 6th repetition by finding distinct encodings (internal representations) for the two situations (compare Figure 3). All this is happening in a completely unsupervised way without any guidance from the outside. The "causal" nature of the STDP rule is at the heart of this mechanism. It allows the network to incorporate predictable input structure into its own dynamics. At the same time, we have shown that STDP needs to be complemented by two homeostatic plasticity mechanisms. Without them the network will lose its favorable learning properties and may even develop seizure-like activity bursts (compare Figure 5).

Our network can be contrasted to recurrent networks without plasticity. Such static networks have received significant attention in the recent past, giving rise to the field of reservoir computing (Jaeger, 2001 ; Maass et al., 2002). The performance of a reservoir network relies on two requirements: (a) that different inputs to the network result in separable outputs based on the reservoir's response (the separation property) and that (b) the network activity states maintain information about recent inputs (the fading memory property). Given the high dimensionality of the reservoir, the separation property is easy to meet. Dockendorf et al. (2009) have confirmed this property for in vitro networks of cortical neurons. The memory property has been addressed in a series of experimental studies, across different brain areas, that compare the neuronal response to a stimulus B vs. the response to B when it was preceded by stimulus A (Brosch and Schreiner, 2000 ; Bartlett and Wang, 2005 ; Broome et al., 2006 ; Nikolic et al., 2006). For example in (Nikolic et al., 2006), the authors analyzed neuronal responses in cat primary visual cortex, area 17, to a sequence of two letter images and were able to recover the identity of the first and second letter reliably using a simple linear classifier.

The most important force shaping the representations in the SORN is STDP. Although the STDP model we used is much simplified, it captures what is arguably the essence of STDP: a "causal" modification of synaptic strengths. In recent years much evidence has accumulated suggesting that the brain's encoding of stimuli is subject to modifications due to STDP-like mechanisms. Several studies showed that repetitive stimulation with temporally patterned inputs causes a rapid STDP-based synaptic reorganization (Yao and Dan, 2001 ; Fu et al., 2002 ; Yao et al., 2007). Specifically, in Yao et al. (2007) a short repeated exposure to natural movies induced a rapid improvement in response reliability in cat visual cortex. Interestingly, the movie stimulation also left a "memory trace" which could be picked up in subsequent spontaneous activity.

It is interesting to note that in all the example tasks we considered the SORNs outperformed optimized versions of recurrent networks without plasticity. We find it unsurprising but rather reassuring that networks that try to discover and incorporate the temporal structure of their inputs into their dynamics outperform static reservoirs. Under repetitive stimulation with temporally structured inputs, SORNs selforganize in efficient ways that boost the network memory and separation properties. In our results, the SORNs could incorporate much longer input sequences as compared to the static reservoirs of similar size (Figure 2). SORNs developed internal representations where each input condition, reflecting both spatial and temporal aspects of the input, produced a tight cluster of network states that was well separated from those of other input conditions. This results in orderly and stereotyped trajectories through the network state space, that can be easily separated by a linear readout.

Reservoir computing architectures are thought to function best when their dynamics are critical (which we also found true for random reservoirs). It has been proposed that self-organization based on neuronal plasticity is able to achieve critical dynamics (Lazar et al., 2007 ; Gómez et al., 2009). Interestingly, the SORNs develop dynamics that are subcritical (compare Figure 4). This raises two questions. First, what is the exact mechanism that gives rise to the subcritical dynamics? Second, why are the subcritical dynamics of SORNs superior to the critical dynamics of static networks? Regarding the latter, we speculate that SORNs' ability to incorporate the *predictable* sequence of inputs into their internal dynamics makes it unnecessary to maintain criticality, which should give the best fading memory for *arbitrary* input sequences. But if there is predictable structure in the input, the recurrent network should try to exploit and use it's resources to model this specific structure rather than striving to have a general purpose fading memory.

The current model is particularly suited for efficient hardware implementation due to the simplicity of the chosen model neurons. In the current design individual neurons do not have any intrinsic memory properties, which makes a strong point that all memory information is maintained by the recurrent dynamics. An open problem is to investigate the generality of these ideas in the context of more elaborate network models based on integrate and fire neurons or conductance based neurons, which also include direct connections between inhibitory units. Future work needs to address if the performance advantage of SORNs over static networks transfers from the simple problems studied here to more difficult engineering problems in time series prediction, speech recognition, etc.

We have shown how the synergistic combination of different local plasticity mechanisms can shape the global structure and dynamics of RNNs in meaningful and adaptive ways. This emergent property could not have been easily predicted on the basis of the individual mechanisms – the whole is more than the sum of its parts. This implies that as we try to understand neural plasticity and how it shapes the brain's representation and processing, it is insufficient to study individual mechanisms in isolation. Only by studying their interaction at the network level, we have a chance to unravel this mystery.

CONFLICT OF INTEREST STATEMENT

The authors declare that the research was conducted in the absence of any commercial or financial relationships that could be construed as a potential conflict of interest.

ACKNOWLEDGMENTS

This work was supported by the Hertie Foundation, grant PLICON (EC MEXT-CT-2006-042484) and GABA Project (EU-04330).

FOOTNOTE

1. Dynamics suitable for computation.

REFERENCES

1. Bartlett, E. L., and Wang, X. (2005). Long-lasting modulation by stimulus context in primate auditory cortex. *J. Neurophysiol.* 94, 83–104.

2. Bertschinger, N., and Natschläger, T. (2004). Real-time computation at the edge of chaos in recurrent neural networks. *Neural Comput.* 16, 1413–1436.

3. Bi, G. Q., and Poo, M. M. (1998). Synaptic modifications in cultured hippocampal neurons: dependence on spike timing, synaptic strength, and postsynaptic cell type. *J. Neurosci.* 18, 10464–10472.

4. Broome, B. M., Jayaraman, V., and Laurent, G. (2006). Encoding and decoding of overlapping odor sequences. *Neuron* 51, 467–482.

5. Brosch, M., and Schreiner, C. E. (2000). Sequence sensitivity of neurons in cat primary auditory cortex.*Cereb. Cortex* 10, 1155–1167.

6. Churchland, M. M., Yu, B. M., Sahani, M., and Shenoy, K. V. (2007). Techniques for extracting single-trial activity patterns from large-scale neural recordings. *Curr. Opin. Neurobiol.* 17, 609–618.

7. Desai, N. S., Rutherford, L. C., and Turrigiano, G. G. (1999). Plasticity in the intrinsic excitability of cortical pyramidal neurons. *Nat. Neurosci.* 2, 515–520.

8. Dockendorf, K. P., Park, I., He, P., Prncipe, J. C., and DeMarse, T. B. (2009). Liquid state machines and cultured cortical networks: the separation property. *Biosystems* 95, 90–97.

9. Durstewitz, D., and Deco, G. (2008). Computational significance of transient dynamics in cortical networks. *Eur. J. Neurosci.* 27, 217–227.

10. Elman, J. (1990). Finding structure in time. *Cogn. Sci.* 14, 179–211.

11. Elman, J. L., and Zipser, D. (1988). Learning the hidden structure of speech. *J. Acoust. Soc. Am.* 83, 1615–1626.

12. Fu, Y., Djupsund, K., Gao, H., Hayden, B., Shen, K., and Dan, Y. (2002). Temporal specificity in the cortical plasticity of visual space representation. *Science* 296, 1999–2003.

13. Gómez, V., Kaltenbrunner, A., López, V., and Kappen, H. J. (2009). Self-organization using synaptic plasticity. *Adv. Neural Inf. Process. Syst.* 22, 513–520.

14. Hopfield, J. (1982). Neural networks and physical systems with emergent collective computational abilities. *Proc. Natl. Acad. Sci. U.S.A.* 79, 2554–2558.

15. Jaeger, H. (2001). The “Echo State” Approach to Analysing and Training Recurrent Neural Networks. GMD Report 148. Bremen, GMD - German National Research Institute for Computer Science.

16. Lazar, A., Pipa, G., and Triesch, J. (2007). Fading memory and time series prediction in recurrent networks with different forms of plasticity. *Neural Netw.* 20, 312–322.

17. Maass, W., Natschläger, T., and Markram, H. (2002). Real-time computing without stable states: a new framework for neural computation based on perturbations. *Neural Comput.* 14, 2531–2560.

18. Markram, H., Lübke, J., Frotscher, M., and Sakmann, B. (1997). Regulation of synaptic efficacy by coincidence of postsynaptic APs and EPSPs. *Science* 275, 213–215.

19. Mazor, O., and Laurent, G. (2005). Transient dynamics versus fixed points in odor representations by locust antennal lobe projection neurons. *Neuron* 48, 661–673.

20. Nikolic, D., Häusler, S., Singer, W., and Maass, W. (2006). Temporal dynamics of information content carried by neurons in the primary visual cortex. *Adv. Neural Inf. Process. Syst.* 19, 1041–1048.

21. Rabinovich, M., Huerta, R., and Laurent, G. (2008). Neuroscience: transient dynamics for neural processing. *Science* 321, 48–50.

22. Schrauwen, B., Wardermann, M., Verstraeten, D., Steil, J. J., and Stroobandt, D. (2008). Improving reservoirs using intrinsic plasticity. *Neurocomputation* 71, 1159–1171.

23. Steil, J. J. (2007). Online reservoir adaptation by intrinsic plasticity for backpropagation-decorrelation and echo state learning. *Neural Netw.* 20, 353–364.

24. Turrigiano, G. G., Leslie, K. R., Desai, N. S., Rutherford, L. C., and Nelson, S. B. (1998). Activity-dependent scaling of quantal amplitude in neocortical neurons. *Nature* 391, 892–896.

25. Yao, H., and Dan, Y. (2001). Stimulus timing-dependent plasticity in cortical processing of orientation.*Neuron* 32, 315–323.

26. Yao, H., Shi, L., Han, F., Gao, H., and Dan, Y. (2007). Rapid learning in cortical coding of visual scenes.*Nat. Neurosci.* 10, 772–778.

27. Zhang, W., and Linden, D. J. (2003). The other side of the engram: experience-driven changes in neuronal intrinsic excitability. *Nat. Rev. Neurosci.* 4, 885–900.

CHAPTER 2

A Review: Artificial Neural Networks as Tool for Control Food Industry Process

Estrella Funes, Yosra Allouche, Gabriel Beltrán, Antonio Jiménez

IFAPA Venta del Llano (Andalusia Regional Government), Jaén, Spain

ABSTRACT

In the last year, interest in using Artificial Neural networks as a modeling tool in food technology is increasing because they have found extensive utilization in solving many complex real world problems. Due to this and as previous step at development of some project, this paper intends to introduce the reader inside neural networks: general characteristics of the ANN, their architectures, their rules of learning, types of networks and ANN's create process. Also this paper presents a comprehensive review of food industrial applications of artificial neural networks in the last year. ANN industrial applications are grouped and tabulated by their main functions and what they actually performed on the referenced papers with except the applications in the olive oil industry that are described with special emphasis.

KEYWORDS

Artificial Neural Networks, Olive Oils, Sensor On-Line, Process Control

1. INTRODUCTION

In the last decades important researches in the area of artificial intelligence were developed. One of the main objectives of the science is the implementation of new technologies with capacities approaching the human being, and even better. Artificial intelligence is a powerful accomplishment in various fields, since it can supplement the human being in various activities, both easy to perform, how complex and even impossible to make as it is carrying out an activity without interruption for long periods of time and where knowledge is required.

In the search on the nature of intelligent mechanisms to try to develop machines that perform complex tasks, it is important to observe certain characteristics of human brain that can be considered crucial to the development of artificial neural networks (ANN).

Artificial neural networks are control systems necessary to solve problems in which the analytical methods are difficult to apply and their results have to be in a specific interval, e.g., in real time. The main objective of ANN is the development of mathematical algorithms that will enable ANNs to learn by mimicking information processing and knowledge acquisition in the human brain. ANN models contain layers of simple computing nodes that operate as nonlinear summing devices. These nodes are richly interconnected by weighted connection lines, and the weights are adjusted when data are presented to the network during a "training" process. Successful training can result in artificial neural networks that perform tasks such as predicting an output value, classifying an object, approximating a function, recognizing a pattern in multifactorial data, and completing a known pattern [1] [2] .

The objective of this paper is to provide a preliminary understanding of ANNs and answer why and when these computational tools are needed, how they are used and their possible applications inside food industry and giving as example the field of the olive grove and olive oil industry. A field of olive oil which is undergoing a constant technological evolution.

2. SIMILARITIES BETWEEN BIOLOGICAL NEURAL NETWORKS (BNN) AND ARTIFICIAL NEURAL NETWORKS (ANN)

2.1. Biological Neuron

The human nervous system consists of billions of neurons of various types and lengths relevant to their location in the body. Every neuron shares many characteristics with other cells of the human body but it has particular and special properties to receive process and transmit an electric signal through all the interconnections of the brain communication system [3] . Some neurons are connected with receivers or effectors and others are connected to transmit and process information. The area of connection is called a synapse and binds the axon of a neuron with dendrites of the adjacent as shown in Figure 1.

The information is transmitted from one neuron to another. This information is transmitted in the form of impulses through the dendrites. Impulses tend to excite the cell. When the excitation accumulated exceeds a threshold value, the neurons send a signal through the axon to other neurons. The majority of the ANN models present the basic operation of the neuron [4] .

2.2. Artificial Neuron

Artificial neurons are models that attempt to simulate the behavior of biological neurons. Each neuron is represented by processing unit that forms a part of one greater entity: the neural network. In general, each neuron sends its output to others by its axon and the axon takes information through potential differences or waves of power, which depends on the potential of the neuron. The input signals are produced by its synapse, adding all excitatory influences [ANN simulates it by positive weights and inhibit by negative weights]. If positive excitatory influences dominate, the neuron gives a positive signal and sends this message to other neurons by its output synapse. According to Figure 2, this unit of process (artificial neuron) consists of a series of entries Xi (x_1, x_2,...x_n), which are equivalent to the dendrites where they receive information in the form of stimulation. The weights that exist in the synapses Wi are equivalent in biological neuron to transmit mechanisms. The union of these values (Xi and Wi) equals the inhibitory and excitatory chemical signals that occur in the synapses and to induce neuron to change their behavior. These values are the gateway of the network weighting function that converts these values in potential.

This potential is equivalent to the total number of signals that arrive on a neuron in the biological neuron by their dendrites. The weighting function is a weighted sum of the inputs and the synaptic weights. The output of weighting function comes from the activation function that transforms this value into another form that the outputs neurons can work. The output network is evaluated in the activation function that gives rise to this neuron signal output to another neighbor. The activation function will be given by the potential of resulting that can be of different ways (Table 1). In summary, the artificial neuron behaves as a biological neuron but in a very sim- plified way [3] .

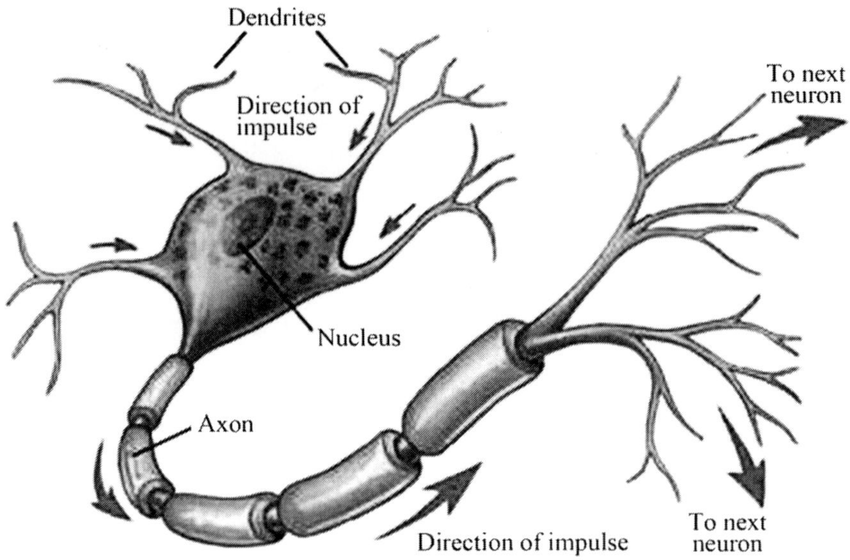

Figure 1. Structure ofa biological neuron. To can compare the performance of artificial neural it must know the structure of a biological neuron [4] .

3. ANN: MODELS, CHARACTERISTICS, ARCHITECTURE, LEARNING

3.1. Models of Artificial Neural Networks

In the Table 2 it can find the principal and known models of artificial neural networks.

3.2. Characteristics of an Artificial Neural Network

Artificial neural networks have a large number of features similar to the brain due to its constitution and its foundations, as it can be to learn from the experience [4] . The artificial neural networks are increasingly used in new technologies in various areas, and present a lot of advantages such as:

- Adaptive learning capacity consists on learning how to perform some tasks by training with illustrative examples. Neural networks are dynamic adaptive systems, due to their process units. The neurons can adjust themselves and change depending on the new conditions that appear. In fact, a neural network can generate its own weight distribution by means of learning and even after this continue learning. When designing a neural network must be given the appropriate structure for its learning and training with pattern. It is necessary to develop a good algorithm of learning that provides the network the ability to discriminate, by training with patterns.

- Self-organization. The neural networks employ their ability to adaptive learning for to autoorganize the information they receive during learning and/or the operation. In this way, neural networks are able to give an answer to a new situation, solve problems in which the input information is not very clear or incomplete.
- Tolerance to failures. Neural networks tolerate failures with respect to the data that extract or abstract the essential features of the learned entries, so they can properly process incomplete or distorted data. Furthermore, networks can continue their function (with some degradation) although destroyed part of the network. This is because the neural networks have their information distributed in the connections between neurons. These connections will have their values according to received stimuli, and an output pattern that represents the information stored that will generate.
- Operation in real time. Neural computations can be performed in parallel. Most of the networks can operate in a real-time environment, the need for change in weights connections or training is minimal.
- Easy insertion into existing technology. A network can be rapidly trained, tested, verified and transferred to a hardware implementation [5].

3.3. Architecture of a Neural Network

- The artificial neuron is an element of simple processing that produces a single output from an inputs vector. When talk about architecture of an ANN, must be taken into account:
- Number of level or layers: the distribution of neurons in the neural network is done forming levels or layers of a determined number of nodes each one. There are input, output and hidden layers.

- Connection patterns depending on the links between the elements of the different layers. The ANN can be classified as totally connected when all the outputs from a level get to all and each one of the nodes in the following level. In this case, there will be more connections than nodes. When some links in the network are lost, the network is partially connected.

- The direction of the information flow through the layers. The connectivity between the nodes of a neural network is related with the way in which the exits of neurons are directed to become into entrances other neurons. The output signal of a node can be one of the entrances another element of process, or even be an entrance to it. When any output of the neurons is input of neurons of the same level or preceding levels, the network is described as feedforward. By the contrary, if there is at least one connected exit as entrance of neurons of previous levels or of the same level, including themselves, the network is denominated of feedback. The feedback networks that have at least a closed loop of back propagation are called recurrent.

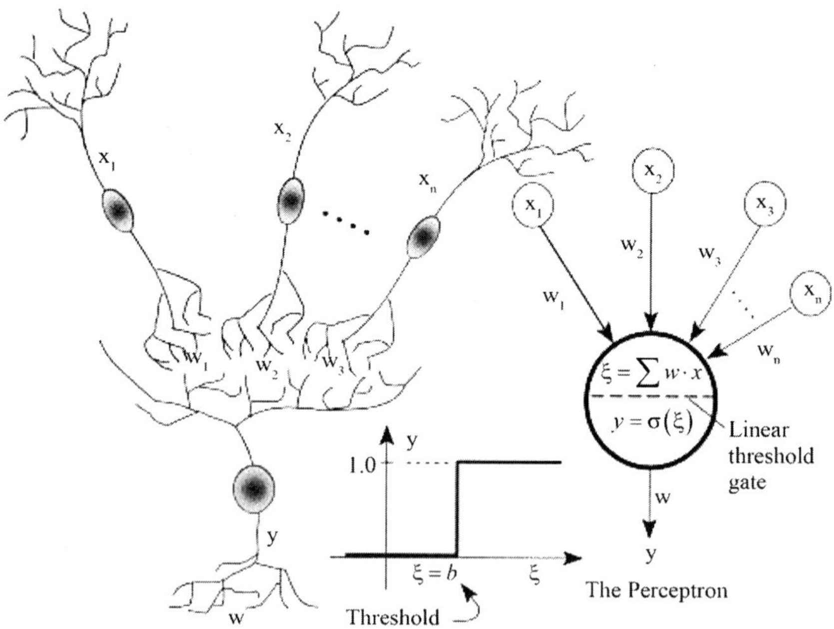

Figure 2. Similarity among a biological neuron and an artificial neuron. In this figure it can observe the difference and similarly between both neuron [4] .

Table 1. Roles of activation of an artificial neural network. These roles are necessary can activate an ANN. According the use of which network the role will be different [4] .

Name	Formula	Range	Graphical
Identity	$y = x$	$[-\infty, +\infty]$	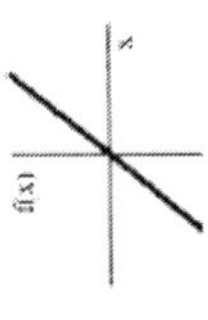
Unit step	$y = sign(x)$ $y = H(x)$	$\{-1, +1\}$ $\{0, +1\}$	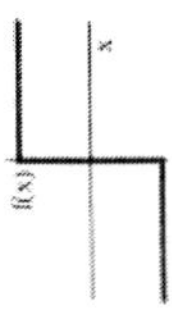
Piecewise linear	$y = \begin{cases} -1, \text{si } x < -l \\ x, \text{ si } +l \le x \le -l \\ +1, \text{si } x > +l \end{cases}$	$[-1, +1]$	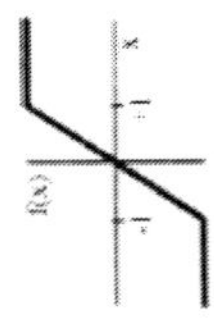
Sigmoid	$y = 1/(1+e^{\wedge}(-x))$ $y = tgh(x)$	$[0, +1]$ $[-1, +1]$	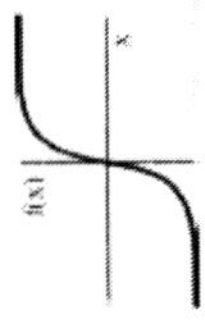
Gaussian	$y = Ae^{-Bx^2}$	$[0, +1]$	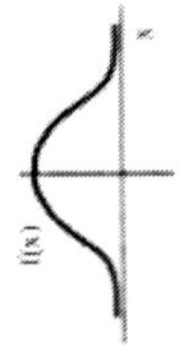
Sinusoidal	$y = A\sin(\omega x + \varphi)$	$[-1, +1]$	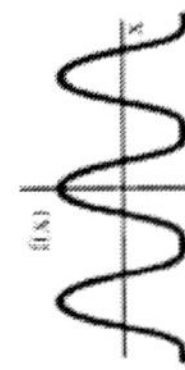

Table 2. Summarize the well-known neural networks models (own elaboration).

Name	Applications	Commentaries	Limitations	Year	Creators
Perceptron (P)	Recognition of characters printed	The oldest network	You may not recognize complex characters	1958	Rosemblatt
Adaline/Madaline	Filtering of signals. Equalizer Adaptive	Quick, easy implement with analog circuits	It is only possible classifying spaces linearly separated	1960	Widrow
Avalanche	Recognition of talks continued. Robot control	No network easy done these tasks	It is not easy to alter the speed or interpolate the movement	1967	Grossberg
Cerebellation	Control of the movement of arms of a robot	Similar to Avalanche	It requires complicated control inputs	1969	Marr, Albus,
Brain-State-Box	Extraction of knowledge database	Possibly best realization that SAS Hopfield networks	Potential realization applications not completely studied	1977	Anderson
Neocognitron	Recognition of characters manuscripts	Easy conceptualize. Insensitive to the translation, rotation and scale	It requires many elements of process, levels and connections	1978	Fukushima
Self-Organizing-Map (SOM)	Recognition of patterns, coding data optimization	Make maps features common of cough learned data	It requires a lot training	1980	Kohonen
Hopfield	Reconstruction patterns and optimization	Can be implemented in VLSI.	Capacity and stability	1982	Hopfield
Machines Boltzmann and Cauchy	Recognition patterns (images, sound and radar). Optimization	Simple networks. Capacity of representation optimal patterns	The machine of Bolzmann need a very long time learning	1985	Hinton
Associative memory bi-directional	Memory heteroasociativa access by content.	Learning and simple architecture	Low capacity storage. The data should be encoded	1985	Kosko
Adaptive resonance theory (ART)	Recognitionpatterns (radar, (sonar, etc.)	Sophisticated, little used	Sensitive to the translation, distortion and scale	1986	Carperrter, Grossberg
Counterpropagation	Understanding images	Combination of Perceptron and TPM	Many neurons and connections	1986	Hectnielsen
Back-propagation (BP)	Voice synthesis from text. Control robots. Forecast Recognition patterns	The most popular network. Numerous applications successfully. Ease of learning. Powerful.	It needs a lot of time for the learning and many examples	1986	Werbos, Parker, Rumelhart

3.4. Process of Learning

Learning ANN consists in determining precise values of weights for all connections, all trained for the efficient resolution of a problem. During the training session, the weights converge gradually to the values that make each entry to produce the desired output vector [3] .

The end of the learning period can be determined:

- using a number fixed cycles,
- when the error falls below a preset amount,
- when the modification of the weights is irrelevant.

Depending on the scheme of learning and the problem to be solved, three types of learning schemes can be distinguished [4] :

- Supervised learning. In the supervised knowledge the training is controlled by an external agent (supervisor, teacher), which watches the answer that the network is supposed to generate from a determined entrance. The supervisor compares the output of the network with the expected one and determines the amount of the modification to be made in the weight. The objective is to decrease the difference between the answer of the network and the desired values. For example the ANN that uses this type of learning is Adaline, MPL, BP, and Associative Memory Bi-directional.
- Unsupervised learning. The ANNs with unsupervised learning (also known as self-supervised) do not require any external element to adjust the weight of the communication links to their neurons. The main problem in the unsupervised classification is to divide the space where the objects are in groups or categories. The ANN that use this type of learning are Hopfield, Machines Boltzmann and Cauchy, networks with competitive learning, such as SOM and ART.

4. ANN: MODELS, CHARACTERISTICS, ARCHITECTURE, LEARNING

The development of a successful ANN project has to be constituted for a number of stages, as illustrated in Figure 3.

4.1. Problem Definition and Formulation

In the first phase, the problem has to definite. There exist engineering problems for which finding the perfect solution requires a practically impossible amount of resources and an acceptable solution would be fine. ANNs can give good solutions for such classes of problems.

4.2. System Design

In order to select a good NN configuration, there are several factors to take into consideration. The major points of interest regarding the ANN topology selection are related to network design, training, and practical considerations. Some of the design considerations include determining the number of input and output nodes to be used, the number of hidden layers in the network and the number of hidden nodes used in each hidden layer. The number of input nodes is typically taken to be the same as the number of state variables. The number of output nodes is typically the number that identifies the general category of the state of the system. Each node constitutes a processing element and it is connected through various weights to other elements. In order to predict with the least possible error, these values must be optimized.

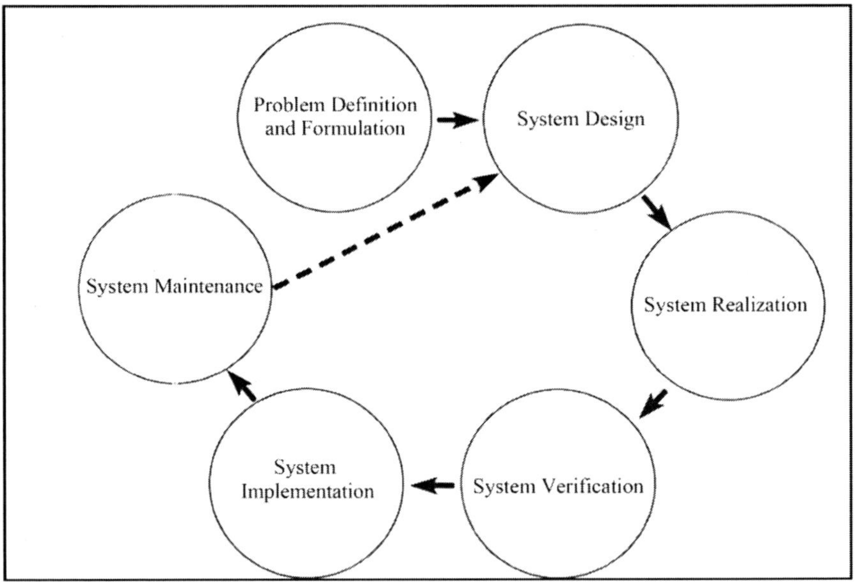

Figure 3. The various phases in an ANN development project. These stages are fundamental to right performance of ANN (own elaboration).

The data must be enough big to cover the possible known variation in the problem domain. This database into three subsets for training-test and validation. The samples that compose the input and output must have the same format and must be normalized in the range 0 or 1 or −1 to 1 depending on the transfer function used. The initial weights of an NN play a significant role in the convergence of the training method. Knowing input data and output data and objects obtain, we can design ANN. First there was to choose to type of ANN and once defined the topology of the ANN it is required to establish algorithm that allow an adequation of the inner parameters of the network.

4.3. System Realization

System realization involves one of two basic steps, training/learning. In this training step, the connection parameters were adjusted with training function showed in Table 1, such that, for the given dataset the neural network-predicted output dataset matched with real output dataset. The training step is influence for various parameters (network size, learning rate, number of training cycles, acceptable error, etc.) can affect the design and error of the final network. For this, we must considered determining the input and output variables, choosing the size of the training data set, initializing network weights, choosing training parameter values (such as learning rate and momentum rate), and selecting training stopping criteria, are important for several network topologies. In function of ANN type we choose of type supervised or not supervised learning. The number of training examples used to train an ANN is sometimes critical to the success of the training process. If the number of training examples is not sufficient, then the network cannot correctly learn the actual input-output relation of the system. If the number of training examples is too large, then the network training time will be longer. For some applications, such as real-time adaptive neural control, training time is a critical.

4.4. System Verification

Verification is intended to confirm the capability of the ANN based model to respond accurately to examples never used in network development. This stage includes comparing the performance of ANN based model to those of other approaches. For each training function, the matrix of weight must be optimized in its respective learning process. In the verification process, the obtained data by ANN must be compared with the real ones to optimize the parameters of the ANN by statistical tools. The testing/validation process is the final stage of the work. When all ANN parameters are optimized, the validation process is only used to test them.

4.5. System Implementation

This phase includes embedding the obtained network in appropriate working system such as hardware controller or computer program, depending of the function have ANN, such as, optimization process or control process or predict on-line. Final testing of the integrated system should also be carried out before its release to the end user.

4.6. System Maintenance

This stage includes update of the medium, changes in variables, new data that influence the development of the network.

5. APPLICATIONS OF NEURAL NETWORKS

The ANNs are an emerging computer technology that can be used in a large number and variety of applications such as, control, monitoring and modeling, recognition, detection an research for patterns, predicts on-line, image processing, optimization and signal processing. This applications can use in several fields as production of manufacturing, agriculture…, business, marketing, medicine, transports, energy, trade the greater, etc… In this paper, we can see some examples, what use has artificial neural network in different food industry.

5.1. Control, Monitoring and Modeling of Industrials Processes

At present, the industry has a number of objectives: improve of quality products, reducing of waste, elimination of toxins, and above all the increase of the benefits. The industry has also to be efficient. For this reason neural networks are implanted in control, modeling and monitoring of existing industrial processes, thus reducing costs process and increasing the outputs, as it can be seen at the various articles referenced in the Table 3.

5.2. Recognition, Detection, Classification and Search for Patterns

The ANNs are used for recognition, detection, classification and search for patterns, e.g., for classification of seeds or apples in real time through digital images, analyzing texture and color. Besides defect detection on cherries or pattern recognition of fruit shape quantitatively between others as the Table 4 incanted.

5.3. Prediction On-Line Parameters

Predicting the behavior of an even which depends from the time, with base in a group of values that are obtained from different moments. In food industries, the ANNs are very used, overall for predict on-line quality and characteristics of food, parameters of elaboration process, prediction temperatures for food storage. The Table 5 shows some applications of ANN in food industry and kind of ANN used.

5.4. Image Processing

More than 200 applications of neural networks in image processing and discuss the present and possible future role of neural networks, especially feed-forward neural networks, Kohonen feature maps and Hop1eld neural networks. The various applications are used in different industries, for example in forest industry or clothe industry to see defects of raw material and final product. In food industry, the image processing method has become very popular and indirectly used to determine the sugar content of fruits. For example, high correlation coefficients have been obtained between the sugar content and the red value in the chromatic coordinates of banana peels [59] or the application of ANN to the color grading of apples [60] .

5.5. Optimization

In general, the objective in optimization is to allocate a limited amount of resources to a set of certain partial tasks such that some objective or costs function is minimized (or maximized). A great number of variables usually enter the problem, and to evaluate and minimize the objective function, a combinatorial problem has to be solved. In the Table 6 we can be observed some examples of optimization of elaboration process conditions of food or optimization of storage conditions of vegetables and fruits between others.

5.6. Signal Processing

From a signal processing perspective, it is imperative to develop a proper understanding of basic neural network structures and how they impact signal processing algorithms and applications. A challenge in surveying the field of neural network paradigms is to identify those neural network structures that have been successfully applied to solve real world problems from those that are still under development or have difficulty scaling up to solve realistic problems. When dealing with signal processing applications, it is critical to understand the nature of the problem formulation so that the most appropriate neural network paradigm can be applied. In addition, it is also important to assess the impact of neural networks on the performance, robustness, and cost-effectiveness of signal processing systems and develop methodologies for integrating neural networks with other signal processing algorithms. Because of the ANN are used in different fields and industries. In food industry, we can find several researches. One of all develops two-way classification models using ANN to classify the stored beef meat samples into two groups: "unspoiled" and "spoiled"; based on the microbial population as alternative to electronic noise [67] .

Table 3. Several examples of application of control, monitoring and modelling of industrial process (own elaboration).

Sample	Object of investigation	Type of ANN	References
Food in general	Modeling and control of a food extrusion process using ANN	BP	Popescu *et al.*, (2001) [5]
	Modeling sterilization process of canned foods ANNs	BP	Gonçalves. *et al.*. (2005) [6]
	Prediction models and for optimization of constant temperature retort (CRT) thermal processing of conduction heating foods using ANN.	Non defined	Chen. and Ramaswamy, (2002) [7]
	Application of ANN in the modelling of the thermal treatment of food	RNN. BP	Luera Peña. Minim. (2001) [8]
Apple	ANN modeling of apple drying process.	MLPFF	Khoshhal. *et al.*, (2010) [9]
Apple juice	Modeling dead-end ultrafiltration of apple juice using ANN	MLPFF, BP	Gökmen. *et al.*, (2009) [10]
Bread	Dynamic model of the kneading process based on ANN that allows predicting the bread dough temperature and the delivered power	RNN	Lamrini, *et al.*, (2012) [11]
Lime	ANN modeling of mass transfer during osmotic dehydration of kaffir lime peel.	MLPFF	Lertworasirikul. Saetan. (2010) [12]
Squid protein	Modeling of squid protein hydrolysis using ANN	MLP	Abakarov, *et al.*, (2011) [13]
Tomato	Modeling of tomato drying using ANN	MLP	Movagharnejad. Nikza, (2007) [14]
Viscous fluids	Neural network modeling of end-over-end thermal processing of particulates in viscous fluids	MLP BP	Meng and Ramaswamy, (2010) [15]
Wheat	Modeling of wheat soaking using two ANNs	MLP. RBF	Kashaninejad, (2009) [16]
Wine	Modeling wine characteristics using ANN	MLPFF, BP	Vlassides, *et al.*, (2001) [17]

Table 4. Several examples of application of recognition, detection, classification and research for patterns (own elaboration).

Sample	Object of investigation	Type of ANN	References
Apple	Develop ANN-based apple classifier	MLP, BP	Bhatt *et al*., (2014) [18]
	Apple classification system based on machine vision and ANN, which classifies apple in real time on the basis of physical parameters of apple such as size, color and external defects.	MLP, BP	Bhatt and Pant, (2013) [19]
	Classification of apple surface features using machine vision and ANN	MLPFF	Yang, (1993) [20]
Beans	Classification of beans using computer vision system and ANNs	MLPFF	Kılıc *et al*., (2007) [21]
Boiled shrimp	Classifications of boiled shrimp's shape using image analysis and ANN model.	MLP	Poonnoy *et al*., (2014) [22]
Cherries	Use of genetic artificial neural networks and spectral imaging for defect detection on cherries	MLPFF, BP	Guyer and Xiukun Yang, (2000) [23]
Food in general	A new pattern recognition method for detecting fouling on stainless steel is presented in food processing.	MLPFF, BP	Wallhäußer *et al*., (2011) [24]
Fruits	Pattern recognition of fruit shape quantitatively with ANN	SOM	Morimoto *et al*., (2000) [25]
Grape	Determination of anthocyanin concentration in whole grape skins using hyperspectral imaging and adaptive boosting ANN	MLPFF	Fernandes *et al*., (2011) [26]
Lamb	Classification according to prediction the lamb carcass grades using features extracted from lamb chop images	MLP	Chandraratne *et al*., (2007) [27]
Pig muscles	Duroc and Iberian pork neural network classification by visible and near infrared reflectance spectroscopy	RBF	Del Moral, *et al*., (2009) [28]
The starch of potato, cassava, corn	Classify the data set and to predict mechanical properties (tensile strength and strain at break of starch-based films using ANN	MLPFF	Dieulot and Skurtys, (2013) [29]
Wine	Prediction of problematic wine fermentations using ANN	MLP, BP	Román *et al*., (2011) [30]

6. APPLICATIONS IN FIELD OF THE OLIVE OIL

In this field, the artificial neural networks have different applications: to classify the olive fruit for their processing, to test the ripeness of fruit, to detect frauds or adulterations in the olive oils or to predict the characteristic of olive oil obtained and to optimize the extraction process of olive oil.

For example, ANN was employed to fruit classification according to olive cultivar, as a tool to guarantee varietal authenticity. In this study, six olive cultivars of Trás-os-Montes region were collected in different groves and during four crop years. Ten biometrical parameters were measured and used for the linear discriminant analysis (LDA) and ANN classification model development. In this work, single MPLFF classification networks, trained using a supervised learning algorithm was used. The MLP methodology demonstrated to be a practical and effective tool for olive cultivar classification, showing high global sensitivity and specificity, even for unknown samples (76% and 78% for the validation data set, respectively). With this work between others, the olive oil producers can use to prevent possible olive oil adulterations with olives of non-allowed cultivars, according to the Protected Designation of Origin (DPO) or monovarietal olive oil's official regulation [68] .

Table 5. Several examples of application prediction on-line parameters (own elaboration).

Sample	Object of investigation	Type of ANN	References
Agricultural products	ANN method to predict the drying characteristics of agricultural products such as hazelnut, bean and chickpea	MLPFF, BP	Topuz (2010) [31]
Carrot	Predicting properties of dried carrot based on composition, drying technique, and microstructural features	P	Kerdpiboon, *et al.*, (2006) [32]
Cheese	Predicting the final process time in the acidification step for cheese production.	P	Jun-ichiHoriuchi, *et al.*, (2004) [33]
Chicken	Predicting the final temperature of chicken carcasses in an industrial scale using ANN	MLPFF, BP	Silveira, *et al.* (2014) [34]
Chicken nuggets	Developing a model to predict the mechanical textural properties of the fried breaded chicken nuggets by using image texture processing	MLPFF	Qiao *et al.*, (2007) [35]
Corn	Artificial neural networks to predict corn yield	MLPFF, BP	Uno *et al.*, (2005) [36]
Crackers	Moisture content and water activity prediction of semi-finished cassava crackers from drying process with ANN	MLPFF	Lertworasirikul and Tipsuwan, (2008) [37]
Fresh outcrop milk cream	Discriminating season of production and feeding regimen of butters based on infrared spectroscopy and ANN	MLP	Gori *et al.*, (2012) [38]
Food in general	Predicting of cold spot temperature in retort sterilization of starch-based foods	BP	Llave *et al.*, (2012) [39]
	Predicting the internal texture characteristics from extrusion food surface images with ANN	MLPFF, BP	Fan *et al.*, (2013) [40]
	Predicting of thermal conductivity of food as a function of moisture content, temperature and apparent porosity.	MLPFF, BP	Sablani, and Rahman, (2003) [41]
Grain	Predicting the moisture content of grain drying process using genetic algorithm	BP, RBF	Liu, *et al.*, (2007) [42]
Ice cream	Predicting total acceptance of ice cream using ANN	MLPFF	Bahramparvar, *et al.*, (2013) [43]
Meal of beef	Predicting shrinkage of ellipsoid beef joints as affected by water immersion cooking using image analysis and ANN	BP	Zheng *et al.*, (2007) [44]
Molasses vinegar	Prediction by ANN of the physicochemical quality of cane molasses vinegar by time-temperature effect of food to flash evaporator-distiller	MLPFF, BP	Vásquez and Lescano, (2010) [45]
Oils	ANN as an alternative to other methods utilized to determine the fatty acid composition of oils.	MLPFF	Yalcin *et al.*, (2012) [46]
Orange juice	Predicting physical properties of orange juice powder as well as process parameters in spray dryers.	MLPFF	Chegini *et al.*, (2008) [47]
Orange, peach, pear	Predicting of viscosity of fruit juice as a function of concentration and temperature.	MLP	Rai *et al.*, (2005) [48]
Peas	Predicting maturity index peas with ANN	MLP	Higginsa *et al.*, (2010) [49]
Peppers	Predicting of storage quality of fresh-cut green peppers using ANN	MLPFF, BP	Meng, *et al.*, (2012) [50]
Processing food in general	Predicting process parameters involved in thermal/pressure food processing using ANN	BP	Torrecilla *et al.*, (2004) [51]
Plum juice	Predicting the permeate flux of red plum juice during membrane clarification	MLPFF	Nourbakhsh *et al.*, (2014) [52]
Seeds	Predicting deoxynivalenol accumulation in barley seeds contaminated with Fusariumculmorum under different conditions	MLP, RBF	Mateo *et al.*, (2011) [53]

Table 6. Several examples of application optimization (own elaboration).

Sample	Object of investigation	Type of ANN	References
Corn germ	Modeling syrup production process from defatted corn germ	MLPFF BP	Millán and Gil, (2011) [61]
Food in general	Design and optimize high-pressure food processes	MLPFF BP	Torrecilla *et al*., (2005), (2007) [62] [63]
Foodstuffs	Optimization the operating conditions on heat and mass transfer during foodstuffs drying	MLPFF	Hernández, (2009) [64]
Fruit	Optimization of heat treatment for fruit during storage using ANN	MLPFF	Morimoto *et al*., (1997) [65]
	Realizing the optimal control of the fruit-storage process.	MLPFF	Morimoto *et al*., (1997) [66]

Other investigation described a novel method for rapid, automatic and objective prediction of the ripening index of an olive lot [69] . The method used integrates a Machine Vision system, capable of performing a color-based raw prediction of ripening index, with an ANN based algorithm to refine it. Such a refinement was based on a set of chemical parameters (oil content, sugar content and phenol content) which were provided as input to the ANN and which can be obtained by historical curves for the region where the ripening index needs to be predicted. The training of ANN was carried out using rule based on the gradient descent BP algorithm with an adaptive learning rate. For other part, FFBP with Levenberg-Marquardt algorithm is used by Ram et al., (2010) to predict the quantity oil together with model linear regressions (MLR). Both based on quality features derived from known image processing algorithms [70] . The ANNs were more accurate than linear regression, resulting in average linear correlations for two varieties of olives.

Due to the most recent changes in purchasing behavior of consumers are markedly affected by the increasing consumer's interest in a reliable geographical declaration of foods. As regards virgin olive oils, this information is deemed an additional warranty of their quality, authenticity, typicality and safety. For this purpose, Garcia- Gonzalez D. et al., (2009) proposed a new methodology based on MLP, a form of ANN that has been applied to the main series of compounds analyzed in oils to identify the geographical origin of virgin olive oil for this time [71] . This work made a classification function of country, region, province and PDO. Torrecilla et al., (2013) joined to classification of extra virgin olive oils (EVOO) used ANN. This classification was made inside PDO with its owns olive oils. For this purpose, they used as input several parameters, such as, acidity, peroxide index, K_{232}, K_{270}, ΔK, moisture and volatile compounds. With this research we can observed the capability of supervised ANN models with smaller topologies to classify very similar EVOO samples, also these techniques can be applied to detect adulterations [72] . These researchers used a self-organizing map (SOM) to classify the aforementioned UV-vis signals from pure EVOO and adulterated samples with different additives and signals of pure EVOO with addition of two different types of noises. This can be a tool to detect adulterations although it still requires an optimization. This type of ANN was used by Marini et al., (2007) together with a MLPFF [73] . Zupan et al., (1994) already used ANN to classify olive oils. The classification was performed on analysis of eight fatty acids of Italian olive oils. In this research, BP and Kohonen were compared with prediction approximated 95% [74] .

ANNs also have other function inside olive oil field, such as, inside extraction process of olive oil. Furferi et al., 2007, 2008, development artificial neural network to predict of the two parameters: acidity level and peroxides of olive oil extracted by a continuous extraction process. Thus, allowing a real-time control of oil quality during the extraction process. BP was used on two different neural network referring to a 3 phases and 2 phases decanting technology which results obtained were acceptable [75] [76] .

On the other hand, ANNs are an adequate tool to estimate the concentration of phenolic compounds, a pollutant with a high environmental impact in olive oil mill wastewater, without any previous phenological knowledge [77] . The type of ANN used in this work is a BP with feed-forward connections between neurons or back connections with a prediction horizon and supervised learning. The ANNs developed and optimized was able to predict the caffeic acid concentration value in a waste stream from olive oil extraction. This research had important results because the difference between the real and the predicted values was about 5%. In other study performed by the same investigators, an integrated artificial neural network (ANN)/laccase biosensor is designed to prediction of cathecol concentration of olive oil mill wastewater. The obtained results gave difference between the real and the predicted values less than 1% [78] .

Other researchers as Bordons and Cueli in (2004) intended to find an application of a predictive controller that deals with measurable disturbances in extraction process in an olive oil mill. This research was focused on the thermal part of the process and it shown the combination of experimental results with validated results. This work didn't use ANN as tool of work but we can observe as if other researchers used this tool for elaboration process optimization [79] . This is the case of Jiménez et al., 2008, 2009 [2] [80] . In this work the authors built an ANN for real-time prediction of the moisture and fat content in olive pomace using two-phase olive oil pro- cessing. With help of AOTF-NIR system that obtained the oil spectrum from horizontal decanter and data analysis, they development an ANN based on the BPFF. These ANN's predictions can be performed quickly during the extraction process, allowing a rapid regulation of technological variables for minimal fat loss.

7. CONCLUSION

In this review the reader can introduce principal characteristics and the ANN's creation process in the ANN's world. Although the fundamental conclusion is numerous applications that ANNs have as we can observe in this review. In some field the ANNs can be used for their enormous advantages, over all for their ability to learn from examples make them efficient problem solving paradigms. Older artificial neural networks have limitations that should not be overlooked, the interest in ANN's field continues growing and optimizing better for solving, controlling or modeling all class of problems that can appear.

ACKNOWLEDGEMENTS

This work is supported by excellence project of Junta de Andalucía, "Modeling and optimization of the process of elaboration of virgin olive oil. Project" I. and it is a preliminary work of E.F.'s PhD thesis.

CONFLICT OF INTEREST

Estrella Funes, Yosra Allouche, Gabriel Beltrán, Antonio Jiménez The authors have declared no conflict of interest.

REFERENCES

1. Basheer, I.A. and Hajmeer, M. (2000) Artificial Neural Networks: Fundamentals, Computing, Design, and Application. Journal of Microbiological Methods, 43, 3-31.

2. Jiménez, A., Aguilera, M.P., Uceda, M. and Beltrán, G. (2009) Neural Network as Tool for Virgin Olive Oil Elaboration Process Optimization. Journal of FoodEngineering, 95, 135-141.

3. BasogainOlabe, X. (2005) Artificial Neural Networks and Their Applications. School of Engineering, Bilbao.

4. Matich, D.J. (2001) Neural Networks: Basic Concepts and Applications. National University of Technology—Faculty Regional Rosario. Group of Applied Research in the Chemical Engineering (GIAIQ).

5. Popescu, O., Popescu, D., Wilder, J. and Karwe, M. (2001) A New Approach to Modelling and Control of a Food Extrusion Process Using Artificial Neural Network and an Expert System. Journal of Food Process Engineering, 24, 17- 36.

6. Gonçalves, E.C., Minim, LA., Coimbra, J.S.R. and Minim, V.P.R. (2005) Modeling Sterilization Process of Canned Foods Using Artificial Neural Networks. Chemical Engineering and Processing, 44, 1269-1276.

7. Chen, C.R. and Ramaswamy, H.S. (2002) Modeling and Optimization of Constant Retort Temperature (CRT) Thermal Processing Using Coupled Neural Networks and Genetic Algorithms. Journal of Food Process Engineering, 25, 351- 379.

8. Luera Peña, W.E. and Minim, L.A. (2001) Application of Neural Networks in the Modelling of the Thermal Treatment of Food. LWT-Food Science and Technology, 3, 81-88.

9. Khoshhal, A., Dakhel, A., Etemadi, A. and Zereshk, S. (2010) Artificial Neural Network Modeling of Apple Drying Process. Journal of Food Process Engineering, 33, 298-313.

10. Gökmen, V., Açar, O.C., Serpen, A. and Süğüt, I. (2009) Modeling Dead-End Ultrafiltration of Apple Juice Using Artificial Neural Network. Journal of FoodProcessEngineering, 32, 248-264.

11. Lamrini, B., Della Valle, G., TreleaPerrot, I.C. and Trystram, G. (2012) A New Method for Dynamic Modeling of Bread Dough Kneading Based on Artificial Neural Network. Food Control, 26, 512-524.

12. Lertworasirikul, S. and Saetan, S. (2010) Artificial Neural Network Modeling of Mass Transfer during Osmotic Dehydration of Kaffir Lime Peel. Journal of Food Engineering, 98, 214-223.

13. Abakarov, A., Teixeira, A., Simpson, R., Pinto, M. and Almonaci, S. (2011) Modeling of Squid Protein Hydrolysis: Artificial Neural Network Approach. Journal of Food Process Engineering, 34, 2026-2046.

14. Movagharnejad, K. and Nikzad, M. (2007) Modeling of Tomato Drying Using Artificial Neural Network. Computers and Electronics in Agriculture, 59, 78-85.

15. Meng, Y. and Ramaswamy, H.S. (2010) Neural Network Modeling of End-over-End Thermal Processing of Particulates in Viscous Fluids. Journal of Food Process Engineering, 33, 23-47.

16. Kashaninejad, M., Dehghani, A.A. and Kashiri, M. (2009) Modeling of Wheat Soaking Using Two Artificial Neural Networks (MLP and RBF). Journal of Food Engineering, 91, 602-607.

17. Vlassides, S., Ferrier, J.G. and Block, D.E. (2001) Using Historical Data for Bioprocess Optimization: Modeling Wine Characteristics Using Artificial Neural Networks and Archived Process Information. Biotechnology and Bioengine- ering, 73, 55-68.

18. Bhat, A.K. (2014) An Analysis of the Performance of Artificial Neural Network Technique for Apple Classification. AI & Society, 29, 103-111.

19. Bhatt, A.K. and Pant, D. (2013) Automatic Apple Grading Model Development Based on Back Propagation Neural Network and Machine Vision, and Its Performance Evaluation. AI & Society, 30, 45-56.

20. Yang, Q. (1993) Classification of Apple Surface Features Using Machine Vision and Neural Networks. Computers and Electronics in Agriculture, 9, 1-12.

21. Kılıç, K., Boyaci, H.I., Koecsel, H. and Kuesmenoglu, I. (2007) A Classification System for Beans Using Computer Vision System and Artificial Neural Networks. Journal of Food Engineering, 78, 897-904.

22. Poonnoy, P., Yodkeaw, P., Sriwai, A., Umongkol, P. and Intamoon, S. (2014) Classification of Boiled Shrimp's Shape Using Image Analysis and Artificial Neural Network Model. Journal of Food Process Engineering, 37, 257-263.

23. Guyer, D. and Yang, X. (2000) Use of Genetic Artificial Neural Networks and Spectral Imaging for Defect Detection on Cherries. Computers and Electronics in Agriculture, 29, 179-194.

24. Wallhäußer, E., Hussein, W.B., Hussein, M.A., Hinrichs, J. and Becker, T.M. (2011) On the Usage of Acoustic Properties Combined with an Artificial Neural Network—A New Approach of Determining Presence of Dairy Fouling. Jour- nal of Food Engineering, 103, 449-456.

25. Morimoto, T., Takeuchi, T., Miyata, H. and Hashimoto, Y. (2000) Pattern Recognition of Fruit Shape Based on the Concept of Chaos and Neural Networks. Computers and Electronics in Agriculture, 26, 171-186.

26. Fernandes, A.M., Oliveira, P., Moura, J.P., Oliveira, A.A., Falco, V., Correia, M.J. and Melo-Pinto, P. (2011) Determination of Anthocyanin Concentration in Whole Grape Skins Using Hyperspectral Imaging and Adaptive Boosting Neural Networks. Journal of Food Engineering, 105, 216-226.

27. Chandraratne, M.R., Kulasiri, D. and Samarasinghe, S. (2007) Classification of Lamb Carcass Using Machine Vision: Comparison of Statistical and Neural Network Analyses. Journal of Food Engineering, 82, 26-34.

28. Del Moral, F.G., Guillén del Moral, L.G., O'Vallea, F. and Martínez del Moral, R.G. (2009) Duroc and Iberian Pork Neural Network Classification by Visible and Near Infrared Reflectance Spectroscopy. Journal of Food Engineering, 90, 540-547.

29. Dieulot, J.-Y. and Skurtys, O. (2013) Classification, Modeling and Prediction of the Mechanical Behavior of Starch- Based Films. Journal of Food Engineering, 119, 188-195.

30. Román, C., Hernández, O.G. and Urtubia, U.A. (2011) Prediction of Problematic Wine Fermentations Using Artificial Neural Networks. Bioprocess and Biosystem Engineering, 34, 1057-1065.

31. Topuz, A. (2010) Predicting Moisture Content of Agricultural Products Using Artificial Neural Networks. Advances in Engineering Software, 41, 464-470.

32. Kerdpiboon, S., Kerr, W. and Devahastin, S. (2006) Neural Network Prediction of Physical Property Changes of Dried Carrot as a Function of Fractal Dimension and Moisture Content. Food Research International, 39, 1110-1118.

33. Horiuchi, J., Shimada, T., Funahashi, H., Tada, K., Kobayashi, M. and Kanno, T. (2004) Artificial Neural Network Model with a Culture Database for Prediction of Acidification Step in Cheese Production. Journal of Food Engine- ering, 63, 459-465.

34. Silveira, C., Belledeli, B.D., Soares, M.B.A., Treichel, H. and Mazutti, M.A. (2014) Use of Artificial Neural Network for Industrial Prediction of Final Temperature of Chicken Carcasses. Journal of Food Process Engineering, 37, 1-9.

35. Qiao, J., Wang, N., Ngadi, M.O. and Kazemi, S. (2007) Predicting Mechanical Properties of Fried Chicken Nuggets Using Image

Processing and Neural Network Techniques. Journal of Food Engineering, 79, 1065-1070.

36. Uno, Y., Prasher, S.O., Lacroix, R., Goel, P.K., Karimi, Y., Viau, A. and Patel, R.M. (2005) Artificial Neural Networks to Predict Corn Yield from Compact Airborne Spectrographic Imager Data. Computers and Electronics in Agriculture, 47, 149-161.

37. Lertworasiriku, S. and Tipsuwan, Y. (2008) Moisture Content and Water Activity Prediction of Semi-Finished Cassava Crackers from Drying Process with Artificial Neural Network. Journal of Food Engineering, 84, 65-74.

38. Gori, A., Cevoli, A., Fabbri, A., Cavoni, M.F. and Losi, G. (2012) A Rapid Method to Discriminate Season of Production and Feeding Regimen of Butters Based on Infrared Spectroscopy and Artificial Neural Networks. Journal of Food Engineering, 109, 525-530.

39. Llave, Y.A., Hagiwara, T. and Sakiyama, T. (2012) Artificial Neural Network Model for Prediction of Cold Spot Tem- perature in Retort Sterilization of Starch-Based Foods. Journal of Food Engineering, 109, 553-560.

40. Fan, F.H., Ma, Q., Ge, J., Peng, Q.Y., William, W. and Riley, T.S.Z. (2013) Prediction of Texture Characteristics from Extrusion Food Surface Images Using a Computer Vision System and Artificial Neural Networks. Journal of Food Engineering, 118, 426-433.

41. Sablani, S.S. and ShafiurRahman, M. (2003) Using Neural Networks to Predict Thermal Conductivity of Food as a Function of Moisture Content, Temperature and Apparent Porosity. Food Research International, 36, 617-623.

42. Liu, X., Chen, X., Wu, W. and Peng, G. (2007) A Neural Network for Predicting Moisture Content of Grain Drying Process Using Genetic Algorithm. Food Control, 18, 928-933.

43. Bahramparvar, M. (2013) Predicting Total Acceptance of Ice Cream Using Artificial Neural Network. Journal of Food Processing and Preservation, 38, 1080-1088.

44. Zheng, C., Sun, D.W. and Zheng, L. (2007) Predicting Shrinkage of Ellipsoid Beef Joints as Affected by Water Immersion Cooking Using Image Analysis and Neural Network. Journal of Food Engineering, 79, 1243-1249.

45. Vásquez, V. and Lescano, C. (2010) Prediction by Artificial Neural Networks of the Physicochemical Quality of Cane Molasses Vinegar by Time-Temperature Effect of Food to Flash Evaporator-Distiller. Scientia Agropecuaria, 1, 63-73.

46. Yalcin, H., Toker, O.S., Ozturk, I., Dogan, M. and Kism, O. (2012) Prediction of Fatty Acid Composition of Vegetable Oils Based on Rheological Measurements Using Nonlinear Models. European Journal of Lipid Science and Techno- logy, 114, 1217-1224.

47. Chegini, G.R., Khazaei, J., Ghobadian, B. and Goudarzi, A.M. (2008) Prediction of Process and Product Parameters in an Orange Juice Spray Dryer Using Artificial Neural Networks. Journal of Food Engineering, 84, 534-543.

48. Rai, P., Majumdar, G.C., DasGupta, S. and De, S. (2005) Prediction of the Viscosity of Clarified Fruit Juice Using Artificial Neural Network: A Combined Effect of Concentration and Temperature. Journal of Food Engineering, 68, 527- 533.

49. Higginsa Di Prestwidge, A., Stirling, D. and Yost, J. (2010) Forecasting Maturity of Green Peas: An Application of Neural Networks. Computers and Electronics in Agriculture, 70, 151-156.

50. Meng, X., Zhang, M. and Adhikari, B. (2012) Prediction of Storage Quality of Fresh-Cut Green Peppers Using Artificial Neural Network. International Journal of Food Science and Technology, 47, 1586-1592.

51. Torrecilla, J.S., Otero, L. and Sanz, P.D. (2004) A Neural Network Approach for Thermal/Pressure Food Processing. Journal of Food Engineering, 62, 89-95.

52. Nourbakhsh, H., Emam-Djomeh, Z., Omid, M., Mirsaeedghazi, H. and Moini, S. (2014) Prediction of Red Plum Juice Permeate Flux during Membrane Processing with ANN Optimized Using RSM. Computers and Electronics in Agricul- ture, 102, 1-9.

53. Mateo, F., Gadea-Gironés, R., Mateo, E.M. and Jiménez, M. (2011) Multilayer Perceptron Neural Networks and Radial-Basis Function Networks as Tools to Forecast Accumulation of Deoxynivalenol in Barley Seeds Contaminated with Fusarium culmorum. Food Control, 22, 88-95.

54. Adhikari, B. and Jindal, V.K. (2000) Artificial Neural Networks: A New Tool for Prediction of Pressure Drops of Non- Newtonian Fluid Foods through Tubes. Journal of Food Engineering, 46, 43-51.

55. Razmi-Rad, E., Ghanbarzadeh, B., Mousavi, S.M., Emam-Djomeh, Z. and Khazae, J. (2007) Prediction of Rheological Properties of Iranian Bread Dough from Chemical Composition of Wheat Flour by Using Artificial Neural Networks. Journal of Food Engineering, 81, 728-734.

56. Abbasi, H., Mahdi, S., Ardabili, S., Emam-djomeh, Z., Mohammadifar, M., Zekri, M. and Aghagholizadeh, A. (2012) Prediction of Extensograph Properties of Wheat-Flour Dough: Artificial Neural Networks and a Genetic Algorithm Approach. Journal of Texture Studies, 43, 326-337.

57. Curvelo, J.C., Dias, G., Rrodrigues de souza, R. and Tambourgi, E. (2010) Applying of Neural Network on the Wine Sensorial Analysis from Barbados Cherry. Journal of Food Process Engineering, 33, 365-378.

58. Rojas Naccha, J. and Vásquez Villalobos, V. (2012) Prediction by Artificial Neural Networks (ANN) of the Diffusivity, Mass, Moisture, Volume and Solids on Osmotically Dehydrated Yacon (Smallantus sonchifolius). Scientia Agrope- cuaria, 3, 201-214.

59. Nakano, K., Kurata, K. and Kaneko, M. (1991) Effects of Alcohol Treatment on Banana Ripening and Application of Image Processing to Colour Grade Measuring. Journal Society Agriculture Structure, 21, 203-208.

60. Nakano, K. (1997) Application of Neural Networks to the Color Grading of Apples. Computers and Electronics in Agriculture, 18, 105-116.

61. Millán, F. and Gil, L. (2011) Design and Optimization of an Artificial Neural Network for Modeling Syrup Production Process from Defatted Corn Germ. Annals of the Metropolitan University, 11, 177-190.

62. Torrecilla, J.S., Otero, L. and Sanz, P.D. (2005) Artificial Neural Networks: A Promising Tool to Design and Optimize High-Pressure Food Processes. Journal of Food Engineering, 69, 299-306.

63. Torrecilla, J.S., Otero, L. and Sanz, P.D. (2007) Optimization of an Artificial Neural Network for Thermal/Pressure Food Processing: Evaluation of Training Algorithms. Computers and Electronics in Agriculture, 56, 101-110.

64. Hernández, J.A. (2009) Optimum Operating Conditions for Heat and Mass Transfer in Foodstuffs Drying by Means of Neural Network Inverse. Food Control, 20, 435-438.

65. Morimoto, T., De Baerdemaeker, J. and Hashimoto, Y. (1997) Optimization of Heat Treatment for Fruit during Storage Using Neural Networks and Genetic Algorithms. Computers and Electronics in Agriculture, 19, 87-101.

66. Morimoto, T., De Baerdemaeker, J. and Hashimoto, Y. (1997) An Intelligent Approach for Optimal Control of Fruit- Storage Process Using Neural Networks and Genetic Algorithms. Computers and Electronics in Agriculture, 18, 205- 224.

67. Balasubramania, S., Panigrahi, S., Loguec, C.M., Gub, H. and Marchello, M. (2009) Neural Networks-Integrated Metal Oxide-Based Artificial Olfactory System for Meat Spoilage Identification. Journal of Food Engineering, 91, 91-98.

68. Peres, A.M., Baptista, P., Malheiro, R., Dias, L.G., Bento, A. and Pereira, J.A. (2011) Chemometric Classification of Several Olive Cultivars from Trás-os-Montes Region (Northeast of Portugal) Using Artificial Neural Networks. Che- mometrics and Intelligent Laboratory Systems, 105, 65-73.

69. Furferi, R., Carfagni, M. and Daou, M. (2008) Real-Time Estimation of Olive Oil Quality Parameters: A Combined Approach Based on ANNs and Machine Vision. New Aspects of Computers Part: II. Proceedings of the 12th WSEAS International Conference on Computers, Iraklion, 23-25 July 2008, 387-392.

70. Ram, T., Wiesman, Z., Parmet, I. and Edan, Y. (2010) Olive Oil Content Prediction Models Based on Image Proces- sing. Biosystems Engineering, 105, 221-232.

71. García-González, D.L., Luna, G., Morales, M.T. and Aparicio, R. (2009) Stepwise Geographical Traceability of Virgin Olive Oils by Chemical Profiles Using Artificial Neural Network Models. European Journal of Lipid Science and Technology, 111, 1003-1013.

72. Torrecilla, J.S., Cancilla, J., Matute, G. and Díaz Rodríguez, P. (2013) Neural Network Models to Classify Olive Oils within the Protected Denomination of Origin Framework. International Journal of Food Science and Technology, 48, 2528-2534.

73. Marini, F., Magrí, A.L., Bucci, R. and Magrí, A.D. (2007) Use of Different Artificial Neural Networks to Resolve Binary Blends of Monocultivar Italian Olive Oils. Analytica Chimica Acta, 599, 232-240.

74. Zupan, J., Novic, M., Li, X. and Gasteiger, J. (1994) Classification of Multicomponent Analytical Data of Olive Oils Using Different Neural Networks. Analytica Chimica Acta, 292, 219-234.

75. Furferi, R., Carfagni, M. and Daou, M. (2007) Artificial Neural Network Software for Real Time Estimation of Olive Oil Qualitative Parameters during Continuous Extraction. Computers and Electronics in Agriculture, 55, 115-131.

76. Furferi, R., Carfagni, M. and Daou, M. (2008) Real-Time Estimation of Peroxides and Acidity Level of Extra-Virgin Olive Oil: An Integrated Approach. International Journal of Mathematics and Computers in Simulation, 2, 206-214.

77. Torrecilla, J.S., Mena, M.L., Yáñez-Sedeño, P. and García, J. (2007) Application of Artificial Neural Network to the Determination of Phenolic Compounds in Olive Oil Mill Wastewater. Journal of Food Engineering, 81, 544-552

78. Torrecilla, J.S., Mena, M.L., Yáñez-Sedeño, P. and García, J. (2008) Field Determination of Phenolic Compounds in Olive Oil Mill Wastewater by Artificial Neural Network. Biochemical Engineering Journal, 38, 171-179.

79. Bordons, C. and Cueli, J. (2004) Predictive Controller with Estimation of Measurable Disturbances. Application to an Olive Oil Mill. Journal of Process Control, 14, 305-315.

80. Jiménez, A., Beltrán, G., Aguilera, M.P. and Uceda, M. (2008) A Sensor-Software Based on Artificial Neural Network for the Optimization of the Olive Oil Elaboration Process. Sensors and Actuators B: Chemical, 129, 985-990.

CHAPTER 3

Real-time biomimetic Central Pattern Generators in an FPGA for hybrid experiments

Matthieu Ambroise[1], **Timothée Levi**[1*],**Sébastien Joucla**[2], **Blaise Yvert**[2] and **Sylvain Saïghi**[1]

[1]Laboratoire IMS, UMR Centre National de la Recherche Scientifique, University of Bordeaux, Talence, France

[2]Laboratoire INCIA (Institute for Cognitive and Integrative Neuroscience), UMR Centre National de la Recherche Scientifique, University of Bordeaux, Talence, France

This investigation of the leech heartbeat neural network system led to the development of a low resources, real-time, biomimetic digital hardware for use in hybrid experiments. The leech heartbeat neural network is one of the simplest central pattern generators (CPG). In biology, CPG provide the rhythmic bursts of spikes that form the basis for all muscle contraction orders (heartbeat) and locomotion (walking, running, etc.). The leech neural network system was previously investigated and this CPG formalized in the Hodgkin–Huxley neural model (HH), the most complex devised to date. However, the resources required for a neural model are proportional to its complexity. In response to this issue, this article describes a biomimetic implementation of a network of 240 CPGs in an FPGA (Field Programmable Gate Array), using a simple model (Izhikevich) and proposes a new synapse model: activity-dependent depression synapse. The network implementation architecture operates on a single computation core. This digital system works in real-time, requires few resources, and has the same bursting activity behavior as the complex model. The implementation of this CPG was initially validated by comparing it with a simulation of the complex model. Its activity was then matched with pharmacological data from the rat spinal cord activity. This digital system opens the way for future hybrid experiments and represents an important step toward hybridization of biological tissue and artificial neural networks. This CPG network is also likely to be useful for mimicking the locomotion activity of various animals and developing hybrid experiments for neuroprosthesis development.

1. INTRODUCTION

Millions of people worldwide are affected by neurological disorders which disrupt connections between brain and body, causing paralysis or affecting cognitive capabilities. The number is likely to increase over the next few years and current assistive technology is still limited. In recent decades, extensive research has been devoted to Brain-Machine Interfaces (BMIs) and neuroprosthesis in general (Hochberg et al., 2006, 2012; Nicolelis and Lebedev, 2009), working toward effective treatment for these disabilities. The development of these devices has had and, hopefully, will continue to have a profound social impact on these patients' quality of life. These prostheses are designed on the basis of our knowledge of interactions with neuronal cell assemblies, taking into account the intrinsic spontaneous activity of neuronal networks and understanding how to stimulate them into a desired state or produce a specific behavior. The long-term goal of replacing damaged neural networks with artificial devices also requires the development of neural network models that match the recorded electrophysiological patterns and are capable of producing the correct stimulation patterns to restore the desired function. The hardware set-up used to interface the biological component is a Spiking Neural Network (SNN) system implementing biologically realistic neural network models, ranging from the electrophysiological properties of a single neuron to large-scale neural networks.

Our study describes the development of a neuromorphic hardware device containing a network of real-time biomimetic Central Pattern Generators (CPG). The main goal of this research is to create artificial CPGs that will be connected to *ex vivo* spinal cord of rats and guinea pigs, thus achieving one main objective of the Brainbow European project (Brainbow, 2012) toward hybridization. Hardware-based SNN systems were developed for hybrid experiments with biological neurons and the description of those pioneer platforms was reported in the literature (Jung et al., 2001; Le Masson et al., 2002; Vogelstein et al., 2006). The Brainbow project will go further by using a large-scale neural network instead of few neurons to substitute the functions of a biological sub-network. The final goal is the development of a new generation of neuro-prostheses capable to restore the lost communication between neuronal circuitries.

Locomotion is one of the most basic abilities of animals. Neurobiologists have established that locomotion results from the activity of half-center oscillators that provides alternating bursts. The first half-center oscillator was proposed by Brown (1914). Pools of interneurons control flexor and extensor motor neurons with reciprocal inhibitory connections. Most rhythmic movements are programmed by central pattern-generating networks consisting of neural oscillators (Marder and Bucher, 2001; Ijspeert, 2008). CPGs are neural networks capable of producing rhythmic patterned outputs without rhythmic sensory or central input. CPGs underlie the production of most rhythmic motor patterns and have been extensively studied as models of neural network function (Hooper, 2000). Half-center oscillators control swimming in xenopus, salamander (Ijspeert et al., 2007), and lamprey (Cohen et al., 1992), as well as leech heartbeat (Cymbalyuk et al., 2002), as described in numerous publications. One key article on modeling the leech heartbeat system is Hill et al. (2001), where the Hodgkin–Huxley formalism is used to reproduce the CPG.

The main novelty of this research was to implement the leech heartbeat system neural network with minimum resources while maintaining its biomimetic activity. Indeed, the final application is a hybrid experiment that requires spike detection, spike sorting, and micro-electrode stimulation. All of these modules are implemented in the same digital board. To achieve this, the Hill et al. (2001) model and results were reproduced using a simpler model (Izhikevich, 2004), implemented in an FPGA (Field Programmable Gate Array) board. This digital board made it possible to design a frugal, real-time network of several CPGs (in this case, a network of 240 CPGs implemented on a Spartan6 FPGA board). For instance, this CPG network is capable of mimicking the activity of a salamander, which requires 40 CPGs (Ijspeert, 2001), or developing hybrid experiments (Le Masson et al., 2002) for neuroprosthesis development (Brainbow, 2012).

The first part of this article describes the biological leech heartbeat system, based on one segmental CPG. The next section focuses on choosing a frugal neuron model to match the same biological behavior. The following section explains the topology of a single neuron and its implementation in the hardware, followed by its extension to a neuron computation core for increasing the size of the neural network. The next stage was to develop a new synaptic model reproducing activity-dependent depression phenomena to fit the biological activity of a leech heartbeat. The architecture of this digital system is then described in full, including the various blocks. Finally, the system was used to design a CPG network, validated by comparing our measurements with *ex vivo* rat spinal cord locomotion results following pharmacological stimulation.

2. MATERIALS AND METHODS

2.1 Description of the Leech Biological Heartbeat System

All leech heartbeat studies agree that the CPG (Figure 1C) responsible for this activity (Figure 1A) requires few neurons, making it an ideal candidate system for elucidating the various biomechanisms governing CPG behavior.

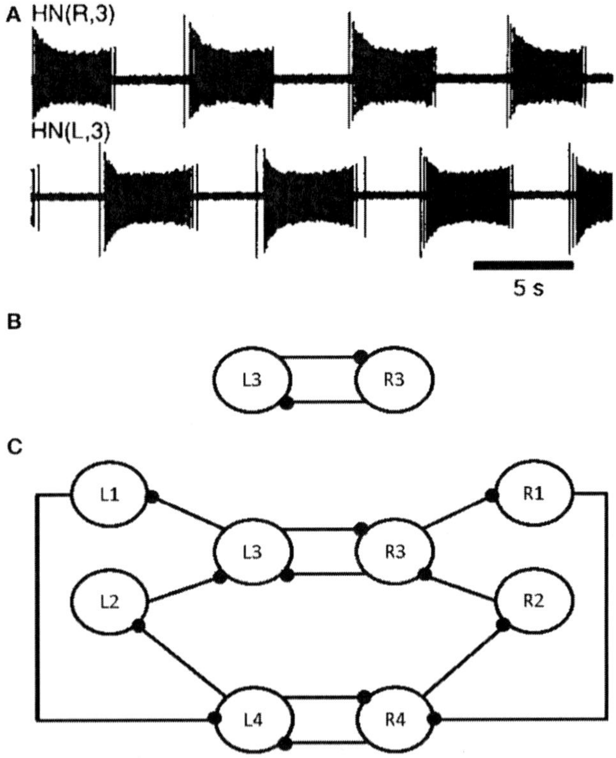

Figure 1. Electrical activity of the leech heartbeat system and diagram of the CPG. Neuron cell bodies are represented by circles. Axons and neurite processes are represented by lines. Inhibitory chemical synapses are represented by small filled dots. **(A)** Electrical activity of two heart interneurons recorded extracellularly from a chain of ganglia (Hill et al., 2001). **(B)** A diagram of the elemental oscillator in the leech heartbeat system. **(C)** A diagram of the segmental oscillator in the leech heartbeat system, including two elemental oscillators, L3/R3 and L4/R4, and two pairs of coordination neurons, L1/R1 and L2/R2.

Modeling studies indicate that the burst duration of a leech heart interneuron in an elemental oscillator is regulated by the interneuron itself and by the opposite interneuron (see L3 and R3 in Figure 1B) (Calabrese, 1995; Nadim et al., 1995; Olsen et al., 1995; Hill et al., 2001; Jezzini et al., 2004; Norris et al., 2007). Figure 1A shows the electrical activity in the leech heartbeat system from extracellular recordings. The pair of neurons is known as an elemental oscillator (Figure 1B), i.e., the smallest unit that capable of producing robust oscillations under normal conditions. These neurons oscillate in alternation with a period of about 10–12 s (Krahl and Zerbst-Boroffka, 1983; Calabrese et al., 1989; Olsen and Calabrese, 1996) demonstrated that the synaptic connections among interneurons and from interneurons to motor neurons were inhibitory. The synaptic interaction between reciprocally inhibitory heart interneurons consists of a graded component in addition to spike-mediated synaptic transmissions (Angstadt and Calabrese, 1991). This kind of synapse is really difficult to implement in hardware as it contains sigmoid functions, differential equations, memory of last spikes, and so on. A description of our synapse model reproducing the same behavior is included below.

Nadim et al. (1995) and Olsen et al. (1995) developed a biophysical model of a pair of reciprocally inhibitory interneurons in the leech heartbeat system. This model included synaptic ionic currents based on voltage-clamp data. Synaptic transmissions between the interneurons consist of spike-mediated and graded synaptic currents. The Hill et al. (2001) model was derived from a previous two-cell, elemental oscillator model (Nadim et al., 1995) by incorporating intrinsic and synaptic current modifications based on the results of a realistic waveform voltage-clamp study (Olsen and Calabrese, 1996). This new, segmental oscillator model behaves more similarly to biological systems. Figure 1C shows a model of the system. The real-time digital segmental oscillator model design will be based on this architecture. The next part will describe the system modeling the leech heartbeat with the goal of implementing it in hardware. The leech heartbeat CPG was chosen for the long duration of the burst.

2.2 System Modeling for Hardware Implementation

2.2.1 State of art

Some previous studies used silicon neurons (Indiveri et al., 2011) to simulate the leech heartbeat system (Simoni et al., 2004; Simoni and DeWeerth, 2007). Sorensen et al. (2004) created a hybrid system of a heart interneuron and a silicon neuron. The silicon neuron provides real-time operation and implements a version of the Hodgkin–Huxley formalism (Hodgkin and Huxley, 1952). However, due to the complexity of the model, it was only possible to use a small number of silicon neurons and, therefore, only one CPG. This study describes the same results using a large CPG network (240 CPGs on a Spartan6 FPGA board), in preparation for future hybrid experiments with different CPGs. For instance, in the salamander model (Ijspeert, 2001), the body CPG consists of 40 interconnected segmental networks.

When a silicon neuron and heart interneuron are connected with reciprocal inhibitory synapses of appropriate strength, they form a hybrid elemental oscillator that produces oscillations remarkably similar to those seen in the living system. Olypher et al. (2006) described the control of burst duration in heart interneurons using a hybrid system, where a living, pharmacologically-isolated, heart interneuron was connected with artificial synapses to a model heart interneuron running in real-time (software). Using an FPGA board will make it possible to operate in real time using a large number of neurons, together with customized systems for various applications (hybrid experiments).

A few studies (Torres-Huitzil and Girau, 2008; Rice et al., 2009; Serrano-Gotarredona et al., 2009;Barron-Zambrano et al., 2010; Barron-Zambrano and Torres-Huitzil, 2013) reported on CPG in FPGA for robotic applications. These studies used simple neuron-models and were more bio-inspired than biomimetic. Guerrero-Riberas et al. (2006) implemented a network of LIF neurons with synapses and plasticity, but not in biological time, so it was impossible to perform hybrid experiments. While multi-legged robots need CPG to move or coordinate their movements, they implement an Amari–Hopfield CPG (Amari, 1972) or basic CPGs (Van Der Pol, 1928), modeled as non-linear oscillators. Those models provide sinusoidal oscillations that are not biorealistic. The ultimate goal of these studies is to create a robot that mimics biological behavior but these systems cannot be used for hybrid experiments. Analog hardware has also been implemented (Linares-Barranco et al., 1993; Still and Tilden, 1998; Lewis et al., 2001; Nakada, 2003; Still et al., 2006; Lee et al., 2007; Wijekoon and Dudek, 2008). However, it is very difficult to tune analog circuits due to parameter mismatch. For these works, they either design bio-inspired oscillators for creating CPG or implement few biomimetic neurons.

2.2.2 Choice and presentation of the izhikevich model

In designing a SNN, the first step is the choice of a biologically realistic model. Indeed, a mathematical model based differential equations is capable of reproducing a behavior quite similar to that of a biological cell. The choice of model was based on two criteria: the family of neurons able to be reproduced and the number of equations. These criteria were used to compare several models, including the Leaky Integrate and Fire model (LIF)

(Indiveri, 2007), the Hodgkin–Huxley model (HH), and the Izhikevich model (IZH).

Hill et al. (2001) used the HH to reproduce the leech heartbeat system with eight neurons (Figure 1C). From the equations defined in this paper, it was established that the eight neurons in the heartbeat leech behaved like regular spiking ones (RS). Indeed, this model was composed of nine voltage-dependent currents with different calcium conductances.

The HH model reproduces all types of neurons with good accuracy (spike timing and shape). Its main drawbacks are the large number of parameters and the equations required. In the heartbeat network, the main focus is on excitatory neurons, like RS. The HH model required 32 parameters for an RS and 26 for a fast-spiking neuron (FS) (Grassia et al., 2011). Furthermore, simulating an RS neuron required four ionic channels (dynamics of potassium and sodium ions, leak current, and slow potassium). In contrast, LIF only involves two equations but is only capable of simulating a few types of neurons.

The IZH represents a good solution, as it is based on two equations and is capable of reproducing many different families of neurons by changing four parameters. Furthermore, according to Izhikevich (2004), this model is resource-frugal, a key advantage when the aim is to design a large CPG network embedded in the same board as other modules required for hybrid experiments (spike detection, spike sorting, stimulation, etc.).

The IZH model depends on four parameters, which make it possible to reproduce the spiking and bursting behavior of specific types of cortical neurons. From a mathematical standpoint, the model is described by a two-dimensional system of ordinary differential equations (Izhikevich, 2003):

$$\frac{dv}{dt} = 0.04v^2 + 5v + 140 - u + I_{\mathrm{Izh}} \tag{1}$$

$$\frac{du}{dt} = a(bv - u) \tag{2}$$

with the after-spike resetting conditions:

$$\text{if} \quad v \geq 30\,\mathrm{mV} \Rightarrow \begin{cases} v \leftarrow c \\ u \leftarrow u + d \end{cases} \tag{3}$$

In equation (3), v is the membrane potential of the neuron, u is a membrane recovery variable, which takes into account the activation of potassium and inactivation of sodium channels, and I_{Izh} describes the input current from other neurons.

The IZH model was chosen to emulate the behavior of the excitatory cells for its simplicity and its capacity to implement various families of neurons. The next step was to determine the network system topology. The next section describes the design of one neuron and its extension to a neuron computation core, then the different synapse models implemented, and, finally, the topology of the network.

2.3 System Topology

2.3.1 Topology of one neuron core: architecture and implementation

In order to make the Izhikevich neural network more biomimetic, the I_{Izh} current from equation (1) was split into three: I_{bias}, I_{exc}, and I_{inh}. I_{bias} is the biasing current, I_{exc} is the positive contribution due to excitatory synapses, and I_{inh} is the negative contribution of inhibitory synapses. Those currents will be detailed in Synapse Model. As suggested in Cassidy and Andreou (2008), equation (1) was multiplied by 0.78125 to make it easier to implement on a digital board. These modifications gave (4), where the $u$coefficient is still 1 thanks to I_{bias} current.

$$\frac{dv}{dt} = {}^{1}\!/_{32}\, v^2 + 4v + 109.375 - u + I_{bias} + I_{exc} + I_{inh}$$

$$\frac{du}{dt} = a \cdot (bv - u) \tag{4}$$

Moreover, $\frac{dv}{dt} = \frac{v[n+1]-v[n]}{\Delta t}$ and, as the time step of the IZH model is equal to one millisecond ($\Delta t = 1$):

$$v[n+1] = {}^{1}\!/_{32}\, v[n]^2 + 5v[n] + 109.375 - u[n] + I_{bias}[n] + I_{exc}[n] + I_{inh}[n]$$

$$u[n+1] = u[n] + a.(b.v[n] - u[n]) \tag{5}$$

One neuron was implemented on the FPGA board according to these equations and specifications. This neuron was then extended into a neuron computation core that updated the u and v values of all neurons in the network. Consequently, the neuron implementation became a neuron computation core. For instance, around 2000 independent neurons could be implemented on our digital board. In this system, the type of neuron is defined by the four Izhikevich parameters: a, b, c, and d from equations (2) and (3). Moreover, the state of a neuron is defined by values u and v, and the three current values. Those 9 values were saved in a RAM for use in the next millisecond in the step computation. By extension, the same process can be used for every neuron in the network.

Each *u* and *v* computation step is run in parallel, using two pipelines based on the architecture presented in [9]. The topology is presented in Figure 2. All parameters from equations (2), (3), and (4), as well as the*u* and *v* values used in the computation are synchronized in one cycle before going through the pipelines (not shown in Figure 2).

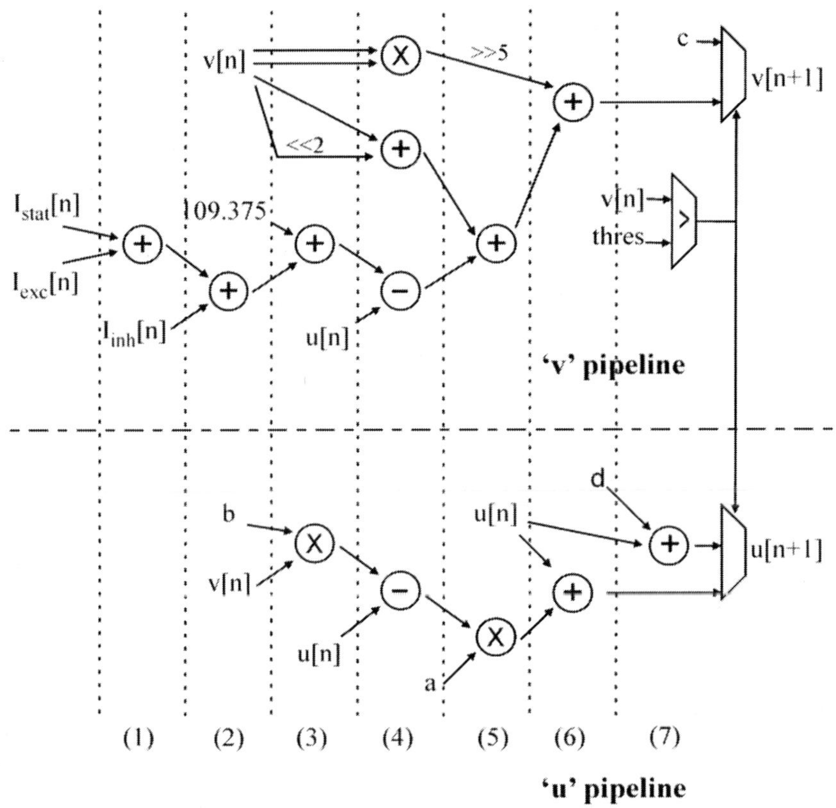

Figure 2. Architecture of the "u" and "v" pipelines in the neural computation core. The computation cycles are separated by dotted lines.

To resume, each neuron is represented by one "v" and "u" value, four Izhikevich coefficients (a, b, c, and d), and three currents (I_{bias}, I_{exc}, and I_{inh}).

I_{exc}, I_{inh}, and I_{bias} are added in two cycles at the beginning of the "v" pipeline, while the "u" pipeline is still inactive (steps 1 and 2). The current sum is added to the constant *109.375* and at the same time as the first multiplication (step 3). By multiplexing operands, the same multiplier is used for the following multiplications in different computation cycles. In step 4, v^2 is obtained by another multiplication. A simple two-bit shift makes it possible to obtain $4v$ and add it to v. At the same time, u is used in two subtractions. Step 5 consists of a 5-bit shift to obtain $(1/32)v^2$, an addition, and the last multiplication. In step 6, the computation of both u and v is completed. In the next step, the v value is tested against the threshold to determine whether the neuron has emitted a spike or not. This test gives the next u and $v$values for this neuron to be stored in the RAM.

An RS neuron with $a =$ 0.002, $b =$ 0.2, $c =$ −65, and $d =$ 8 was used to implement the CPG.

Once the neuron computation core was implemented, the synaptic model was chosen and implemented.

2.3.2 Synapse model

A network is defined by a group of neurons and a group of synapses. Once the neuron model had been chosen, it was obviously necessary to choose a synapse model. Like the neuron model, this model had to be biomimetic but frugal in its use of resources. In biology, synapses are described as links between neurons that transmit different types of synaptic currents to each other to either excite or inhibit neuron activity. In our implementation, a synaptic weight (W_{syn}) was added to the synaptic current. When W_{syn} was positive, it was added to I_{exc} (excitatory synaptic current) and when W_{syn} was negative, it was added to I_{inh}(inhibitory synaptic current).

Thanks to AMPA and GABA effects, all synaptic current excitations or inhibitions, respectively decay exponentially (Ben-Ari et al., 1997). AMPA is an excitatory neurotransmitter that depolarizes the neuron membrane whereas GABA is an inhibitory neurotransmitter with a hyperpolarizing effect. Depolarization or hyperpolarization are represented by a positive or negative contribution on the synaptic current.

The synaptic current I_{syn} was implemented with a time constant τ_{syn} for the exponential decay, as follows:

$$I_{syn}(t) = -\tau_{syn} \cdot I'_{syn}(t) = -\tau_{syn} \cdot \frac{I_{syn}(t+T) - I_{syn}(t)}{T} \tag{6}$$

$$I_{syn}(t+T) = \left(1 - \frac{T}{\tau_{syn}}\right) \cdot I_{syn}(t) \tag{7}$$

When computation step T equals one millisecond and τ_{syn} is in ms:

$$I_{syn}(t+1) = \left(1 - \frac{1}{\tau_{syn}}\right) \cdot I_{syn}(t) \tag{8}$$

$$I_{syn}[n+1] = I_{syn}[n] - \frac{1}{\tau_{syn}} \cdot I_{syn}[n] \tag{9}$$

Adding the synaptic weight to the synaptic current, the new equation is:

$$I_{syn}[n+1] = I_{syn}[n] - \frac{1}{\tau_{syn}} \cdot I_{syn}[n] + W_{syn}[n] \tag{10}$$

The synaptic computation core implementation is based on the same principle as the neuron computation core. However, this model is not adequate to fit biological data. It was, therefore, decided to implement an activity-dependent depression, where the new synaptic weight, W_s, was dependent on W_{syn}.

2.3.3 Activity-dependent depression

As the synaptic behavior described in Hill et al. (2001) requires too many resources to be implemented on FPGA, the method chosen to fit overall biological behavior was activity-dependent depression (Tabak et al., 2000). Activity-dependent depression of synapses is another biological phenomenon consisting of reducing a synaptic weight after a spike. In biology, each synapse contribution is provided by a synaptic vesicle. These vesicles contain ions that empty out at each spike and then regenerate, following an exponential rule. According to Matsuoka (1987), four methods provide a stable rhythm within a network (regulation of stimulus intensity, change in input, alteration of stimuli, and change in synaptic weight). The phenomenon, known as activity-dependent depression changes the synaptic weight depending on the activity of the network.

This phenomenon has been reported in neurobiology literature but no model had been devised. This paper proposes a model of this activity-dependent depression that was implemented in digital hardware to improving our CPG network.

As previously explained, each time a neuron emits a spike; the synapse adds a synaptic weight (W_{syn}) to the synaptic current. At the same time, the factor (δ_{syn}) indicating the level of depression on a synaptic weight increases. Furthermore, δ_{syn} regulates W_{syn}. The value of δ_{syn} is between 0 and 1. Consequently, when δ_{syn} equals zero there is no depression on W_{syn} and when δ_{syn} equals one there is maximum depression on W_{syn} and the synapse is exhausted.

W_s was used instead of W_{syn} as the synaptic weight for each synapse. Then, according to the activity-dependent depression effect, when there is a spike, W_s is added to the synaptic current:

$$W_s[n] = W_{syn} - \delta_{syn}[n] \cdot W_{syn} \tag{11}$$

The other effect of activity-dependent depression is to increase δ_{syn} after each spike, thanks to the percentage dissipation (P).

$$\delta_{syn}[n+1] = \delta_{syn}[n] + P \cdot (1 - \delta_{syn}[n]) \tag{12}$$

The regeneration or reloading of synaptic vesicles is represented by δ_{syn} decreasing to zero. Thus, δ_{syn}decays exponentially when no spike is emitted. So, using the method described in Synapse Model:

$$\delta_{syn}[n+1] = \delta_{syn}[n] - \frac{1}{\tau_{reg}} \cdot \delta_{syn}[n] \tag{13}$$

To summarize, all synapses are now represented by (12), (13), (14) and:

$$I_{syn}[n+1] = I_{syn}[n] - \frac{1}{\tau_{syn}} \cdot I_{syn}[n] + W_s[n] \tag{14}$$

The main parameters are: synaptic weight, W_{syn}; level of depression, δ_{syn}; and percentage dissipation, P. All these parameters are stored in the RAM on the digital board. Furthermore, this computation required greater precision due to the sensitivity of the parameters. The 26-bit signed fixed representation chosen had 1-bit for the sign, 9-bits for the whole numbers, and 16 for the decimals.

Once the neuron and synapse models had been designed, it was possible to develop the neural network topology.

2.3.4 Network topology

Three elementary blocks. The architecture was based on three main blocks: the neuron implemented (or neuron computation core), a synapse, and the RAM. The connectivity between those blocks is shown in Figure 3.

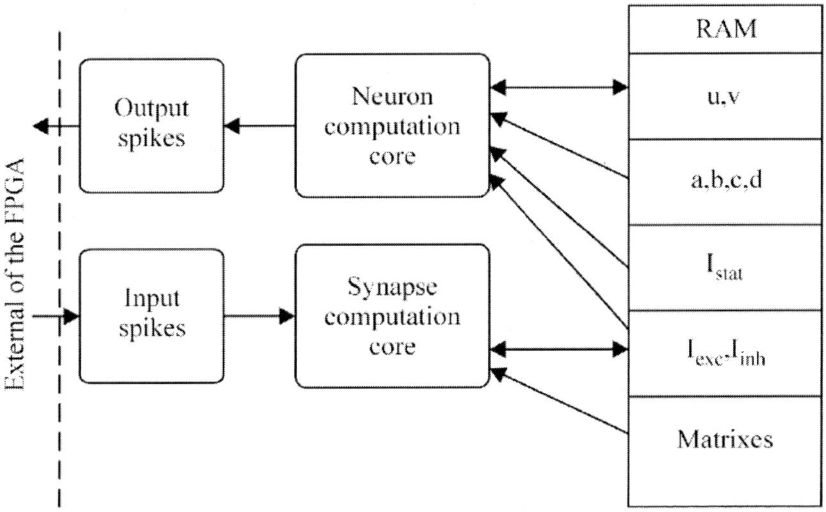

Figure 3. Global architecture of the spiking neural network.

So far, the neuron computation core can update the state ("u" and "v" variables) of each neuron. In the digital network, the role of the synapse is to update all synaptic currents and weights related to the activity of all neurons, so the synapse block exhibits two behaviors (spiking or not). These two behaviors are summarized in Table 1.

The IZH model has a time step of one millisecond, so the other computation was synchronized with this time step. The new values of u and v, the exponential decay of I_{syn}, and the new values of each synaptic current are computed in the same millisecond.

Moreover, a biological neural network is composed of *Nn* neurons and *Ns* synapses. To define which neuron is connected to which and with which kind of synapse (excitatory or inhibitory), the network is described using two matrixes: connectivity and synaptic weight (see Figure 3). To save RAM, both matrixes are implemented as sparse matrixes with *Nn* lines. The *i*th line in the connectivity matrix corresponds to the connectivity of presynaptic neuron N_i to the other neurons. The synapses are identified by the postsynaptic neuron addresses. For example, the connection to neuron N_j is identified by the number j on the ith line. In the worst case, each neuron is connected to itself and all the others, giving Nncolumns. Each matrix line ends with a virtual neuron (address $Nn + 1$). This implementation is not optimum for the worst case, but the gain is significant for biologically plausible networks, where the total number of synapses is at least four times smaller. Marom and Shahaf (2002) and Garofalo et al. (2009)estimated the average connectivity level of neural networks at their mature phase each neuron is mono-synaptically connected to 10–30% of all the other neurons.

Table 1. Description of the equations for synaptic currents and activity-dependent depression.

When a spike is emitted	**When no spike is emitted**
Synaptic current	Synaptic current
$I_{exc}[n+1] = I_{exc}[n] + W_s[n]$ or $I_{inh}[n+1] = I_{inh}[n] + W_s[n]$	$I_{exc}[n+1] = I_{exc}[n]$ and $I_{inh}[n+1] = I_{inh}[n]$
Activity-dependent depression	Activity-dependent depression
$\delta_{syn}[n+1] = \delta_{syn}[n] + P(1 - \delta_{syn}[n])$	$\delta_{syn}[n+1] = \delta_{syn}[n] - \dfrac{\delta_{syn}[n]}{\tau_{reg}}$

There is a direct link between the matrixes: the synaptic weight matrix is the same size as the connectivity matrix, i.e., the same number of lines and columns, with the virtual neurons in the same position (Figure4). The connectivity between two neurons described by the coordinates (k, l) in the connectivity matrix has the weight shown in box (k, l) in the synaptic weight matrix. A third matrix based on the same principle completes the system: the percentage efficiency matrix, which gives the percentage dissipation, P, of each synapse in a network, as defined in the previous section on activity-dependent depression. We will describe now the state machine of the neural network.

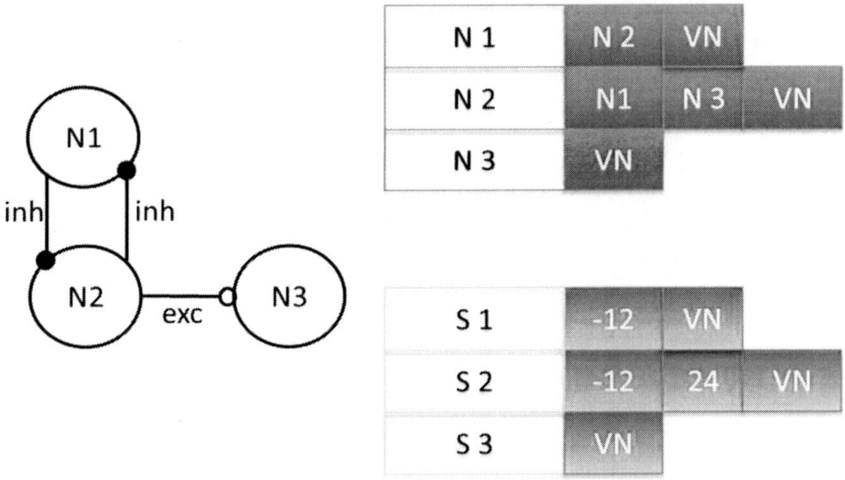

Figure 4. Example of matrix design depending on the neural network. Neuron 2 (N2) is connected by an inhibitory synapse to neuron 1 (N1) and by an excitatory synapse to neuron 3 (N3). Then, on line 2 in the connectivity matrix, N2 is connected to N1, N3, and a virtual neuron (VN), indicating the end of the connection. In the synaptic weight matrix, the synapses for neuron 2 (S2) have a negative weight for the inhibitory synapse and a positive weight for the excitatory synapse. Note the correspondence of its position in both matrixes.

Network machine states. The synaptic current is computed in three successive steps:

- EXT state: for closed-loop experiments, we implement this state in which external feedback can interact with the artificial neural network. This first state consists of using the synaptic block to update the synaptic current. In this case, presynaptic spikes are external events (see Figure 3), such as stimulation from biological neurons in the case of neuroprosthesis. This state makes it possible to stimulate each neuron.

- NEUR state: during this step, the neuron membrane ("*u*" and "*v*" from Figure 2) and all exponential decay values are computed in parallel.

- SYN state: the last step consists of updating the synaptic current to reflect the presynaptic spikes computed in the NEUR state. These updated current values are used in the EXT state during the next cycle.

The EXT, NEUR, and SYN states must be completed within a one millisecond time step. If the computation of all three states is completed in less than 1 ms, an IDLE state is implemented until the end of the cycle.

Moreover, the blocks (neuron computation core and synapse computation core) described in Figure 3 are multiplexed in time to reduce the implementation area in large-scale neural networks.

Our architecture has two main limits: the number of available cycles (Nc) in one millisecond and the size of the RAM used to save all parameters. Two equations derived from these limits determine the maximum size of the implementable neural network, in terms of number of neuron (Nn) and synapses (Ns).

In the EXT state, all synaptic currents are updated in 10 cycles for each neuron, i.e., 10· Nn cycles. Each neuron requires 11 cycles to compute the NEUR state, i.e., 11· Nn cycles. The synaptic current update during state SYN requires 10 cycles per synapse, i.e., 10· Ns. Figure 2 describes 7 cycles for the neuron computation core, but 4 more cycles are required to read and save the various parameters in the RAM.

This leads to the following equation for computing the maximum number of neurons that may be implemented, depending on the number of cycles available:

$$10 \cdot Nn + 11 \cdot Nn + 10 \cdot Ns \leq Nc \tag{15}$$

Having built all the component parts of this real-time, biomimetic digital system, it was possible to validate it by several experiments, presented in the following section.

3. RESULTS

A CPG is defined by the number of neurons and the families of neurons and synapses. The leech heartbeat neural network was simulated by an appropriate CPG configuration.

Hill et al. (2001) presented an elemental oscillator, based on two excitatory neurons linked by inhibitory synapses. A segmental oscillator may consist of 4–10 neurons. A two-neuron network (elemental oscillator from Figure 1B) was chosen to validate our topology, followed by an eight-neuron neural network (segmental oscillator from Figure 1C). The activity of our system was then compared with that of an *ex vivo* rat spinal cord, stimulated with pharmacological solutions. It was also demonstrated that the period of bursting activity could be modified depending on one parameter. This will be useful in future closed-loop hybrid experiments.

Biological CPGs provide specifications concerning their behavior. Indeed, their activity is characterized by periodic long bursts (lasting many seconds). Each burst begins by a quick rise in spike frequency to a maximum and ends with a low final spike frequency.

3.1 Comparison of Biological/Digital Elemental Oscillator

The first example of a CPG was the elemental oscillator (with only two neurons). To reproduce activity accurately, it was necessary to obtain the following values: τ_{ampa} (time constant of the inhibitory synaptic current exponential decay), τ_{reg} (time constant of the recovery of synaptic vesicles), and P (percentage dissipation). These values will be the same for each synapse. The following values were chosen to match biological behavior: $\tau_{current} = 100$ ms and $\tau_{reg} = 4444$ ms (so $1/\tau_{current} = 0.01$ and $1/\tau_{reg} = 0.0002$). TheI_{bias} current was equal to 8 for both neurons. The synaptic weights are −5.1 and the percentages of dissipation are 1.49.

This model was validated by comparing its implementation with the complex model in Hill et al. (2001)(see Figure 5).

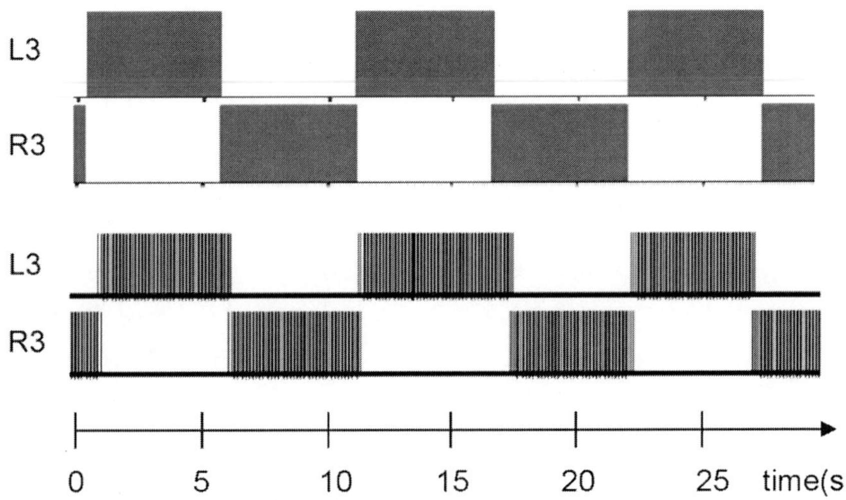

Figure 5. Comparison between elemental oscillator (Figure **1B**) bursting activity in the complex model simulated by scilab, as described in **Hill et al. (2001)** and the elemental oscillator presented above thanks to a logic analyzer. The time scale is the same.

In this case, the activity of one neuron inhibits the second neuron. Due to activity-dependent depression and the GABA effect, the inhibition ends and lets the second neuron fire again. In both cases (biological modeling system and digital system), the bursting activity was similar in terms of period and duty cycle, thus validating the simplified elemental oscillator with the complex one. The next step was to validate the segmental oscillator and compare its implementation with biological data.

3.2 Comparison of Biological/Digital Segmental Oscillator

Keeping the time constant, the biological behavior of the eight-neuron network was duplicated using the following parameters. This time, an eight-neuron CPG was implemented using the same values for $\tau_{current}$and τ_{reg} than as those used for the elemental oscillator. The use of 8 neurons made it possible to maintain the period without variation (see Table 2) by slowing down the two pairs of oscillators with coordination neurons (De Schutter, 2000).

Table 2. Comparison of burst characteristics in the two digital implementations and the biological system.

	Biological system (Hill et al., 2001)	**Elemental oscillator (digital)**	**Segmental oscillator (digital)**
Mean period	10–12 s	12.6 ± 1.4 s	11.2 ± 1 s
Mean duty cycle	57.2 ± 2.9%	54.7 ± 6%	46.1 ± 6%
Mean spike frequency	11.9 ± 2.1 Hz	12.1 ± 1 Hz	11.2 ± 1 Hz
Initial spike frequency	4.3 ± 0.7 Hz	8.5 ± 0.2 Hz	8.6 ± 0.4 Hz
Peak spike frequency	17.5 ± 3.2 Hz	13 ± 0 Hz	12.5 ± 0 Hz
Final spike frequency	5.8 ± 1.0 Hz	8.1 ± 0.2 Hz	9.3 ± 3 Hz

In Figures 1C, 6, L3/R3 and L4/R4 correspond to the two elemental oscillators and are coupled to the L1/R1 and L2/R2 coordination neurons. The connectivity between each neuron is following Figure 1C. The synaptic weights are −7 and the percentage of dissipation is 2.65.

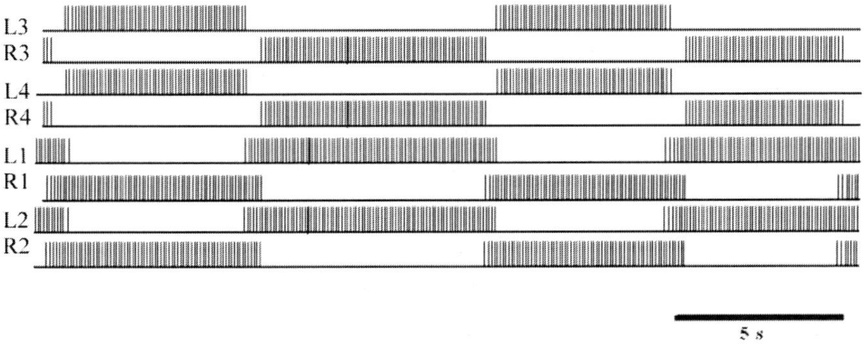

Figure 6. Logic analyzer measurements of the digital eight-neuron CPG. L3 and R3 show the activity of the first oscillator. L4 and R4 show the activity of the second oscillator. L1/R1 and L2/R2 are the coordination neurons.

The mean period, duty cycle, and variations in spike frequency depending on their position in the burst were measured to quantify the overlap of bursting activity (Table 2). The mean period of this digital implementation was similar to biological values. Note that the segmental oscillator exhibited less variation than the elemental system, thanks to its coordination neurons.

Also, in general, the spike frequency of our implementation was similar to that of the biological system. Due to our synapse model, the frequency reached a maximum in each spike burst but remained on a plateau instead of decreasing to the minimum frequency immediately. In the biological system, the behavior described is due to the enhancement and attenuation of variations in conductance. However, the IZH model does not include conductance, so it cannot be as biomimetic as the HH model. This highlights a weak point of the implementation presented here, but even the HH model, Hill et al. (2001) was unable to mimic this biological variation in spike frequency in a single burst. One discrepancy between the model and the biological system is that the initial and final spike frequencies of a burst were consistently lower in the biological system. In both implementations, the most inconvenient drawback was the variation in the duty cycle, explained by the stability of the IZH model. One perspective of this work to ensure stability is described in the discussion section.

These experiments validated the implementation of our elemental and segmental oscillators. This table also confirms that designing a biomimetic system was a good choice. Indeed, the variations of the duty cycle and the period for the bursting activity could not be reproduced by bio-inspired oscillators. The next step was to identify one parameter that would modify the bursting activity period, which would be useful in closed-loop applications.

3.3 Variation in the Mean Period Depending on One Parameter

A CPG is defined here by the number of neurons and the type synapses involved, the static currents of each neuron, the percentage dissipation, and the synaptic efficiency time constant.

Changing the synaptic efficiency time constant τ_{reg} modifies the period of each spike burst (Table 3). The variation in τ_{reg} affects the period and duration of each burst, as well as the duty cycle and the variability of these parameters: the greater the value of $1/\tau_{reg}$, the longer the mean period of bursting activity.

Table 3. Variation in the mean period depending on the τ_{reg} parameter.

$1/\tau_{reg}(ms^{-1})$	Mean period (s)
0.09	4.4 ± 1.6
0.15	7.2 ± 1.2
0.20	11.2 ± 1
0.22	12.9 ± 1.1

The possibility of modifying the period using a single parameter is very useful and was applied in a closed-loop hybrid experiment concerning locomotion behaviors.

3.4 FPGA Resources

Originally, a CPG consisted of 8 neurons and 12 synapses but 2 additional synapses per CPG were required to create a network of CPG, by connecting CPG to another one. Thus, each CPG consisted of 8 neurons and 14 synapses. In terms of cycles and available memory, this implementation was capable of running 240 CPGs on a Spartan 6 digital board [see equation (15) and Table 4]. The power consumption of one CPG is 8 mW and for CPGs is 20 mW. We could reduce it in the future by designing a custom ASIC. For neuroprosthesis application, the power consumption should be lower than 80 mW/cm2 chronic heat dissipation level considered to prevent tissue damage (Zumsteg et al., 2005).

Table 4. Resources required for one CPG on a Spartan 6 digital board.

Resources	Total available	Used for one CPG	Used for 240 CPGs
Slice FF's	184304	1,093 (0.6%)	1,459 (0.8%)
Slice LUT	92152	1,037 (1.2%)	1,756 (1.9%)
DSP48A1	180	10 (5.6%)	10 (5.6%)
RAMB16BWER	268	1	42 (756 kb)
Total RAM	4824 kb	9 kb (0.2%)	765 kb (16%)

3.5 Comparison with *Ex-Vivo* Rat Spinal Cord Results Using Pharmacological Stimulation

The final validation of this system consisted of comparing the CPG output with *ex vivo* physiological data obtained from the spinal cord of newborn rat [postnatal day (P)1–2]. Bursting locomotor-like activity was induced by bath-application of aCSF (artificial cerebrospinal fluid) mixed with N-methyl-DL-aspartate (NMA; 10 μM), serotonin (5HT; 5 μM), and dopamine (DA; 50 μm) (all purchased from Sigma-Aldrich, France).

For the elemental oscillator. Neuron N1 (corresponds to neuron L3 in Figure 1B) is connected to neuron N2 (corresponds to neuron R3 in Figure 1B) by an inhibitory synapse with a synaptic weight of −7 and a percentage dissipation of 12%. The I_{bias} current is equal to 7 for both neurons.

Figure 7 shows that the digital system fits the biological recordings of the newborn rat spinal cord. The period and duty cycle of the bursting activity are the same, confirming that the digital system was suitable for hybrid experiments. Instead of using pharmacological stimulation, the digital board will be used in the near future to create a hybrid experiment involving the *ex vivo* spinal cord and the digital CPGs. A closed-loop is also possible thanks to the possibility of changing the mean period of bursting activity by modifying a single parameter (τ_{reg}).

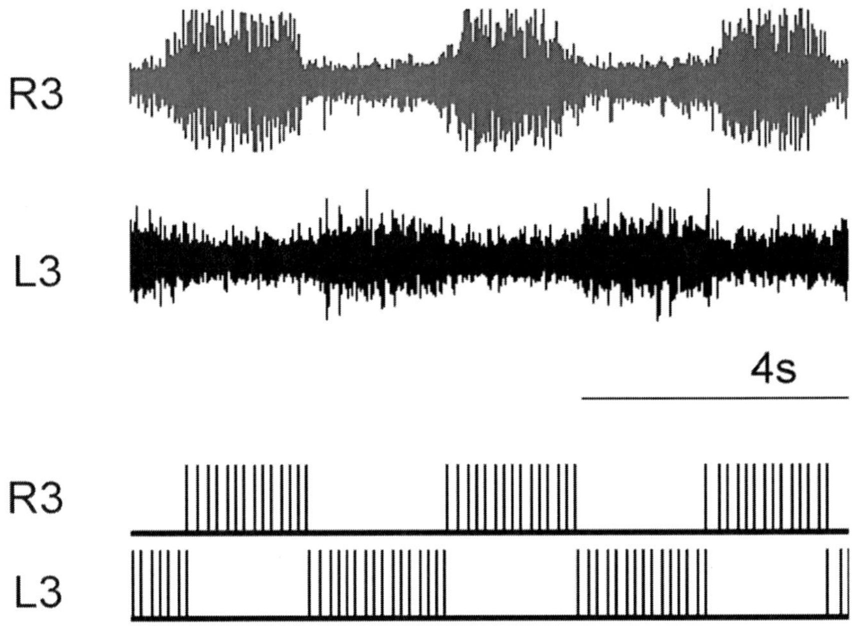

Figure 7. Comparison of pharmacological *in-vitro* spinal cord with digital CPG.

4. DISCUSSION

One key step in designing a neuroprosthesis is to produce a large, resource-frugal biomimetic SNN. A biologically realistic CPG (i.e., the leech heartbeat system neural network) was implemented with a minimum resource cost in terms of neuron model, while maintaining its biomimetic activity, as shown in the Results. The first step was to model the biological leech heartbeat system using a single, segmental CPG. The next stage was to choose an efficient neuron model that required few resources for its digital implementation but remained biorealistic enough to match the behavior of biological cells. The topology and hardware implementation of a single neuron were then extended to form a neuron computation core built into a large-scale neural network: 240 CPGs on a Spartan6 FPGA board. Furthermore, the new synaptic model proposed reproduced the activity-dependent depression phenomenon, which had only previously been described in biology literature. The architecture of the entire real-time systemwas described in detail. Finally, the system was validated by several experiments comparing both elemental and segmental oscillators with biological data, and comparing the segmental oscillator with *ex vivo* rat spinal cord stimulated by pharmacological solutions.

The short-term prospect of this work is to improve the stability of the system using another neuron model. Currently our work is focused on the quartic model (Touboul, 2009), which is more stable than the Izhikevich one and also requires few resources. As described in Table 2, this system is subject to variations in duty cycle and mean period, likely to be reduced by using the new model. However, these variations also exist in biology, so it is necessary to study the actual effect of these variations in the biological system to determine whether they should be eliminated or not.

In the medium term, this system will be included in a hybrid experiment using an *ex vivo* rat spinal cord. The experiment board includes several modules, including an MEA (Micro-Electrode Array) and spike detection block, to detect and record neural activity in the spinal cord. All these modules, together with the CPG network, will be implemented in the same FPGA. Our neurophysiologist colleagues will identify the best spinal cord sites to stimulate and record bursting activity. These sites will be hybridized to the output of the artificial CPG described in this paper and, in turn, its activity will drive the various ventral root outputs of the spinal cord into full locomotor-like activity. These future experiments aim to demonstrate that hybrid artificial/biological networks provide possible solutions for rehabilitating lost central nervous system function.

Our CPG network could be also used to study the locomotion of different animals. Indeed, according toIjspeert (2001), the locomotion activity of a salamander requires 40 CPGs, so the 240 CPGs implemented on the Spartan 6 digital board would be suitable for studying more complex locomotion. Our system will be used in a closed-loop system with different sensors and actuators.

CONFLICT OF INTEREST STATEMENT

The authors declare that the research was conducted in the absence of any commercial or financial relationships that could be construed as a potential conflict of interest.

ACKNOWLEDGMENTS

This work is supported by the European Union's Seventh Framework Programme (ICT-FET FP7/2007-2013, FET Young Explorers scheme) under grant agreement n° 284772 BRAIN BOW (www.brainbowproject.eu).

REFERENCES

1. Amari, S. (1972). Characteristic of the random nets of analog neuron-like elements. *IEEE Trans. Syst.Man Cybern.* 2, 643–657.

2. Angstadt, J. D., and Calabrese, R. L. (1991). Calcium currents and graded synaptic transmission between heart interneurons of the leech.*J. Neurosci.* 11, 746–759.

3. Barron-Zambrano, J. H., and Torres-Huitzil, C. (2013). FPGA implementation of a configurable neuromorphic CPG-based locomotion controller. *Neural Netw.* 45, 50–61.

4. Barron-Zambrano, J. H., Torres-Huitzil, C., and Girau, B. (2010). FPGA-based circuit for central pattern generator in quadruped locomotion. *Aust. J. Intell. Inform. Process. Syst.* 12, 24–29.

5. Ben-Ari, Y., Khazipov, R., Leinekugel, X., Caillard, O., and Gaiarsa, J. L. (1997). GABAA, NMDA and AMPA receptors: a developmentally regulated 'ménage à trois'. *Trends Neurosci.* 20, 523–529.

6. Brainbow. (2012). *Brainbow Project European Union's Seventh Framework Programme (ICT-FET FP7/2007-2013, FET Young Explorers scheme) Under Grant Agreement n° 284772*. Available online at: www.brainbowproject.eu

7. Brown, T. (1914). On the nature of the fundamental activity of the nervous centres; together with an analysis of the conditioning of

rhythmic activity in progression and a theory of the evolution of function in the nervous system. *J. Physiol.* 48, 18–46.

8. Calabrese, R. L. (1995). "Half-center oscillators underlying rhythmic movements," in *The Handbook of Brain Theory And Neural Networks*, ed M. A. Arbib (Cambridge, MA: MIT Press), 444–447.

9. Calabrese, R. L., Angstadt, J., and Arbas, E. (1989). A neural oscillator based on reciprocal inhibition. *Perspect. Neural Syst. Behav.* 10, 33–50.

10. Cassidy, A., and Andreou, A. G. (2008). "Dynamical digital silicon neurons," in *IEEE Biomedical Circuits and Systems Conference, BioCAS 2008*, 289–292.

11. Cohen, A. H., Ermentrout, G. B., Kiemel, T., Kopel, N., Sigvardt, K. A., and Williams, T. L. (1992). Modelling of intersegmental coordination in the lamprey central pattern generator for locomotion. *Trends Neurosci.* 15, 434–438.

12. Cymbalyuk, G. S., Gaudry, Q., Masino, M. A., and Calabrese, R. L. (2002). Bursting in leech heart interneurons: cell-autonomous and network-based mechanisms. *J. Neurosci.* 22, 10580–10592.

13. De Schutter, E. (ed.) (2000). *Computational Neuroscience: Realistic Modeling for Experimentalists*. Boca Raton, FL: CRC Press.

14. Garofalo, M., Nieus, T., Massobrio, P., and Martinoia, S. (2009). Evaluation of the performance of information theory-based methods and cross-correlation to estimate the functional connectivity in cortical networks. *PLoS ONE* 4:e6482.

15. Grassia, F., Buhry, L., Levi, T., Tomas, J., Destexhe, A., and Saighi, S. (2011). Tunable neuromimetic integrated system for emulating cortical neuron models. *Front. Neurosci.* 5:134.

16. Guerrero-Riberas, R., Morrison, A., Diesmann, M., and Pearce, T. (2006). Programmable logic construction kits for hyper-real-time neuronal modeling. *Neural Comput.* 18, 2651–2679.

17. Hill, A. A., Lu, J., Masino, M. A., Olsen, O. H., and Calabrese, R. L. (2001). A model of a segmental oscillator in the leech heartbeat neuronal network. *J. Comput. Neurosci.* 10, 281–302.

18. Hochberg, L. R., Bacher, D., Jarosiewicz, B., Masse, N. Y., Simeral, J. D., Vogel, J., et al. (2012). Reach and grasp by people with tetraplegia using a neurally controlled robotic arm. *Nature* 485, 372–375.

19. Hochberg, L. R., Serruya, M. D., Friehs, G. M., Mukand, J. A., Saleh, M., Caplan, A. H., et al. (2006). Neuronal ensemble control of prosthetic devices by a human with tetraplegia. *Nature* 442, 164–171.

20. Hodgkin, A. L., and Huxley, A. F. (1952). A quantitative description of membrane current and its applications to conduction and excitation in nerve. *J. Physiol.* 117, 500–544.

21. Hooper, S. (2000). Central pattern generators. *Curr. Biol.* 10, 176–177.

22. Ijspeert, A. (2001). A connectionist central pattern generator for the aquatic and terrestrial gaits of a simulated salamander. *J. Biol. Cybern.* 84, 331–348.

23. Ijspeert, A. (2008). Central pattern generators for locomotion control in animals and robots: a review. *J. Neural Netw.* 21, 642–653.

24. Ijspeert, A., Crespi, A., Ryczko, D., and Cabelguen, J. (2007). From swimming to walking with a salamander robot driven by a spinal cord model. *Science* 315, 1416–1420.

25. Indiveri, G. (2007). Synaptic plasticity and spike-based computation in VLSI networks of integrate-and-fire neurons. *Neural Inform. Process. Lett. Rev.* 11, 135–146.

26. Indiveri, G., Linares-Barranco, B., Hamilton, T., Van Schaik, A., Etienne-Cummings, R., and Delbruck, T. (2011). Neuromorphic silicon neuron circuits. *Front. Neurosci.* 5:73.

27. Izhikevich, E. M. (2003). Simple model of spiking neurons. *IEEE Trans. Neural Netw.* 14, 1569–1572.

28. Izhikevich, E. M. (2004). Which model to use for cortical spiking neurons. *IEEE Trans. Neural Netw.* 15, 1063–1070.

29. Jezzini, S., Hill, A. A., Kuzyk, P., and Calabrese, R. L. (2004). Detailed model of intersegmental coordination in the timing network of the leech heartbeat central pattern generator. *J. Neurophysiol.* 91, 958–977.

30. Jung, R., Brauer, E. J., and Abbas, J. J. (2001) Real-time interaction between a neuromorphic electronic circuit and the spinal cord. *IEEE Trans. Neural Syst. Rehabil. Eng.* 9, 319–326.

31. Krahl, B., and Zerbst-Boroffka, I. (1983). Blood pressure in the leech. *J. Exp. Biol.* 107, 163–168.

32. Le Masson, G., Renaud-Le Masson, S., Debay, D., and Bal, T. (2002). Feedback inhibition controls spike transfer in hybrid thalamic circuits. *Nature* 417, 854–858.

33. Lee, Y. J., Lee, J., Kim, K. K., Kim, Y. B., and Ayers, J. (2007). Low power cmos electronic central pattern generator design for a biomimetic underwater robot. *Neurocomputing* 71, 284–296.

34. Lewis, M. A., Hartmann, M. J., Etienne-Cummings, R., and Cohen, A. H. (2001). Control of a robot leg with an adaptive VLSI CPG chip.*Neurocomputing* 38–40, 1409–1421.

35. Linares-Barranco, B., Sánchez-Sinencio, E., Rodríguez-Vázquez, A., and Huertas, J. L. (1993). "CMOS Analog Neural Network Systems Based on Oscillatory Neurons," in *Silicon Implementation of Pulse Coded Neural Networks*, eds M. Zaghloul, J. Meador, and R. Newcomb (Boston: Kluwer Academic Publishers), 199–247.

36. Marder, E., and Bucher, D. (2001). Central pattern generators and the control of rhythmic movements. *Curr. Biol.* 11, 986–996.

37. Marom, S., and Shahaf, G. (2002) Development, learning and memory in large random networks of cortical neurons: lessons beyond anatomy. *Q. Rev. Biophys*. 35: 63–87.

38. Matsuoka, K. (1987). Mechanism of frequency and pattern control in the neural rhythm generators. *Biol. Cybern*. 56, 345–353.

39. Nadim, F., Olsen, O. H., De Schutter, E., and Calabrese, R. L. (1995). Modeling the leech heartbeat elemental oscillator. *J. Comput. Neurosci.* 2, 215–235.

40. Nakada, K. (2003). An analog cmos central pattern generator for interlimb coordination in quadruped locomotion. *IEEE Tran. Neural Netw*. 14, 1356–1365.

41. Nicolelis, M. A. L., and Lebedev, M. A. (2009). Principles of neural ensemble physiology underlying the operation of brain-machine interfaces. *Nat. Rev. Neurosci.* 10, 530–540.

42. Norris, B., Weaver, A., Wenning, A., Garcia, P., and Calabrese, R. L. (2007). A central pattern generator producing alternative outputs: phase relations of leech hear motor neurons with respect of premotor synaptic input. *J. Neurophysiol*. 98, 2983–2991.

43. Olsen, O. H., and Calabrese, R. L. (1996). Activation of intrinsic and synaptic currents in leech heart interneurons by realistic waveforms.*J. Neurosci.* 16, 4958–4970.

44. Olsen, O. H., Nadim, F., and Calabrese, R. L. (1995). Modeling the leech heartbeat elemental oscillator: II. Exploring the parameter space.*J. Comput. Neurosci.* 2, 237–257.

45. Olypher, A., Cymbalyuk, G., and Calabrese, R. L. (2006). Hybrid systems analysis of the control of burst duration by low-voltage-activated calcium current in Leech heart interneurons. *J. Neurophysiol.* 96, 2857–2867. doi: 10.1152/jn.00582.2006

46. Rice, K. L., Bhuiyan, P. A., Taha, T. M., Vutsinas, C. N., and Smith, M. C. (2009). “FPGA Implementation of Izhikevich Spiking Neural Network for Character Recognition,” in *International Conference on Reconfigurable Computing and FPGAs*, (Cancun), 451–456.

47. Serrano-Gotarredona, R., Oster, M., Lichtsteiner, P., Linares-Barranco, A., Paz-Vicente, R., Gómez-Rodríguez, F., et al. (2009). CAVIAR: A 45k-Neuron, 5M-Synapse, 12G-connects/sec AER Hardware Sensory-Processing-Learning-Actuating System for High Speed Visual Object Recognition and Tracking. *IEEE Trans. Neural Netw*. 20, 1417–1438.

48. Simoni, M., and DeWeerth, S. (2007). Sensory feedback in a half-center oscillator model. *IEEE Trans. Biomed. Eng*. 54, 193–204.

49. Simoni, M., Cymbalyuk, G., Sorensen, M., R., Calabrese, R. L., and DeWeerth, S. (2004). A multi-conductance silicon neuron with biologically matched conductances. *IEEE Trans. Biomed. Eng*. 51, 342–354.

50. Sorensen, M., DeWeerth, S., Cymbalyuk, G., and Calabrese, R. L. (2004). Using a hybrid neural system to reveal regulation of neuronal network activity by an intrinsic current. *J. Neurosci*. 24, 5427–5438.

51. Still, S., and Tilden, M. W. (1998). “Controller for a four legged walking machine,” in *Neuromorphic Systems: Engineering Silicon from Neurobiology*, eds L. S. Smith and A. Hamilton, (World Scientific Publishing Co Pte Ltd), 138–148

52. Still, S., Hepp, K., and Douglas, R. J. (2006). Neuromorphic walking gait control. *IEEE Trans. Neural Netw*. 17, 496–508.

53. Tabak, J., Senn, W., O'Donovan, M., and Rinzel, J. (2000). Modeling of spontaneous activity in developing spinal cord using activity-dependent depression in an excitatory network. *J. Neurosci*. 20, 3041–3056.

54. Torres-Huitzil, C., and Girau, B. (2008). Implementation of central pattern generator in an FPGA-based embedded system. *18th International Conference on Artificial Neural Networks, Vol.* 5164, 179–187.

55. Touboul, J. (2009). Importance of the cutoff value in the quadratic adaptive integrate-and-fire model. *Neural Comput*. 21, 2114–2122.

56. Van Der Pol, B. (1928). The heartbeat considered as a relaxation oscillation, and an electrical model of the heart. *Philos. Mag*. 6, 763–775.

57. Vogelstein, R. J., Tenore, F., Etienne-Cummings, R., Lewis, M. A., and Cohen, A. H. (2006). Dynamic control of the central pattern generator for locomotion. *Biol. Cybern*. 95, 555–566.

58. Wijekoon, J., and Dudek, P. (2008). Compact silicon neuron circuit with spiking and bursting behavior. *Neural Netw*. 21, 524–534.

59. Zumsteg, Z., Kemere, C., O'Driscoll, S., Santhanam, G., Ahmed, R. E., Shenoy, K. V., et al. (2005). Power feasibility of implantable digital spike sorting circuits for neural prosthetic systems. *IEEE Trans. Neural Syst. Rehabil. Eng*. 13, 272–279.

CHAPTER 4

A New Data Mining Scheme Using Artificial Neural Networks

S. M. Kamruzzaman [1,*] and A. M. Jehad Sarkar [2]

[1]Department of Electronics Engineering, Hankuk University of Foreign Studies, 89 Wangsan-ri, Mohyeon-myon, Yongin-si, Kyonggi-do, 449-791, Korea

[2]Department of Digital Information Engineering, Hankuk University of Foreign Studies, 89 Wangsan-ri, Mohyeon-myon, Yongin-si, Kyonggi-do, 449-791, Korea

ABSTRACT

Classification is one of the data mining problems receiving enormous attention in the database community. Although artificial neural networks (ANNs) have been successfully applied in a wide range of machine learning applications, they are however often regarded as black boxes, *i.e.*, their predictions cannot be explained. To enhance the explanation of ANNs, a novel algorithm to extract symbolic rules from ANNs has been proposed in this paper. ANN methods have not been effectively utilized for data mining tasks because how the classifications were made is not explicitly stated as symbolic rules that are suitable for verification or interpretation by human experts. With the proposed approach, concise symbolic rules with high accuracy, that are easily explainable, can be extracted from the trained ANNs. Extracted rules are comparable with other methods in terms of number of rules, average number of conditions for a rule, and the accuracy. The effectiveness of the proposed approach is clearly demonstrated by the experimental results on a set of benchmark data mining classification problems.

KEYWORDS

data mining; neural networks; symbolic rules; weight freezing; constructive algorithm; pruning; clustering; rule extraction; symbolic rules

1. INTRODUCTION

Data mining, also popularly known as knowledge discovery in databases, refers to the process of automated extraction of hidden, previously unknown and potentially useful information from large databases. It is the process of finding and interpreting the valuable information by using the knowledge of multidisciplinary fields such as statistics, artificial intelligence, machine learning, database management and so on [1,2]. While the predictive accuracy obtained by artificial neural networks (ANNs) is often higher than that of other methods or human experts, it is generally difficult to understand how ANNs arrive at a particular conclusion due to the complexity of the ANN architectures [3,4]. It is often said that an ANN is practically a "black box". Even for an ANN with only single hidden layer, it is generally impossible to explain why a particular pattern is classified as a member of one class and another pattern as a member of another class, due to the complexity of the network [5]. This may cause problems in some cases. To solve this problem, researchers are interested in developing a humanly understandable representation for ANNs.

ANNs have the ability of distributed information storage, parallel processing, reasoning, and self-organization. It also has the capability of rapid fitting of nonlinear data, so it can solve many problems which are difficult for other methods [6]. Initially, the application of the ANN in data mining was not positive, and the main reasons were that the ANN has the defects of complex structure, poor interpretability and long training times. But its advantages such as high affordability to the noise data with low error rate, and the continuously advancing and optimization of various network training, pruning, and rule extraction algorithms, make the application of the ANNs in the data mining increasingly favored by the overwhelming majority of users [7–9]. In machine learning and data mining research, rule extraction has become an increasingly important topic, and a growing number of researchers and practitioners have applied ANNs for machine learning in a variety of real world applications [10–14]. An inherent defect of ANNs is that the learned knowledge is masked in a large amount of connections, which leads to the poor transparency of knowledge and poor explanation ability [15]. In order to compensate this defect, developing algorithms to extract symbolic rules from trained neural networks has been a hot topic in recent years.

In many applications, it is highly desirable to extract symbolic rules from these networks. Unlike a collection of weights, symbolic rules can be easily interpreted and verified by human experts. They can also provide new insights into the application problems and the corresponding data [16,17]. A number of works are available in the literature to explain the functionality of ANNs by extracting rules from trained ANNs. The main problem of existing works is that they determine the number of hidden neurons in ANNs manually. Thus, the prediction accuracy and rules extracted from trained ANNs may not be optimal since the performance of ANNs is greatly dependent on their architectures. Furthermore, rules extracted by existing algorithms are not simple; as a result it is difficult to understand by the users.

In this paper we have proposed a new data mining scheme; referred to as ESRNN (Extraction of Symbolic Rules from ANNs) to extract symbolic rules from trained ANNs. A four-phase training algorithm is proposed by using backpropagation learning. In the first and second phases, appropriate network architecture is determined using weight freezing based constructive and pruning algorithms. In the third phase, the continuous activation values of the hidden nodes are discretized by using an efficient heuristic clustering algorithm. Finally, in the fourth phase, symbolic rules are extracted using the frequently occurred pattern based rule extraction algorithm by examining the discretized activation values of the hidden nodes.

The rest of the paper is organized as follows. Section 2 describes the related work. The proposed data mining scheme is presented in Section 3. We discuss the performance evaluation in Section 4. Finally, in Section 5 we conclude the paper.

2. RELATED WORK

A neural network-based approach to mining classification rules from given databases has been proposed in [18]. The network is first trained to achieve some required accuracy rate. Redundant connections of the network are then removed by a network pruning algorithm. The activation values of the hidden nodes in the network are analyzed, and classification rules are generated using the result of this analysis. Two classes of approaches for data mining with ANNs have been proposed in [19]. The first approach, often called rule extraction, involves extracting symbolic models from trained neural networks. The second approach is to directly learn simple, easy-to-understand networks. Data mining using pruned artificial neural network tree (ANNT) has been proposed in [20]. ANNT pruning approach consists of three phases: training, pruning and rule extraction. It improved the generalization ability of the network and the number of rules extracted is reduced. The key technology and ways to achieve the data mining based on neural networks is researched in [7]. The combination of data mining method and neural network model can greatly improve the efficiency of data mining techniques, and has been widely used. How to apply ANN in data mining techniques has reviewed in [2]. Given the current state of the art, neural network deserves a place in the tool boxes of data mining specialists.

In the literature, there are many different approaches for the rule extraction from ANNs. A number of algorithms for extracting rules from trained ANNs have been developed in the last two decades [21–30]. Saito and Nakano proposed a medical diagnosis expert system based on a multilayer ANN in [21]. They treated the network as a black box and used it only to observe the effects on the network output caused by change the inputs. Two methods for extracting rules from ANN are described by Towell and Shavlik in [22]. The first method is the subset algorithm [23], which searches for subsets of connections to a node whose summed weight exceeds the bias of that node. The major problem with subset algorithms is that the cost of finding all subsets increases as the size of the ANNs increases. The second method, the MofN algorithm [24], is an improvement of the subset method that is designed to explicitly search for M-of-N rules from knowledge based ANNs. Instead of considering an ANN connection, groups of connections are checked for their contribution to the activation of a node, which is done by clustering the ANN connections.

Liu and Tan proposed X2R in [25], a simple and fast algorithm that can be applied to both numeric and discrete data, and generate rules from datasets. It can generate perfect rules in the sense that the error rate of the rules is not worse than the inconsistency rate found in the original data. The problem of the rules generated by X2R, are order sensitive, *i.e.*, the rules should be fired in sequence. Liu described a family of rule generators in [26] that can be used to extract rules in various applications. It includes versions that can handle noise in data, produce perfect rules, and can induce order independent or dependent rules. The basic idea of the algorithm is simple: using first order information in the data to determine shortest sufficient conditions in a pattern that can differentiate the pattern from patterns belonging to other classes.

Setiono presented MofN3, a new method for extracting M-of-N rules from ANNs, in [27]. The topology of the ANN is the standard three-layered feedforward network. Nodes in the input layer are connected only to the nodes in the hidden layer, while nodes in the hidden layer are also connected to nodes in the output layer. Given a hidden node of a trained ANN with *N* incoming connections, show how the value of *M* can be easily computed. In order to facilitate the process of extracting M-of-N rules, the attributes of the dataset have binary values −1 or 1. Kamruzzaman and Islam proposed an algorithm, REANN in [28] to extract rules from trained ANNs for medical diagnosis problems. This paper investigates the rule extraction process for only 3 medical datasets.

Jin and Sendhoff provide an up-to-date yet not necessarily complete review of the existing research on Pareto-based multiobjective machine learning (PMML) algorithms in [29]. They illustrate, on three benchmark problems (breast cancer, iris, and diabetes), how can address important topics in machine learning, such as generating interpretable models, model selection for generalization, and ensemble extraction, using the Pareto-based multiobjective approach. They compare three Pareto-based approaches to the extraction of neural ensembles and indicate that the method by trading off accuracy and complexity can provide reliable results. Finally, Wang *et al.* proposed a novel algorithm of regression rules extraction from ANN in [30], which is based on linear intelligent insertion. The linear function and symbolic rules are used to the ANN, and the rules are generated by the decision tree.

The limitations of the existing rule extraction algorithms are summarized as follows:

- Use predefined and fixed number of hidden nodes that require human experience and prior knowledge of the problem to be solved,
- Clustering algorithms used to discretize the output values of hidden nodes are not efficient,
- Computationally expensive,
- Could not produce concise rules, and
- Extracted rules are order sensitive.

To overcome these limitations we have proposed a scheme for data mining by extracting symbolic rules from trained ANNs. The proposed system successfully solves a number of data mining classification problems in the literature and described in detail in the next section.

3. PROPOSED DATA MINING SCHEME USING ANNS

Developing algorithms and applications that are able to gain knowledge of their experience and previous examples, and that show intelligent behavior is the domain of machine learning and ANNs. Data mining on the other hand deals with the analysis of large and complex databases in order to discover new, useful and interesting knowledge using techniques from machine learning and statistics. The data mining process using ANNs with the emphasis on symbolic rule extraction is described in this section. The proposed data mining scheme is composed of two steps: data preparation and rule extraction, as shown in Figure 1 and explained further as follows:

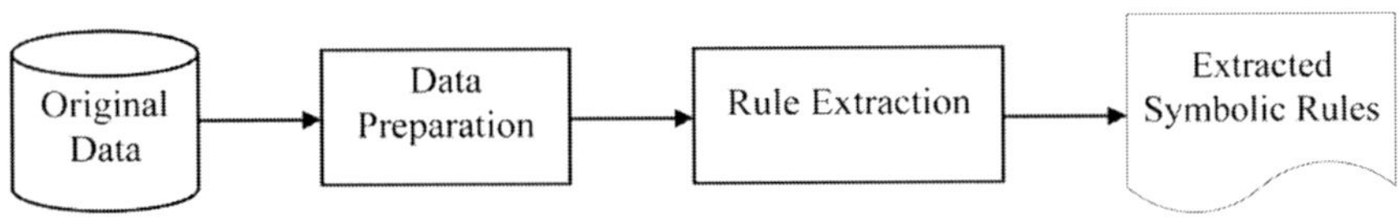

Figure 1. Data mining technique using ANNs.

3.1. Data Preparation

In many fields of artificial intelligence, such as pattern recognition, information retrieval, machine learning, and data mining, one needs to prepare quality data by pre-processing the raw data. The input to the data mining algorithms is assumed to be nicely distributed, containing no missing or incorrect values where all features are important. The real-world data may be incomplete, noisy, and inconsistent, which can disguise useful patterns. Data preparation is a process of the original data to make it fit to a specific data mining method. Data preparation is the first important step in the data mining and plays an important role in the entire data mining process.

The data mining using ANNs can only handle numerical data. How to represent the input and output attributes of a learning problem in a neural network is one of the key decisions influencing the quality of the solutions one can obtain. Depending on the kind of problem, there may be several different kinds of attributes that must be represented. For all of these attribute kinds, multiple reasonable methods of neural network representation exist. We will now discuss each attribute kind and some common methods to represent such an attribute.

- **Real-valued attributes** are usually rescaled by some function that maps the value into the range 0…1 or −1…1 in a way that makes a roughly even distribution within that range.

- **Integer-valued attributes** are most often handled as if they were real-valued. If the number of different values is only small, one of the representations used for ordinal attributes may also be appropriate. Note that often attributes whose values are integer numbers are not really integer-valued but are ordinal or cardinal instead. We consider all integer-valued attributes as real-valued.

- **Ordinal attributes** with m different values are either mapped onto an equidistant scale making them pseudo-real-valued or are represented by $m-1$ inputs of which the leftmost k have value 1 to represent the k-th attribute value while all others are 0. A binary code using only $[\log_2 m]$ inputs can also be used.

- **Nominal attributes** with m different values are usually either represented using a 1-of-m code or a binary code.

- **Missing attribute** values can be replaced by a fixed value (e.g., the mean of the non-missing values of this attribute) or can be represented explicitly by adding another input for the attribute that is 1 if the attribute value is missing.

3.2. Rule Extraction: The ESRNN Algorithm

It is becoming increasingly apparent that without some form of explanation capability, the full potential of ANNs may not be realized. The rapid and successful proliferation of applications incorporating ANNs in many fields, such as commerce, science, industry, medicine *etc.*, offers a clear testament to the capability of ANN paradigm. Extracting symbolic rules from trained ANN is one of the promising areas that are commonly used to explain the functionality of ANNs. The aim of this subsection is to introduce a new algorithm, referred to as ESRNN (extraction of symbolic rules from ANNs), to extract symbolic rules from trained ANNs. We now describe below each of the components of ESRNN in further detail.

A standard three-layer feedforward ANN is the basis of the proposed ESRNN algorithm. The hyperbolic tangent function, which can take any value in the interval [–1, 1], is used as the hidden node activation function. Rules are extracted from near optimal ANN by using a new rule extraction algorithm. The aim of ESRNN is to search for simple rules with high predictive accuracy. The major steps of ESRNN are summarized in Figure 2 and explained further as follows:

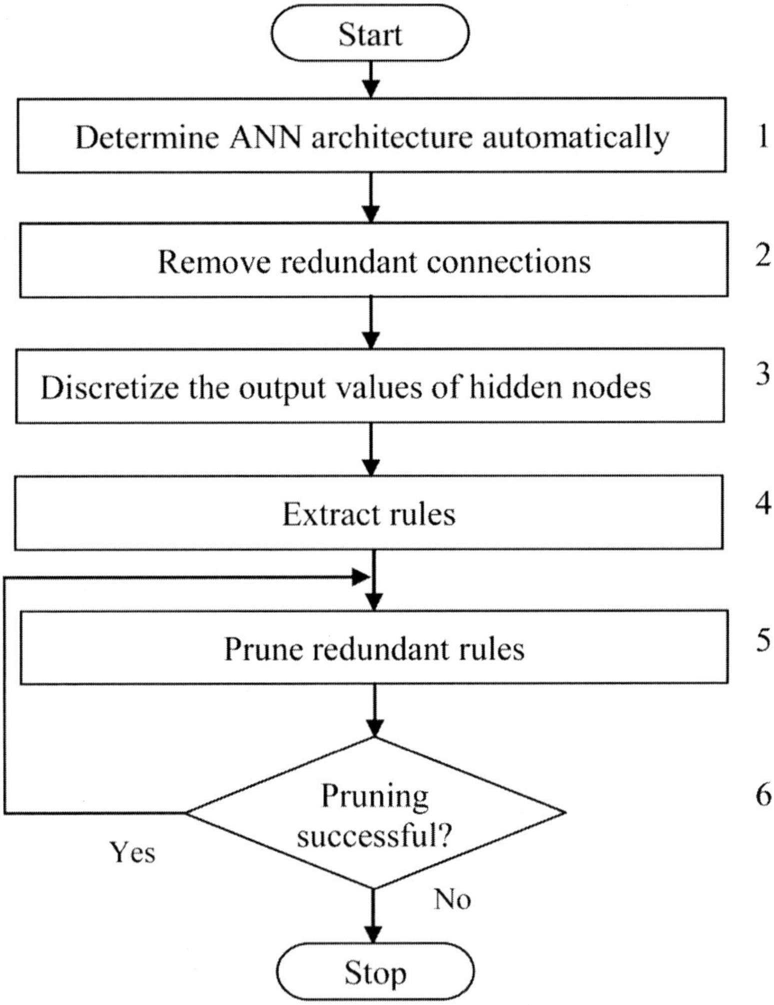

Figure 2. Flow chart of the proposed ESRNN algorithm.

Step 1 Create an initial ANN architecture. The initial architecture has three layers, including an input, an output, and a hidden layer. The number of nodes in the input and output layers is the same as the number of attributes and the classes of the problem. Initially, the hidden layer contains only one node. The number of nodes in the hidden layer is automatically determined by using the weight freezing based constructive algorithm, explained in subsection *A*.Initialize all connection weights randomly within a certain small range.

Step 2 Remove redundant input nodes and connections between input nodes and hidden nodes and between hidden nodes and output nodes by using a basic pruning algorithm, explained in subsection *B*. When pruning is completed, the ANN architecture contains only important nodes and connections. This architecture is saved for the next step.

Step 3 Discretize the outputs of hidden nodes by using an efficient heuristic clustering algorithm, explained in subsection *C*. The reason for discretization is that the outputs of hidden nodes are continuous, and thus the rules can not be readily extractable from the ANN.

Step 4 Extract the rules that map the inputs and outputs relationships. The task of the rule extraction is accomplished in three phases. In the first phase, rules are extracted by using the rule extraction algorithm, explained in subsection *D*, to describe the outputs of ANN in terms of the discretized output values of the hidden nodes. In the second phase, rules are extracted to describe the discretized output values of the hidden nodes in terms of the inputs. Finally in the third phase, combine the rules extracted in the first and second phases.

Step 5 Prune redundant rules extracted in **Step 4** by replacing specific rules with more general ones.

Step 6 Check the classification accuracy of the network. If the accuracy falls below an acceptable level, *i.e.*, rule pruning is not successful then stop. Otherwise go to **Step 5**.

The rules extracted by ESRNN are compact and comprehensible, and do not involve any weight values. The accuracy of the rules from pruned networks is as high as the accuracy of the original networks. The important features of the ESRNN algorithm are the rules extracted by rule extraction algorithm is recursive in nature and is order insensitive, *i.e.*, the rules need not to be required to fire sequentially.

3.2.1. Weight Freezing Based Constructive Algorithm

One drawback of the traditional backpropagation algorithm is the need to determine the number of nodes in the hidden layer prior to training. To overcome this difficulty, many algorithms that construct a network dynamically have been proposed in [31–33]. The most well known constructive algorithms are dynamic node creation (DNC) [34], feedforward neural network construction (FNNC) algorithm, and the cascade correlation (CC) algorithm [35]. The constructive algorithm used in the ESRNN algorithm is based on the FNNC algorithm proposed in [36]. In FNNC algorithm, the training process is stopped when the classification accuracy on the training set is 100% [37]. However, it is not possible to get 100% classification accuracy for most of the benchmark classification problems. In addition, higher classification accuracy on the training set does not guarantee the higher generalization ability *i.e.*, classification accuracy on the testing set.

The training time is an important issue in designing ANNs. One approach for reducing the number of weights to be trained is to train few weights rather than all weights in a network and keep remaining weights fixed, commonly known as weight freezing. The idea behind the weight freezing-based constructive algorithm is to freeze input weights of a hidden node when its output does not change much in the successive few training epochs. Theoretical and experimental studies reveal that some hidden nodes of an ANN maintain almost constant output after some training epochs, while others continuously change during the whole training period.

In our algorithm, it has been proposed that the output of a hidden node can be frozen when its output does not change much in the successive training epochs. This weight freezing method can be considered as combination of the two extremes: for training all the weights of ANNs and for training the weights of only the newly added hidden node of ANNs [38]. The major steps of our weight freezing based constructive algorithm are summarized in Figure 3 and explained further as follows:

Step 1 Create an initial ANN consisting of three layers, *i.e.*, an input, an output, and a hidden layer. The number of nodes in the input and output layers is the same as the number of inputs and outputs of the problem. Initially, the hidden layer contains only one node *i.e.*, $h = 1$, where h is the number of hidden nodes in the network. Randomly initialize all connection weights within a certain range.

Step 2 Train the network on the training set by using backpropagation algorithm until the error E is almost constant for a certain number of training epochs, τ, is specified by the user.

Step 3 Compute the ANN error E. If E is found unacceptable (*i.e.*, too large), then assume that the ANN has inappropriate architecture, and go to the next step. Otherwise stop the training process. The ANN error E is calculated according to the following equations:

$$E(w, v) = \frac{1}{2}\sum_{i=1}^{k}\sum_{p=1}^{C}(S_{pi} - t_{pi})^2 \tag{1}$$

where k is the number of patterns, C is the number of output nodes, and t_{pi} is the target value for pattern x_i at output node p. S_{pi} is the output of the network at output node p.

$$S_{pi} = \sigma(\sum_{m=1}^{h} \delta((x_i)^T w_m) v_{pm}) \tag{2}$$

where h is the number of hidden nodes in the network, x_i is an n-dimensional input pattern, $i = 1, 2, \ldots, k$, w_m is a p-dimensional vector weights for the arcs connecting the input layer and the m-th hidden node, $m = 1, 2, \ldots, h$, v_m is a C-dimensional vector of weights for the arcs connecting the m-th hidden node and the output layer. The activation function for the output layer is sigmoid function $\sigma(y) = 1/(1 + e^{-y})$ and for the hidden layer is hyperbolic tangent function $\delta(y) = (e^y - e^{-y}) / (e^y + e^{-y})$.

Step 4 Compare each hidden node's output $X(n)$ at training epoch n with its previous value $X(n-\tau)$. If $X(n) \cong X(n-\tau)$, freeze the input weights of that node.

Step 5 Add one hidden node to the hidden layer. Randomly initialize the weights of the arcs connecting this new hidden node with input nodes and output nodes. Set $h = h + 1$ and go to **Step 2**.

3.2.2. Pruning Algorithm

Pruning offers an approach for dynamically determining an appropriate network topology. Pruning techniques begin by training a larger than necessary network and then eliminate weights and nodes that are deemed redundant [38,39].

As the nodes of the hidden layer are determined automatically by weight freezing based constructive algorithm in ESRNN, the aim of this pruning algorithm used here is to remove as many unnecessary nodes and connections as possible. A node is pruned if all the connections to and from the node are pruned. Typically, methods for removing weights from the network involve adding a penalty term to the error function. It is hoped that by adding a penalty term to the error function, unnecessary connections will have small weights, and therefore pruning can reduce the complexity of the network significantly. The simplest and most commonly used penalty term is the sum of the squared weights.

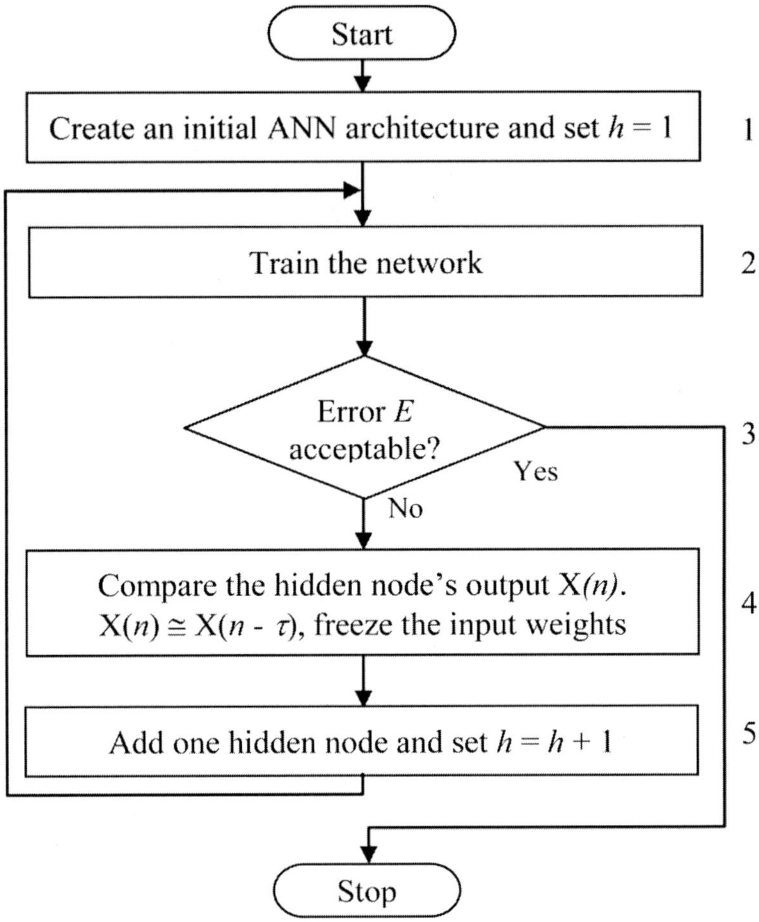

Figure 3. Flow chart of the weight freezing based constructive algorithm.

Given a set of input patterns $x_i \in \Re^n$, $i = 1, 2, \ldots, k$, let w_m is a p-dimensional vector weights for the arcs connecting the input layer and the m-th hidden node, $m = 1, 2, \ldots, h$. The weight of the connection from the l-th input node to the m-th hidden node is denoted by w_{ml}, v_m is a C-dimensional vector of weights for the arcs connecting the m-th hidden node and the output layer. The weight of the connection from the m-th hidden node to the p-th output node is denoted by v_{pm}. It has been suggested that faster convergence can be achieved by minimizing the cross entropy function instead of squared error function [40].

The backpropagation algorithm is applied to update the weights (w, v) and minimize the following error function:

$$\theta(w, v) = F(w, v) + P(w, v) \tag{3}$$

where $F(w, v)$ is the cross entropy function:

$$F(w, v) = -\sum_{i=1}^{k}\sum_{p=1}^{o}\left(t_{pi}\log S_{pi} + (1-t_{pi})\log(1-S_{pi})\right) \tag{4}$$

where S_{pi} is the output of the network:

$$S_{pi} = \sigma\left(\sum_{m=1}^{h}\delta((x_i)^T w_m)v_{pm}\right) \tag{5}$$

where x_i is an n-dimensional input pattern, $i = 1, 2, \ldots, k$, and $(x_i)^T w_m$ denotes the scalar product of the vectors x_i and w_m, $\delta(.)$ is the hyperbolic tangent function and $\sigma(.)$ is the logistic sigmoid function. $P(w, v)$ is a penalty term used for weight decay:

$$P(w, v) = \varepsilon_1\left(\sum_{m=1}^{h}\sum_{l=1}^{n}\frac{\beta(w_{ml})^2}{1+\beta(w_{ml})^2} + \sum_{m=1}^{h}\sum_{p=1}^{o}\frac{\beta(v_{pm})^2}{1+\beta(v_{pm})^2}\right) + \varepsilon_2\left(\sum_{m=1}^{h}\sum_{l=1}^{n}(w_{ml})^2 + \sum_{m=1}^{h}\sum_{p=1}^{o}(v_{pm})^2\right) \tag{6}$$

The values for the weight decay parameters $\varepsilon_1,\varepsilon_1 > 0$ must be chosen to reflect the relative importance of the accuracy of the network verses its complexity. More weights may be removed from the network at the cost of a decrease in the network accuracy with larger values of these two parameters. They also determine the range of values where the penalty for each weight in the network is approximately equal to ε_1. The parameter $\beta > 0$ determines the steepness of the error function near to the origin.

The value of the function $f(w)\, w^2 / (1 + w^2)$ is small when w is close to zero and approaches to 1 as w becomes large. In addition, the derivative function $f'(w) = w^2 / (1 + w^2)^2$ indicates that the backpropagation training will be very little affected with the addition of the penalty function for weights having large values. Consider the plot of the function $f(w) = \varepsilon_1\beta w^2 / (1 + \beta w^2) + \varepsilon_2 w^2$ shown in Figure 4, where $\varepsilon_1 = 0.1$, $\varepsilon_2 = 10^{-5}$, and $\beta = 10$ in Figure 4(a). This function intersects the horizontal line $f = \varepsilon_1$ at $w \approx \pm 5.62$. By decreasing the value of ε_2, the interval over which the penalty value is approximately equal to ε_1 can be made wider as shown in Figure 4(c), where $\varepsilon_2 = 10^{-6}$. A weight is prevented from taking too large value, since the quadratic term becomes dominant for the larger values of w. The derivative of the function $f(w), f'(w)$ near zero is relatively large as shown in Figure 4(b,d). This will give a small weight w stronger tendency to decay to zero.

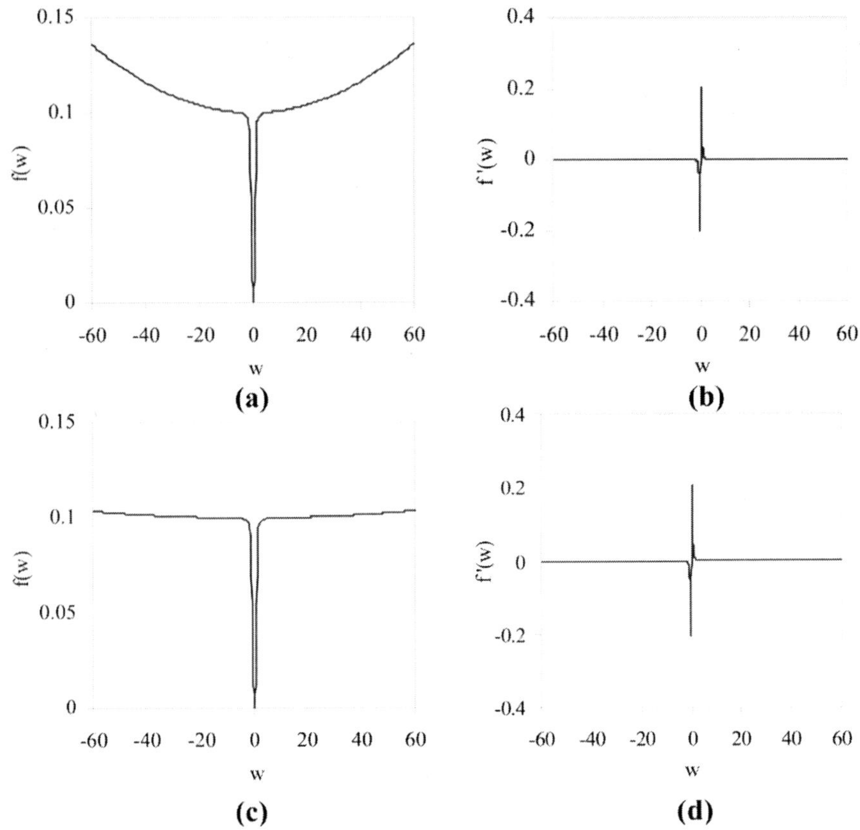

Figure 4. Plots of the function $f(w) = \varepsilon_1\beta w^2 / (1 + \beta w^2) + \varepsilon_2 w^2$ and its derivative $f(w) = 2\varepsilon_1\beta w / (1 + \beta w^2)^2 + 2\varepsilon_2 w$, where $\varepsilon_1 = 0.1$, $\varepsilon_2 = 10^{-5}$, and $\beta = 10$ for **(a,b)** and $\varepsilon_1 = 0.1$, $\varepsilon_2 = 10^{-6}$, and $\beta = 10$ for **(c,d)**.

This pruning algorithm removes the connections of the ANN according to the magnitudes of their weights. As the eventual goal of the ESRNN algorithm is to get a set of simple rules that describe the classification process, it is important that all unnecessary nodes and connections must be removed. In order to remove as many connections as possible, the weights of the network must be prevented from taking values that are too large [41]. At the same time, weights of irrelevant connections should be encouraged to converge to zero. The penalty function is found to be particularly suitable for these purposes.

The steps of the pruning algorithm are explained as follows:

Step 1 Train the network to meet a prespecified accuracy level with the condition (7) satisfied by all correctly classified input patterns.

$$\max_p |e_{pi}| = \max_p |S_{pi} - t_{pi}| \leq \eta_1,\ p = 1, 2, ..., C. \tag{7}$$

Let η_1 and η_2 be positive scalars such that $(\eta_1 + \eta_2) < 0.5$ (η_1 is the error tolerance, η_2 is a threshold that determines if a weight can be removed), where $\eta_1 \in [0, 0.5)$. Let (w, v) be the weights of this network.

Step 2 Remove connections between input nodes and hidden nodes and between hidden nodes and output nodes. This task is accomplished in two phases. In the first phase, connections between input nodes and hidden nodes are removed. For each w_{ml} in the network, if

$$\max_p |v_{pm} w_{ml}| \leq 4\eta_2, \tag{8}$$

then remove w_{ml} from the network.

In the second phase, connections between hidden nodes and output nodes are removed. For eachv_{pm} in the network, if

$$|v_{pm}| \leq 4\eta_2 \leq 4\eta_2, \tag{9}$$

then remove v_{pm} from the network.

Step 3 Remove connections between input nodes and hidden nodes further. If no weight satisfies condition (8) or condition (9), then for each w_{ml} in the network, compute $w_{ml} = \max_p |v_{pm} w_{ml}|$ Remove w_{ml} with smallest w_{ml}. Continue, otherwise stop.

Step 4 Retrain the network and calculate the classification accuracy of the network.

Step 5 If classification accuracy of the network falls below an acceptable level, then stop and use the previous setting of the network weights. Otherwise, go to **Step 2**.

The pruning algorithm used in the ESRNN algorithm intended to reduce the amount of training time. Although it can no longer be guaranteed that the retrained pruned ANN will give the same accuracy rate as the original ANN, the experiments show that many weights can be eliminated simultaneously without deteriorating the performance of the ANN. The two conditions (8) and (9) for pruning depends on the weights for connections between input and hidden nodes and between hidden and output nodes. It is imperative that during the training phase these weights be prevented from getting too large values. At the same time, small weights should be encouraged to decay rapidly to zero.

3.2.3. Heuristic Clustering Algorithm

The process of grouping a set of physical or abstract objects into classes of similar objects is called clustering. A cluster is a collection of data objects that are similar within the same cluster and are dissimilar to the objects in other clusters. A cluster of a data objects can be treated collectively as one group in many applications [42]. There exist a large number of clustering algorithms in the literature, such as, k-means, k-medoids [43,44]. The choice of clustering algorithm depends both on the type of data available and on the particular purpose and applications.

After applying pruning algorithm in ESRNN, the ANN architecture produced by the weight freezing based constructive algorithm contains only important nodes and connections. Nevertheless, rules are not readily extractable because the hidden node activation values are continuous. The discretization of these values paves the way for rule extraction. It is found that some hidden nodes of an ANN maintain almost constant output while other nodes change continuously during the whole training process [45]. Figure 5 shows output of three hidden nodes where a hidden node maintains almost constant output value after some training epochs but output value of other nodes are changing continually. In ESRNN, no clustering algorithm is used when hidden nodes maintain almost constant output value. If the outputs of hidden nodes do not maintain constant value, a heuristic clustering algorithm is used.

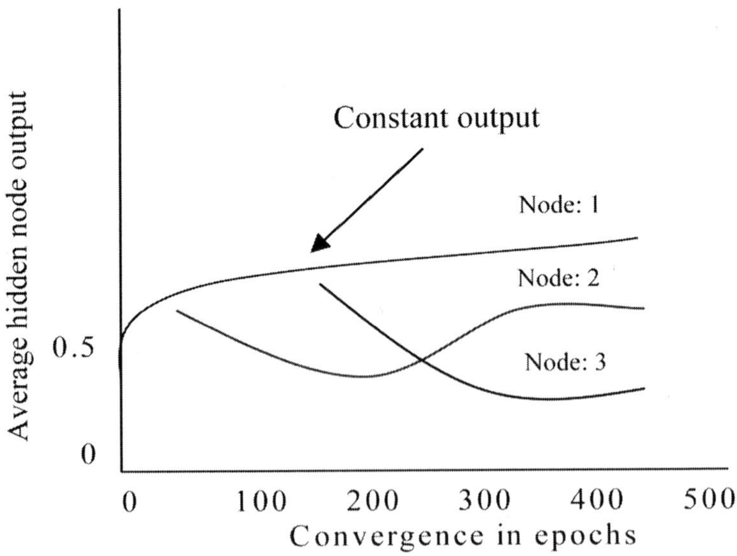

Figure 5. Output of the hidden nodes.

The aim of the clustering algorithm is to discretize the output values of the hidden nodes. Consider that the number of hidden nodes in the pruned network is H. Clustering the activation values of the hidden node is accomplished by a simple greedy algorithm that can be summarized as follows:

1 Find the smallest positive integer d such that if all the network activation values are rounded to d decimal places, the network still retains its accuracy rate.

2 Represent each activation value α by the integer closest to $\alpha \times 10^d$. Let $H_i = <h_{i,1}, h_{i,2}, \ldots, h_{i,k}>$ be the k-dimensional vector of these representations at hidden node i for patterns $x_1, x_2, \ldots, x_k$ and let $H = (H_1, H_2, \ldots, H_H)$ be the $k \times H$ matrix of the hidden representations of all patterns at all H hidden nodes.

3 Let P be a permutation of the set $\{1, 2, \ldots, H\}$ and set $m = 1$.

4 Set $i = P(m)$.

5 Sort the values of the ith column (H_i) of matrix H in increasing order.

6 Find a pair of district adjacent values $h_{i,j}$ and $h_{i,j+1}$ in H_i such that if $h_{i,j+1}$ is replaced by $h_{i,j}$ no conflicting data will be generated.

7 If such a pair of values exists, replace all occurrences of $h_{i,j+1}$ in H_i by $h_{i,j}$ and repeat **Step 6**. Otherwise, set $m = m + 1$. If $m \leq H$, go to **Step 4**, else stop.

In our scheme, the activation value of an input pattern at hidden node m is computed as the hyperbolic tangent function, it will have a value in the range of [−1, 1]. Steps 1 and 2 of the clustering algorithm find integer representations of all hidden node activation values. A small value for d in step 1 indicates that relatively few distinct values for the activation values are sufficient for the network to maintain its accuracy. For example, when $d = 2$, then there could be up to 201 distinct values: −1.00, −0.99, −0.98, …, 0.99, 1.00. For the results reported in this paper, we set the value of $d = 2$.

The array P contains the sequence in which the hidden nodes of the network are to be considered. Different ordering sequences usually result in different clusters of activation values. Once a hidden node is selected for clustering, the discretized activation values are sorted in step 5 such that the activation values are in increasing order. The values are clustered based on their distance. We implemented step 6 of the algorithm by first finding a pair of adjacent distinct values with the shortest distance. If these two values can be merged without introducing conflicting data, they will be merged. Otherwise, a pair with the second shortest distance will be considered. This process is repeated until there are no more pairs of values that can be merged. The next hidden node as determined by the array *P* will then be considered.

3.2.4. Rule Extraction (RE) Algorithm

Classification rules are sought in many areas from automatic knowledge acquisition [46] to data mining [47,48] and ANN rule extraction because some of their attractive features. They are explicit, understandable and verifiable by domain experts, and can be modified, extended and passed on as modular knowledge. The proposed rule extraction (RE) algorithm, can be applied to both numeric and discrete data, is composed of three major functions:

a) Rule Extraction: This function first initializes the extracted rule list to be empty, and sorts the examples according to example frequency. Then it picks the most frequent occurring example as the base to generate a rule and adds the rule to the list of extracted rules. It then finds all the examples that are covered by the rule and removes them from the example space. It repeats the above process iteratively and continuously adds the extracted rules to the rule list until the examples space becomes empty because all data examples have been covered by the rules extracted and they have all been removed.

b) Rule Clustering: Rules are clustered in terms of their class levels. Rules of the same class are clustered together as one group of rules.

c) Rule Pruning: Redundant or more specific rules in each cluster are removed. In each of these clusters, more than one rule may cover the same example. For examples, the rule “if (color = green) and (height < 4) then grass” is already contained in a more general rule “if (color = green) then grass”, and thus the rule “if (color = green) and (height < 4) then grass” is redundant. RE eliminates these redundant rules in each cluster to further reduce the size of the best rule list.

A default rule should be chosen to accommodate possible unclassifiable patterns. If rules are clustered, the choice of the default rule is based on clusters of rules. The steps of the rule extraction algorithm are explained as follows:

Step 1 Extract Rule:

Sort-on-frequency (data-without-duplicates);

$i = 0$;

while (data-without-duplicates is NOT empty){

extract R_i to cover the pattern occurred most frequently;

remove all the patterns covered by R_i ;

$i = i+1$;}

The core of this step is a greedy algorithm that finds the shortest rule based on the first order information, which can differentiate the pattern under consideration from the patterns of other classes. It then iteratively extracts shortest rules and remove the patterns covered by each rule until all patterns are covered by the rules.

Step 2 Cluster Rule:

Cluster rules according to their class levels. Rules extracted in **Step 1** are grouped in terms of their class levels.

Step 3 Prune Rule:

Replace specific rules with more general ones;

Remove noise rules;

Eliminate redundant rules;

Step 4 Check whether all patterns are covered by any rules. If yes then stop, otherwise continue.

Step 5 Determine a default rule. A default rule is chosen when no rule can be applied to a pattern.

RE exploits the first order information in the data and finds shortest sufficient conditions for a rule of a class that can differentiate it from patterns of other classes. It can extract concise and perfect rules in the sense that the error rate of the rules is not worse than the inconsistency rate found in the original data. The novelty of RE is that the rule extracted by it is order insensitive, *i.e.*, the rules need not be required to fire sequentially.

4. PERFORMANCE EVALUATION

This section evaluates the performance of the ESRNN algorithm on a set of well-known benchmark classification problems including diabetes, iris, wine, season, golf playing, and lenses that are widely used in machine learning and data mining research. The datasets representing all the problems were real world data are obtained from [49,50].

4.1. Dataset Description

This subsection briefly describes the datasets used in this study. The characteristics of the datasets are summarized inTable 1. The detailed descriptions of the datasets are available in [49,50].

Table 1. Characteristics of datasets.

Sl No.	Datasets	No. of Examples	Input Attributes	Output Classes
1	Diabetes	768	8	2
2	Iris	150	4	3
3	Wine	178	13	3
4	Season	11	3	4
5	Golf Playing	14	4	2
6	Lenses	24	4	3

The diabetes dataset: The Pima Indians Diabetes data consists of 768 data pairs with eight attributes normalized between 0 and 1. The eight attributes are number of pregnancies (A_1), plasma glucose concentration (A_2), blood pressure (A_3), triceps skin fold thickness (A_4), Two hour serum insulin (A_5), body mass index (A_6), diabetes pedigree function (A_7), and age (A_8). In this database, 268 instances are positive (output equals 1) and 500 instances are negative (output equals 0).

The iris dataset: This is perhaps the best-known database to be found in the pattern recognition literature. The dataset contains three classes of 50 instances each, where each class refers to a type of Iris plant. Four attributes are used to predict the iris class, *i.e.*, sepal length (A_1), sepal width (A_2), petal length (A_3), and petal width (A_4), all in centimeters. Among the three classes, class 1 is linearly separable from the other two classes, and classes 2 and 3 are not linearly separable from each other. To ease knowledge extraction, we reformulate the data with three outputs, where class 1 is represented by {1, 0, 0}, class 2 by {0, 1, 0}, and class 3 by {0, 0, 1}.

The wine dataset: In a classification context, this is a well-posed problem with "well behaved" class structures. A good dataset for first testing of a new classifier, but not very challenging. These data are the results of a chemical analysis of wines grown in the same region in Italy but derived from three different cultivars. The analysis determined the quantities of 13 constituents found in each of the three types of wines. Number of instances 178, number of attributes 13. All attributes are continuous. This was a three-class problem.

The season data: The season dataset contains discrete data only. There are 11 examples in the dataset, each of which consisted of three-elements. These are weather, tree, and temperature. This was a four-class problem.

The golf playing data: The golf playing dataset contains both numeric and discrete data. There are 14 examples in the dataset, each of which consisted of four-elements. These are outlook, temperature, humidity and wind. This is a two-class problem.

The lenses data: The dataset contains 24 examples and are complete and noise free. The examples highly simplified the problem. The attributes do not fully describe all the factors affecting the decision as to which type, if any, to fit. Number of Instances: 24. Number of Attributes: 4; age, spectacle prescription, astigmatic and tear production rate. All attributes are nominal. This was three-class problem: hard contact lenses, soft contact lenses and not contact lenses.

4.2. Experimental Setup

In all experiments, one bias node with a fixed input 1 was used for the hidden and output layers. The learning rate was set between [0.1, 1.0] and the weights were initialized to random values between [−1.0, 1.0]. A hyperbolic tangent function $\delta(y) = (e^y - e^{-y})/(e^y + e^{-y})$ was used as the hidden node activation function and a logistic sigmoid function $\sigma(y) = 1/(1 + e^{-y})$ as the output node activation function.

In this study, all datasets representing the problems were divided into two sets: the training set and the testing set. The numbers of examples in the training set and the testing set was chosen to be the same as those in other works, in order to make the comparison with those works possible. The sizes of the training and testing datasets used in this study are given inTable 2.

4.3. Experimental Results

Tables 3–8 show the ANN architectures produced by the ESRNN algorithm and the training epochs over 10 independent runs on a set of benchmark data mining classification problems. The initial architecture has selected before applying the constructive algorithm, which was used to determine the number of nodes in the hidden layer. The intermediate architecture was the outcome of the constructive algorithm, and the final architecture was the outcome of pruning algorithm used in the ESRNN algorithm. It has been seen that ESRNN can automatically determine compact ANN architectures.

Table 2. Sizes of the training and the testing datasets.

Sl No.	Datasets	Training Examples	Testing Examples
1	Diabetes	384	384
2	Iris	75	75
3	Wine	89	89
4	Season	6	5
5	Golf Playing	7	7
6	Lenses	12	12

Table 3. ANN architectures and the training epochs for the diabetes dataset.

	Initial Architecture		Intermediate Architecture		Final Architecture		No. of Epochs
	No. of Nodes	No. of Connections	No. of Nodes	No. of Connections	No. of Nodes	No. of Connections	
Mean	11 (8-1-2)	10	13.1	31	12.1	19.7	306.4
Min	11 (8-1-2)	10	12.3	23	11.9	15	283
Max	11 (8-1-2)	10	13.9	38	13.2	24.4	329

Table 4. ANN architectures and the training epochs for the irish dataset.

	Initial Architecture		Intermediate Architecture		Final Architecture		No. of Epochs
	No. of Nodes	No. of Connections	No. of Nodes	No. of Connections	No. of Nodes	No. of Connections	
Mean	8 (4-1-3)	7	9	13	9	10.2	198.2
Min	8 (4-1-3)	7	8	8	8	7	185
Max	8 (4-1-3)	7	11	22	10	13.8	220

Table 5. ANN architectures and the training epochs for the wine dataset.

	Initial Architecture		Intermediate Architecture		Final Architecture		No. of Epochs
	No. of Nodes	No. of Connections	No. of Nodes	No. of Connections	No. of Nodes	No. of Connections	
Mean	17 (13-1-3)	16	18.6	38	17	24.8	215
Min	17 (13-1-3)	16	17.2	18	16	22	198
Max	17 (13-1-3)	16	21	62	21	42	240

Table 6. ANN architectures and the training epochs for the season dataset.

	Initial Architecture		Intermediate Architecture		Final Architecture		No. of Epochs
	No. of Nodes	No. of Connections	No. of Nodes	No. of Connections	No. of Nodes	No. of Connections	
Mean	8 (3-1-4)	7	8.9	13.1	8.8	11	87
Min	8 (3-1-4)	7	8	7	8	9.1	74
Max	8 (3-1-4)	7	10	14.2	10.5	15	105

Table 7. ANN architectures and the training epochs for the golf playing dataset.

	Initial Architecture		Intermediate Architecture		Final Architecture		No. of Epochs
	No. of Nodes	No. of Connections	No. of Nodes	No. of Connections	No. of Nodes	No. of Connections	
Mean	7 (4-1-2)	6	8.2	13	7.9	10.5	95.2
Min	7 (4-1-2)	6	7.3	6.1	7.1	6.2	88
Max	7 (4-1-2)	6	9.1	18.2	9.2	13.8	103

Table 8. ANN architectures and the training epochs for the lenses dataset.

	Initial Architecture		Intermediate Architecture		Final Architecture		No. of Epochs
	No. of Nodes	No. of Connections	No. of Nodes	No. of Connections	No. of Nodes	No. of Connections	
Mean	8 (4-1-3)	7	9.1	13.2	8.7	12	106
Min	8 (4-1-3)	7	8.3	7	8.2	7.8	99
Max	8 (4-1-3)	7	10.4	20.8	11	16	126

Figure 6 shows the smallest of the pruned networks over 10 runs for the diabetes problem. The pruned network was only 2 hidden nodes. No input nodes were pruned by pruning algorithm. One hidden node was pruned, as all the connections to and from this node was pruned. The accuracies on the training data and the testing data have reached 76.30% and 75.52%, respectively. The weight of the connection from the first hidden node to the first output node is −1.172 and to the second output node is 1.172 and the weight of the connection from the second hidden node to the first output node is −31.06 and to the second output node is 32.04. The discrete values found by the heuristic clustering algorithm were −0.968, 0.004 and 0.976.

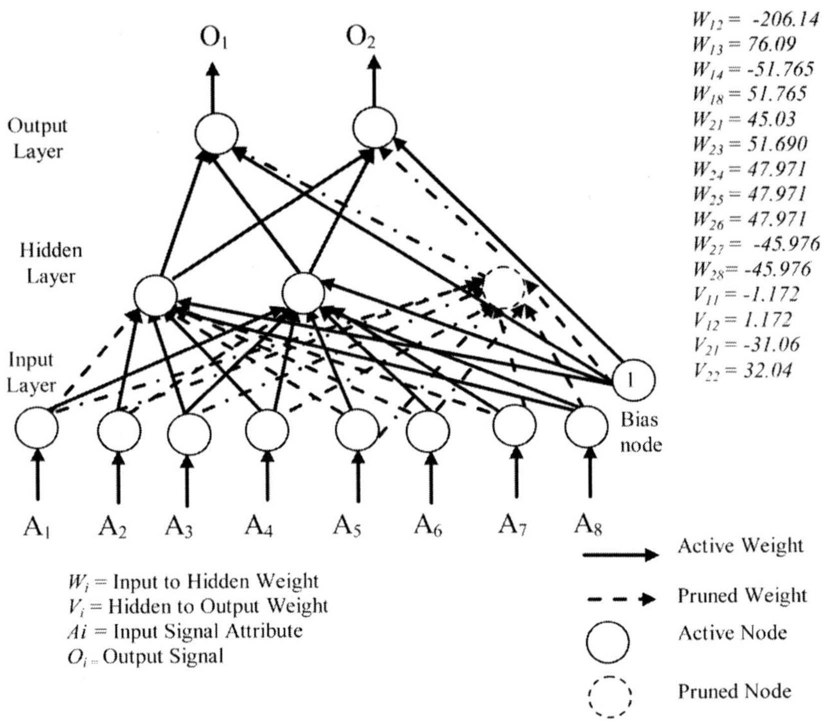

Figure 6. A pruned network for the diabetes data.

Figures 7 and 8 show the training time error for diabetes problem. From the Figure 7, it was observed that the training time error decreased and maintained almost constant after some training epochs, it was further decreased when additional hidden nodes were added. The fluctuation was observed due to the connection pruning and finally maintained almost constant value in account of retraining the pruned network.

The training time error for diabetes data with weight freezing is shown in Figure 8. When error is become constant then weight freezing is done. The effects of hidden node addition with increasing the training epochs for diabetes a problem is shown in Figure 9.

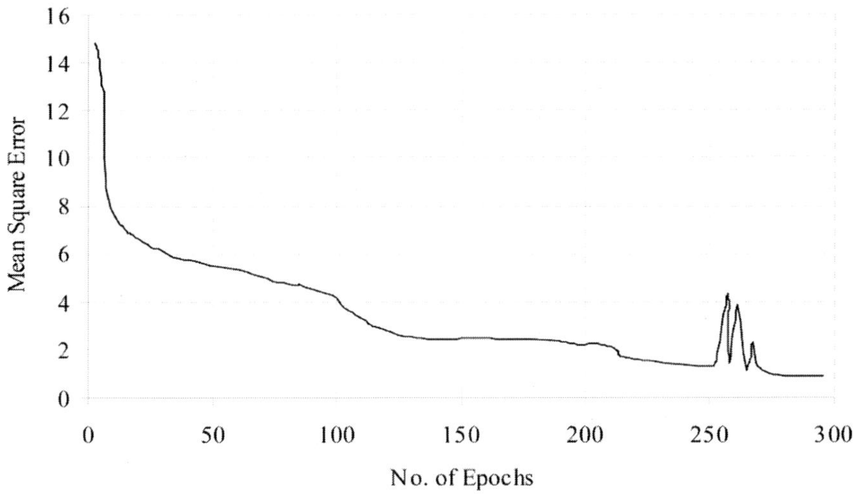

Figure 7. Training time error for the diabetes data.

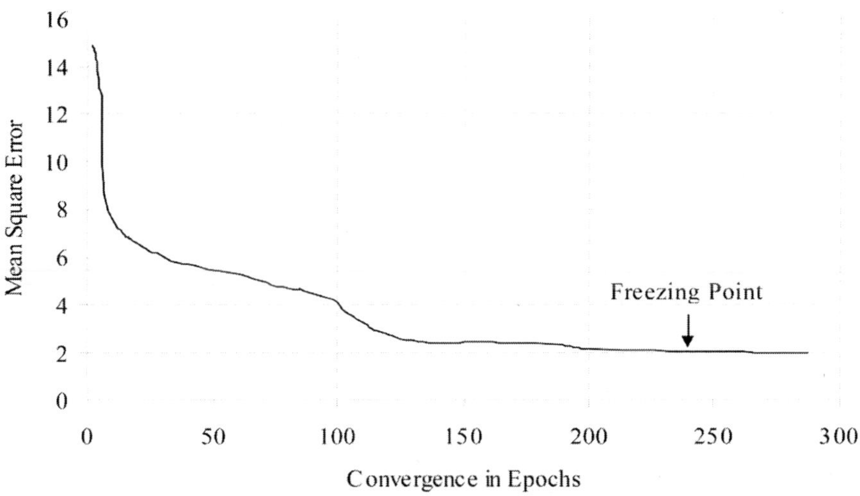

Figure 8. Training time error for the diabetes data with weight freezing.

4.4. Extracted Rules

The number of rules extracted by the ESRNN algorithm and the accuracy of the rules is presented in Table 9, but the visualization of the rules in terms of the original attributes were not discussed. This subsection discusses the rules extracted by ESRNN in terms of the original attributes. The number of conditions per rule and the number of rules extracted have also visualized here.

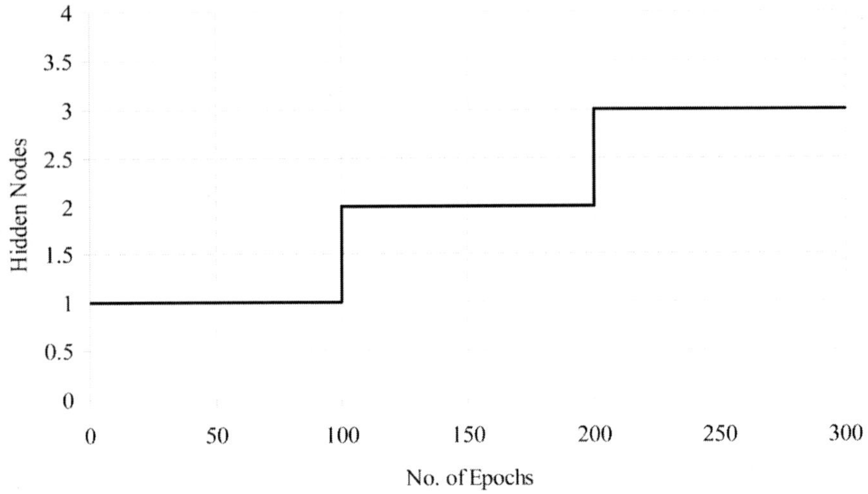

Figure 9. Hidden node addition for the diabetes data.

Table 9. Number of extracted rules and rules accuracies.

Sl No.	Datasets	No. of Extracted Rules	Accuracy
1	Diabetes	2	76.56%
2	Iris	3	98.67%
3	Wine	3	91.01%
4	Season	4	100%
5	Golf Playing	3	100%
6	Lenses	8	100%

The diabetes data

Rule 1: If Plasma glucose concentration (A_2) <= 0.64 and Age (A_8) <= 0.69 then tested negative

Default Rule: tested positive.

The iris data

Rule 1: If Petal-length (A_3) <= 1.9 then iris setosa

Rule 2: If Petal-length (A_3) <= 4.9 and Petal-width (A_4) <= 1.6 then iris versicolor

Default Rule: iris virginica.

The wine data

Rule 1: If Input 10 (A_{10}) <= 3.8 then class 2

Rule 2: If Input 13 (A_{13}) >= 845 then class 1

Default Rule: class 3.

The season data

Rule 1: If Tree (A2) = yellow then autumn

Rule 2: If Tree (A2) = leafless then autumn

Rule 3: If Temperature (A3) = low then winter

Rule 4: If Temperature (A3) = high then summer

Default Rule: spring.

The golf playing data

Rule 1: If Outlook (A1) = sunny and Humidity >=85 then don't play

Rule 2: Outlook (A1) = rainy and Wind= strong then don't play

Default Rule: play.

The lenses data

Rule 1: If Tear Production Rate (A4) = reduce then no contact lenses

Rule 2: If Age (A1) = presbyopic and Spectacle Prescription (A2) = hypermetrope and Astigmatic (A3) = yes then no contact lenses

Rule 3: If Age (A1) = presbyopic and Spectacle Prescription (A2) = myope and Astigmatic (A3) = no then no contact lenses

Rule 4: If Age (A1) = pre-presbyopic and Spectacle Prescription (A2) = hypermetrope and Astigmatic (A3) = yes and Tear Production Rate (A4) = normal then no contact lenses

Rule 5: If Spectacle Prescription (A2) = myope and Astigmatic (A3) = yes and Tear Production Rate (A4) = normal then hard contact lenses

Rule 6: If Age (A1) = pre-presbyopic and Spectacle Prescription (A2) = myope and Astigmatic (A3) = yes and Tear Production Rate (A4) = normal then hard contact lenses

Rule 7: If Age (A1) = young and Spectacle Prescription (A2) = myope and Astigmatic (A3) = yes and Tear Production Rate (A4) = normal then hard contact lenses

Default Rule: soft contact lenses.

Table 9 shows the number of extracted rules and the rules accuracy for a set of benchmark data mining problems. In most of the cases ESRNN produces fewer rules with better accuracy. It was observed that two to three rules were sufficient to solve the problems. The accuracies were 100% for three datasets including season, golf playing, and lenses classification. These datasets have a lower number of examples.\

4.5. Performance Comparisons

This section compares experimental results of the ESRNN algorithm with the results of other works. The primary aim of this work is not to evaluate ESRNN in order to gain a deeper understanding of rule generation without an exhaustive comparison between ESRNN and all other works. Table 10 compares ESRNN results of the diabetes data with those produced by PMML [29], NN RULES [14], C4.5 [46], NN-C4.5 [51], OC1 [51], and CART [52] algorithms. ESRNN achieved 76.56% accuracy although NN-C4.5 was closest second with 76.4% accuracy. Due to the high noise level, the diabetes problem is one of the most challenging problems in our experiments. ESRNN has outperformed all other algorithms. Table 11 compares ESRNN results of the iris data with those produced by PMML [29], NN RULES [14], DT RULES [14], BIO RE [24], Partial RE [24], and Full RE [24] algorithms. ESRNN achieved 98.67% accuracy although NN RULES was closest second with 97.33% accuracy. Here number of rules extracted by ESRNN and NN RULES are equal. Table 12 shows ESRNN results of the wine data. ESRNN achieved 91.01% accuracy by generating 3 rules. No detailed previous work have found for showing comparison of this dataset.

Table 10. Performance comparison of the ESRNN with other algorithms for the diabetes data.

Dataset	Feature	ESRNN	PMML	NN RULES	C4.5	NN-C4.5	OC1	CART
Diabetes	No. of Rules	2	2	4	–	–	–	–
	Avg. No. of Conditions	2	1	3	–	–	–	–
	Accuracy (%)	76.56	75	76.32	70.9	76.4	72.4	72.4

Table 11. Performance comparison of the ESRNN algorithm with other algorithms for the irish data.

Dataset	Feature	ESRNN	PMML	NN RULES	DT RULES	BIO RE	Partial RE	Full RE
Irish	No. of Rules	3	3	3	4	4	6	3
	Avg. No. of Conditions	1	1	1	1	3	3	2
	Accuracy (%)	98.67	91.3	97.33	94.67	78.67	78.67	97.33

Table 12. Performance of the ESRNN algorithm for the wine data.

Dataset	Feature	ESRNN
Wine	No. of Rules	3
	Avg. No. of Conditions	3
	Accuracy (%)	91.01

Table 13 compares the ESRNN results of the season data with those produced by RULES [53] and X2R [25]. All three algorithms achieved 100% accuracy. This is possible because the number of examples is low. ESRNN extracted five rules, whereas RULES extracted seven and X2R six.

Table 13. Performance comparison of ESRNN with other algorithms for season data.

Dataset	Feature	ESRNN	RULES	X2R
Season	No. of Rules	5	7	6
	Avg. No. of Conditions	1	2	1
	Accuracy (%)	100	100	100

Table 14 compares ESRNN results of golf playing data with those produced by RULES [53], RULES-2 [54], and X2R [25]. All four algorithms achieved 100% accuracy because the lower number of examples. Number of extracted rules by ESRNN are three whereas these were eight for RULES and 14 for

RULES-2. Finally, Table 15 compares ESRNN results of lenses data with those produced by PRISM [55]. Both algorithms achieved 100% accuracy because the lower number of examples. Number of extracted rules by ESRNN are eight whereas they were nine for PRISM.

Table 14. Performance comparison of ESRNN with other algorithms for golf playing data.

Dataset	Feature	ESRNN	RULES	RULES-2	X2R
Golf Playing	No. of Rules	3	8	14	3
	Avg. No. of Conditions	2	2	2	2
	Accuracy (%)	100	100	100	100

Table 15. Performance comparison of ESRNN with other algorithm for lenses data.

Dataset	Feature	ESRNN	PRISM
Lenses	No. of Rules	8	9
	Avg. No. of Conditions	3	–
	Accuracy (%)	100	100

5. CONCLUSIONS

In this paper we have presented a neural network based data mining scheme to mining classification rules from given databases. This work is an attempt to apply the connectionist approach to data mining by extracting symbolic rules similar to that of decision trees. An important feature of the proposed rule extraction algorithm is its recursive nature. A set of experiments was conducted to test the proposed approach using a well defined set of data mining problems. The results indicate that, using the proposed approach, high quality rules can be discovered from the given data sets. The extracted rules are concise, comprehensible, order insensitive, and do not involve any weight values. The accuracy of the rules from the pruned network is as high as the accuracy of the fully connected networks. Experiments showed that this method helped a lot to reduce the number of rules significantly without sacrificing classification accuracy. In almost all cases ESRNN outperformed the others. With the rules extracted by the method introduced here, ANNs should no longer be regarded as black boxes.

ACKNOWLEDGMENTS

This work was supported by Hankuk University of Foreign Studies Research Fund of 2011.

REFERENCES

1. Wang, L; Sui, TZ. Application of data mining technology based on neural network in the engineering. Proceedings of International Conference on Wireless Communications, Networking and Mobile Computing, Shanghai, China, 21–25 September 2007; pp. 5544–5547.

2. Nirkhi, S. Potential use of artificial neural network in data mining. Proceedings of International Conference on Computer and Automation Engineering, Singapore, 26–28 February 2010; pp. 339–343.

3. Andrews, R; Diederich, J; Tickle, AB. Survey and critique of techniques for extracting rules from trained artificial neural networks. *Knowl. Based Syst* **1995**, *8*, 373–389.

4. Setiono, R. Extracting rules from neural networks by pruning and hidden-unit splitting. *Neural Comput* **1997**, *9*, 205–225.

5. Darbari, A. Rule Extraction from Trained ANN: A Survey; Technical Report;. Department of Computer Science, Dresden University of Technology: Dresden, Germany, 2000.

6. Biryulev, C; Yakymiv, Y; Selemonavichus, A. Research of artificial neural networks usage in data mining and semantic integration. Proceedings of International Conference on Perspective Technologies and Methods in MEMS Design, Lviv, Ukraine, 20–23 April 2010; pp. 144–149.

7. Ni, X. Research of data mining based on neural networks. *World Academy Sci Eng Tech* **2008**, *39*, 381–384.

8. Vesely, A. Neural networks in data mining. *Agric. Econ* **2003**, *49*, 427–431.

9. Singh, Y; Chauhan, AS. Neural networks in data mining. *J. Theo. and App. Inf. Tech* **2009**, *5*, 37–42.

10. Baesens, B; Setiono, R; Mues, C; Vanthienen, J. Using neural network rule extraction and decision tables for credit-risk evaluation. *Manage. Sci* **2003**, *49*, 312–329.

11. Jacobsson, H. Rule extraction from recurrent neural networks: A taxonomy and review. *Neural Comput* **2005**, *17*, 1223–1263.

12. Kahramanli, H; Allahverdi, N. Rule extraction from trained adaptive neural networks using artificial immune systems. *Expert Syst. Appl* **2009**, *36*, 1513–1522.

13. Setiono, R; Baesens, B; Mues, C. A note on knowledge discovery using neural networks and its application to credit screening. *Eur. J. Operation. Res* **2009**, *192*, 326–332.

14. Tickle, AB; Andrews, R; Golea, M; Diederich, J. The truth will come to light: Directions and challenges in extracting the knowledge embedded within trained artificial neural networks. *IEEE Trans. Neural Netw* **1998**, *9*, 1057–1067.

15. Setiono, R; Leow, WK. FERNN: An algorithm for fast extraction of rules from neural networks. *Appl. Intell* **2000**, *12*, 15–25.

16. Setiono, R; Liu, H. Symbolic representation of neural networks. *IEEE Comput* **1996**, *29*, 71–77.

17. Odajima, K; Hayashi, Y; Gong, T; Setiono, R. Greedy rule generation from discrete data and its use in neural network rule extraction. *Neural Netw* **2008**, *21*, 020–1028.

18. Setiono, R; Liu, H; Lu, H. Effective data mining using neural networks. *IEEE Trans. Knowl. Data En* **1996**, *8*, 957–961.

19. Craven, M; Shavlik, J. Using neural networks for data mining. *Future Gener. Comput. Syt* **1997**, *13*, 211–229.

20. Anbananthen, SK; Sainarayanan, G; Chekima, A; Teo, J. Data mining using pruned artificial neural network tree (ANNT). Proceedings of IEEE International Conference on Information & Communication Technologies, Damascus, Syria, 24–28 April 2006; pp. 1350–1357.

21. Saito, K; Nakano, R. Medical diagnosis expert system based on PDP model. Proceedings of IEEE International Conference on Neural Networks, San Diego, CA, USA, 24–27 July 1988; pp. 1255–1262.

22. Towell, GG; Shavlik, JW. Extracting refined rules from knowledge-based neural networks. *Mach. Learn* **1993**, *13*, 71–101.

23. Fu, L. Rule learning by searching on adapted nets. Proceedings of National Conference on Artificial Intelligence, Anaheim, CA, USA; 1991; pp. 590–595.

24. Towell, GG; Shavlik, JW. Knowledge-based artificial neural networks. *Artif. Intell* **1994**, *70*, 119–165.

25. Liu, H; Tan, ST. X2R: A fast rule generator. Proceedings of IEEE International Conference on Systems, Man and Cybernetics, Vancouver, BC, Canada, 22–25 October 1995; pp. 1631–1635.

26. Liu, H. A family of efficient rule generators. In *Encyclopedia of Computer Science and Technology*; Marcel Dekker Inc: New York, NY, USA, 1998; Volume 39, pp. 15–28.

27. Setiono, R. Extracting M-of-N rules from trained neural networks. *IEEE Trans. Neural Netw* **2000**, *11*, 512–519.

28. Kamruzzaman, SM; Islam, MM. An algorithm to extract rules from artificial neural networks for medical diagnosis problems. *Int. J. Inf. Tech* **2006**, *12*, 41–59.

29. Jin, Y; Sendhoff, B. Pareto-based multiobjective machine learning: An overview and case studies. *IEEE Trans. Syst. Man Cybern. C Appl. Rev* **2008**, *38*, 397–415.

30. Wang, J; Zhang, W; Qin, B; Shi, W. Research on rules extraction from neural network based on linear insertion. Proceedings of WASE International Conference on Information Engineering, Beidaihe, Hebei, China, 14–15 August 2010; pp. 33–36.

31. Kwok, TY; Yeung, DY. Constructive algorithms for structured learning in feedforward neural networks for regression problems. *IEEE Trans. Neural Netw* **1997**, *8*, 630–645.

32. Islam, MM; Yao, X; Murase, K. A constructive algorithm for training cooperative neural network ensembles. *IEEE Trans. Neural Netw* **2003**, *14*, 820–834.

33. Parekh, R; Yang, J; Honavar, V. Constructive neural network learning algorithms for pattern classification. *IEEE Trans. Neural Netw* **2000**, *11*, 436–451.

34. Ash, T. Dynamic node creation in backpropagation networks. *Connect. Sci* **1989**, *1*, 365–375.

35. Fahlman, SE; Lebiere, C. The cascade-correlation learning architecture. *Adv. Neural Inf. Proc. Syst* **1990**, *2*, 524–532.

36. Setiono, R; Hui, LCK. Use of quasi-Newton method in a feedforward neural network construction algorithm. *IEEE Trans. Neural Netw* **1995**, *6*, 273–277.

37. Kamruzzaman, SM; Hasan, AR; Siddiquee, AB; Mazumder, MEH. Medical diagnosis using neural network. Proceedings of International Conference on Electrical and Computer Engineering, Dhaka, Bangladesh, 28–30 December 2004; pp. 537–540.

38. Islam, MM; Akhand, MAH; Rahman, MA; Murase, K. Weight freezing to reduce training time in designing artificial neural networks. Proceedings of International Conference on Computer and Information Technology, Dhaka, Bangladesh, 27–28 December 2002; pp. 132–136.

39. Sietsma, J; Dow, RJF. Neural net pruning-why and how? Proceedings of IEEE International Conference on Neural Networks, San Diego, CA, USA, 24–27 July 1988; pp. 325–333.

40. Ooyen, A; Nienhuis, B. Improving the convergence of backpropagation algorithm. *Neural Netw* **1992**, *5*, 465–471.

41. Reed, R. Pruning algorithms-A survey. *IEEE Trans. Neural Netw* **1993**, *4*, 740–747.

42. Jiawei, H; Kamber, M. *Data Mining: Concepts and Techniques*; Morgan Kaufmann: San Fransisco, CA, USA, 2001.

43. Kaufman, L; Rousseeuw, PJ. *Finding Groups in Data: An Introduction to Cluster Analysis*; John Wiley & Sons: New York, NY, USA, 2005.

44. Raymond, TN; Han, J. Efficient and effective clustering methods for spatial data mining. Proceedings of International Conference on Very Large Data Bases, Santiago de Chile, Chile, 12–15 September 1994; pp. 144–155.

45. Islam, MM; Murase, K. A new algorithm to design compact two hidden-layer artificial neural networks. *Neural Netw***2001**, *4*, 1265–1278.

46. Quinlan, JR. *C45: Programs for Machine Learning*; Morgan Kaufmann: San Fransisco, CA, USA, 1993.

47. Agrawal, R; Imielinski, T; Swami, A. Database mining: A performance perspective. *IEEE Trans. Knowl. Data Eng* **1993**,*5*, 914–925.

48. Yen, S-J; Chen, ALP. An efficient algorithm for deriving compact rules from databases. Proceedings of International Conference on Database Systems for Advanced Applications, Singapore, 11–13 April 1995; pp. 364–371.

49. Murphy, PM; Aha, DW. *UCI Repository of Machine Learning Databases, (Machine-Readable Data Repository)*; Department of Information and Computer Science, University of California: Irvine, CA, USA, 1998. Available online: http://archive.ics.uci.edu/ml/ (accessed on 12 April 2011).

50. Prechelt, L. *PROBEN1-A Set of Neural Network Benchmark Problems and Benchmarking Rules*; Technical Report 21/94; Fakultat fur Informatik, Universitat Karlsruhe: Karlsruhe, Germany, 1994.

51. Setiono, R. Techniques for extracting rules from artificial neural networks. Proceedings of International Conference on Soft Computing and Information Systems, Iizuka, Japan, 16–20 October 1998.

52. Breiman, L; Friedman, J; Olshen, R; Stone, C. Classification and Regression Trees; Wadsworth and Brooks: Monterey, CA, USA, 1984.

53. Pham, DT; Aksoy, MS. RULES: A simple rule extraction system. *Expert Syst. Appl* **1995**, *8*, 59–65.

54. Pham, DT; Aksoy, MS. An algorithm for automatic rule induction. *Artif. Intell. Eng* **1994**, *8*, 277–282.

55. Cendrowska, J. PRISM: An algorithm for inducting modular rules. *Int. J. Man Mach. Stud* **1987**, *27*, 349–370.

CHAPTER 5

Biological Inspiration—Theoretical Framework Mitosis Artificial Neural Networks Unsupervised Algorithm

Lácides Pinto Mindiola, Gelvis Melo Freile, Carlos Socarras Bertiz

Universidad de La Guajira, Riohacha, Colombia

ABSTRACT

The modified approach to conventional Artificial Neural Networks (ANN) described in this paper represents an essential departure from the conventional techniques of structural analysis. It has four main distinguishing features: 1) it introduces a new simulation algorithm based on the biology; 2) it performs relatively simple arithmetic as massively parallel, during analysis of a structure; 3) it shows that it is possible to use the application of the modified approach to conventional ANN to solve problems of any complexity in the field of structural analysis; 4) the Neural Topologies for Structural Analysis (NTSA) system are recurrent networks and its outputs are connected to its inputs [1] and [2] . In NTSA system the DNA of the neuron mother and daughters would be defined by: 1) the same entry, from the corresponding neuron in the previous layer; 2) the same trend vector; 3) the same transfer function (purelin). The mother's neuron and her daughter's neuron differ only in the connection weight and its output signal.

KEYWORDS

Mitosis, Artificial Neuron, Node, Structural Analysis, Neural Networks, Output, Layer, Simulation

1. INTRODUCTION

The ADALINE network is a fairly well known ANN and is very similar to the perceptron except that its transfer function is linear. Since its invention by Bernard Widrow and his graduate student Marcian Hoff in 1960, we have considered a lineal ANN of several layers with simple processors also called "neuron", "node", and "pro- cessing elements". Each layer has its own weight matrix W, its own bias vector b (the bias is much like a weight except that it has a constant input of 1), a net input vector n and a output vector a$(a = purelin(wp + b))$. Note that the scalar input p is multiplied by the scalar weight w to form wp.

A NTSA is an information processing system which operates on inputs to extract information, and produces outputs corresponding to the extracted information. NTSA means that the "distribution factors" in the nodes of the structures are the components of the input vector to the first layer. After a complete presentation of the training data, a new set of weights and biases are obtained, and new outputs are again evaluated in a feed-forward manner until a specific tolerance for error is obtained. Unsupervised training uses unlabeled training data and requires no external teaching.

A NTSA model is composed of simple processors, each having a local memory. Processing elements are con- nected by unidirectional links that carry discriminating data; the linear feed-forward net has been found to be a suitable one for training techniques. Outputs of neurons in one layer are transferred to their corresponding neuron in another layer through a link that amplifies or inhibits such outputs through weighting factors [3] .

This paper evaluates a neural network approach on frame analysis using an unsupervised algorithm. The results are obtained programming the entire formulation of the algorithm using MATLAB. The aim of the study is to estimate the rotational end moment.

The modified neuronal structural analysis requires the addition of a set of new concepts but simple addition of traditional, also called classic. All of them eventually make themselves the nature of a new approach, whose theoretical and conceptual frameworks then present them in a concise form, but as clear as possible, so that it is accessible to all and everyone disposing of the basic conceptual tools tries to address that this new theoretical method is practical.

Artificial Neural Networks (ANN) is able to perform relatively simple arithmetic as massively parallel, during analysis of a structure. To this end, an urgent need is to define the terms: matrices and vectors weights trends, using the parameters derived from the physical and mechanical members.

We do not intend here to discuss the basic principles of analysis of statically indeterminate structures, nor the plausibility of traditional approaches. We simply want to show that it is possible to use the application of the modified approach or conventional artificial neural networks to solve problems of any complexity in the field of structural analysis.

Electronic neural networks generally exist as computer simulations. And that they are typically designed as very large. The ability to simulate is limited by the speed and storage capacity of digital computers available. Some researchers have developed hardware in order to increase processing speed, but conditions are not given to facilitate this event parallel [4] . However, there are two key similarities between biological and artificial neural networks. Firstly, the building blocks of the two kinds of networks are simple computational devices, although artificial neurons are much simpler, so that biological neurons are highly interconnected. Secondly, the connection among the processing elements determines the function of the network. The main objective of this paper is to determine the application of the modified approach or conventional ANN to solve problems of any complexity in the field of structural analysis. The work elaborated the building of the NTSA models as well as its features with the appropriate connections to solve particular problems [1] .

2. NOTATION AND TERMINOLOGY

In this work, figures, mathematical equations and text discussing both figures and mathematical equations will use the following notation.

Basic Concepts

Scalars: Small italic letters a, b, c

Vectors: Small bold nonitalic letters a, b, c

Matrices: Capital bold nonitalic letters A, B, C

A. Structures

We append the number of the layer as a superscript to the rotational moments. Thus, the rotational for the first layer and node four on the right end of the beam is written as $M^1 4_{vd}$, and the rotational moment for the second layer and node five on the bottom column is written as $M^2 5_{ci}$.Figure 1 and Figure 2 show this notation.

End Moments

End Rotational Moments

$M^L n_{sub}$

L—layer, n—node, sub—viz, vd, cs, ci

Viz—left beam

Vd—right beam

Cs—top column

Ci—bottom column

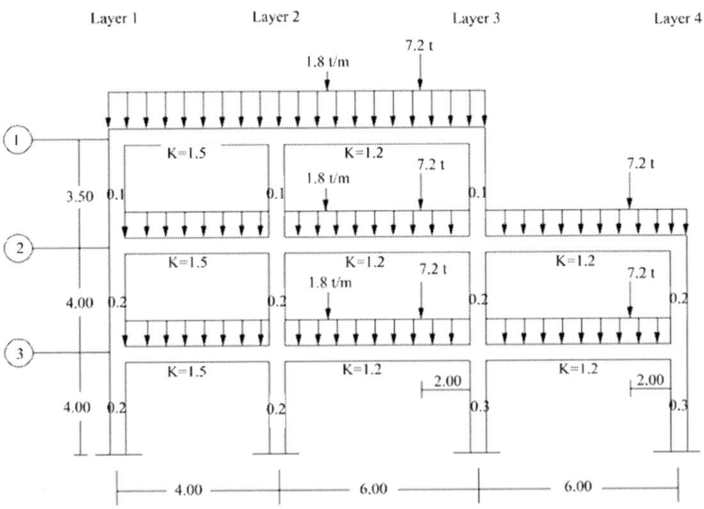

Figure 1. Primitive structure.

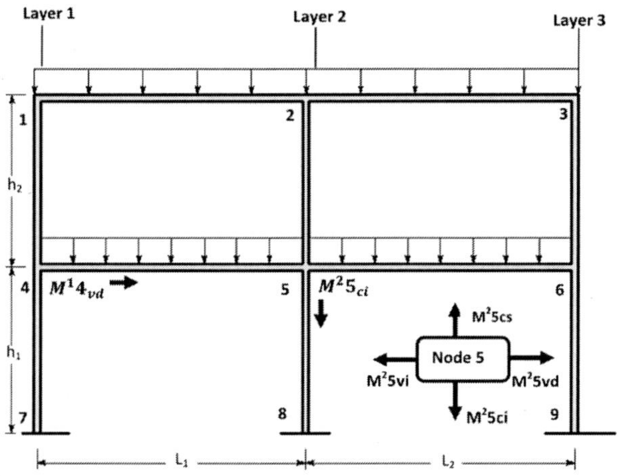

Figure 2. Structure—notation.

Horizontal Displacements

$M^L \Delta H_{i,j}$

ΔH —horizontal displacement, i—top column, j—bottom column

Moments of Perfect Embedding

$ME^L_{i,j}$

$ME^L_{j,i}$

i, j —Extremes

Sum of the Moments of Perfect Embedding

$\sum_{(n)} ME$

B. Neural Networks Model

We need to introduce some additional notation concerning the network architectures. See Figure 3and Figure 4.

Weight Matrices:

Scalar Element

$$w_{i,j}^{L}$$

i—row, j—column, L—layer

Matrix

$$W^{L}$$

Column Vector

$$w_{j}^{L}$$

Row Vector

$${}_{i}w^{L}$$

Bias Vector:

Scalar Element

$$b_{i}^{L}$$

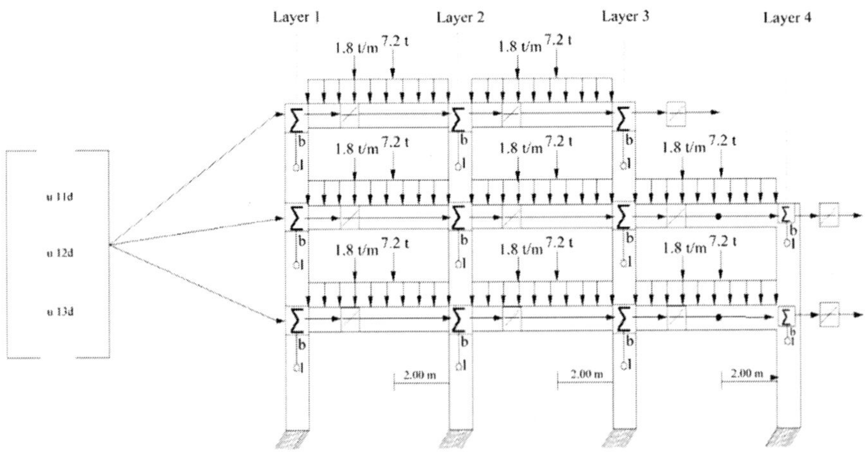

Figure 3. Biological inspiration NTSA system primitive.

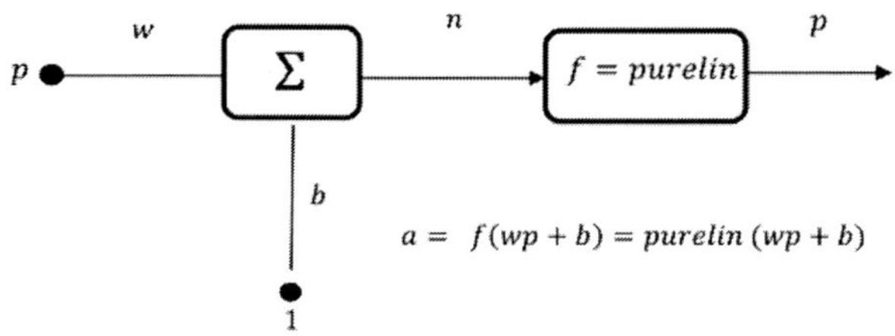

Figure 4. Single-input neuron.

Vector

$\boldsymbol{b}^L$

Input Vector:

Scalar Element

p_i

Net Input Vector:

Scalar Element

n_i^L

Vector

$\boldsymbol{n}^L$

Output Vector:

Scalar Element

a_i^L

Vector

$\boldsymbol{a}^L$

Transfer Function

Scalar Element

$$a_i^L$$

Vector

$$\boldsymbol{a}^L = f^L\left(\boldsymbol{n}^L\right) = purelin^L\left(\boldsymbol{n}^L\right)$$

Multi Layers of Neural networks

Layers Superscript

Input Vector: p

Output Vector: a

Bias Vector: b

Net input Vector: n

Weight Matrix: W

Input Number: R

Neuron Numbers for Layer: S

2. GENERAL ARCHITECTURE NTSA

A NTSA is an information processing system which operates on inputs to extract information, and produces outputs corresponding to the extracted information. NTSA model is composed of simple processors, each having a local memory. Processing elements are connected by unidirectional links that carry discriminating data the linear feed-forward net has been found to be a suitable one for training techniques. Outputs of neurons in one layer are transferred to their corresponding neuron in another layer through a link that amplifies or inhibits such outputs through weighting factors. Except for the processing elements of the input layer, the input of each neuron is the sum of the weighted outputs of the node in the prior layer and a bias. Each neuron is activated according to its input, transfer function, and threshold value [5] .

There exist a variety of ANN models and learning procedures. Feed-forward networks are well known approaches for prediction and database processing applications. In this type (NTSA), the weighted and biases links feed activation functions from the input layer to the output layer in forward direction.

Figure 3 shows the general feed-forward multilayer networks model, including two hidden layers. The "distribution factors" beams [6] , of the input layer constitutes the neurons inputs in layer (L_1), representing a set of variables.

$M^1 4_{vd}$ Rotational Moment Node4 Layer1, Right Hand Beam

$M^1 5_{vd}$ Rotational Moment Node5 Layer2, Down Column

A single-input neuron is shown in Figure 2. The scalar input p is multiplied by the scalar weight w to form wp, one of the terms that are sent to the summer. The neuron has a bias b, which is summed with the weighted input to form the net input n, goes into transfer function f (linear function), which produces the scalar neuron output a.

If we related this simple model of the neuron with biological neuron, then the weight w corresponds to the strength of a synapse, the neuron body is represented by the summation and the transfer function, and the neuron output a represents the signal on the axon [5] .

$$(x_1, x_2, x_3)^T$$

The inputs and outputs for the i_{th} neuron are:

$$n_i^L = W_i^L a_i^{(L-1)} + b_i^L \tag{1}$$

$$a_i^L = f_i\left(n_i^L\right) = f_i^L\left(W_i^L a_i^{(L-1)} + b_i^L\right) \tag{2}$$

where f_i constitutes an activation function linear, its behavior is that of a threshold function, in which the output of the neuron is generated if a threshold level, is reached. The net input and output o the j_{th} neuron are similarity treated as in (1) and (2).

Typically, the "activation function" (purelin) is chosen by the designer and then the parameters w and b will be adjusted by some learning algorithm so that the neuron input/output relationship meets some specific goal.

2.1. Multiple Layers of Neuron

Generally, one neuron with many inputs may not be sufficient. We might need three, four or seven, operating in parallel, in what we will call a layer.

Yet, can be considered a network with several layers. Each layer has its own weight matrix $\boldsymbol{W}$, its own bias vector $\boldsymbol{b}$, a net vector $\boldsymbol{n}$ and an output vector $\boldsymbol{a}$.

The first layer has $\boldsymbol{S}^1$ neurons, $\boldsymbol{S}^2$ neurons in the second layer, etc. As noted, different layers can have different numbers of neurons. Thus, the weight matrix for the first layer is written as $\boldsymbol{W}^1$, and the weight matrix for the second layer is written as $\boldsymbol{W}^2$. This notation is used in all network models of the NTSA system. The outputs of the layers one, two, and three are the inputs for layers two, three and four. Thus, layer three can be viewed as a one-layer network with $\boldsymbol{S}^3$ inputs.

There is, one correspondence between the S^1 neurons of the first layer, S^2 of the second layer neurons, the neurons S^3 of the third layer as long as $S^1 = S^2 = S^3 = \cdots$ and so on, so that the output of a neuron of the first layer is the input to the corresponding neuron in the second layer and so on, that is, the output of the L_{th} layer or output layer is $\boldsymbol{a}^L$ vector:

$$
\begin{array}{l}
a_1^1 \to a_1^2 \to a_1^3 \to \cdots \to a_1^L \\
a_2^1 \to a_2^2 \to a_2^3 \to \cdots \to a_2^L \\
a_3^1 \to a_3^2 \to a_3^3 \to \cdots \to a_3^L \\
\qquad\quad \vdots \\
a_N^1 \to a_N^2 \to a_N^3 \to \cdots \to a_N^L ,
\end{array}
$$

A layer whose output is the network output is called an output layer. The others layers are called “hidden layers”. The network of Figure 3 has an output (layer 4) and three hidden layers (layers 1, 2, and 3).

2.2. Recurrent Networks

The NTSA system is a recurrent network with feedback; some of its outputs are connected to the inputs subsequent network, put it in some way. This is quite different from the networks models that we have studied thus far, which were strictly feed-forward with no backward connection. The NTSA is of the forward-backward class by the system architecture.

3. ARTIFICIAL NEURAL MITOSIS (ANM)

In general, natural biological organisms are much more complicated than the automatic devices. However, some peculiarities that we observe in the organization of the first and the way they perform certain operations or behavior can serve as a reference to the inspiration of an approach required to solve a family of common problems in real life. The set of experiences and difficulties we face when we operate certain automatic devices can be the source of the interpretations of the physiological systems, mainly human and some vertebrate in particular.

The man has been inspired by the central nervous system of humans, which is actually the most difficult of all; which it has been the source of a biological inspiration that led to the creation of some automatic devices such as artificial neural networks. Such devices sometimes have certain limitations in its design and operation, when it comes to solve a certain class of problems. Consequently, we could ensure that such systems are still evolving. This means that the system is open to change, or equivalently; network models support new paradigms; without this, it's essential, such as learning and massive parallelism goals are violated.

The central nervous system of humans is of unlimited complexity. For their study, it is taken to biological neurons as independent units. Thus, it is isolated as the first stage of the problem; the structure and workings of such individual elemental processing units. The second part of the problem aims to understand how neurons are organized into a whole; and how that operation of all expressed, starting logically based on these individual elements of information processing.

Neuron is considered as automatic physiological devices as a "black box", which react to the presence of certain stimuli and issues a response as independent functional units. They have clearly defined characteristics, i.e., neurons that receive signals can be of two types excitatory or inhibitory. To stimulate a neuron is necessary that this receive excitatory stimuli. After a certain time, the neuron will issue "one and only one" output pulse (signal).

What would you do if a neuron of the output layer of a network, like a node of a structure, had to pass over a different signal to other neurons or to the outside world? Another question, why, if the artificial neuron is a real operation of a biological neuron very large simplification, cannot inspire and create an artificial neuron that go further; and generate neurons daughters with the same DNA? The DNA of the neuron mother and daughters would be defined by: 1) the same entry, from the corresponding neuron in the previous layer; 2) the same trend vector; 3) the same transfer function. The mother and her daughter's neuron differ only in the connection weight and its output signal.

In this evolutionary scenario, it is another question. Is it possible to think of a neuronal structure and function, to incorporate the concept of neuronal multiplication? Yes, it's a positive response. Then, the artificial neural mitosis could be a feature attributable to the artificial neurons, though, does not share this property essentially biological equivalent.

There are a variety of kinds of designs and learning techniques that are enriched with its own peculiarities, which can generate or manufacture under its own power users and meeting their own needs. Let us not forget that the field of neural networks is in the throes of evolution and growth; and therefore, for their development, much remains to be done.

The synthesis of biological neurons is important and difficult because it is born the human brain. In it, we were inspired to produce electronic devices or artificial neural networks, which are the synthesis of artificial neurons. Artificial neural networks are characterized by their autonomy and logic flexibility. Running, at least part of the functions of the central nervous system [3] .

Figure 5(a) and Figure 5(b) illustrate about the kinds of biological cells, which serve as inspiration to neuro- structural analysis. The first is a biological neuron. The second is a cell of the class that forms the intestinal epithelium and the epidermis. The latter cells are able to multiply, to generate through a stem cell mitosis two or more identical to the mother daughters cells genetically. In other words, mitosis is a process of equitable sharing of genetic material, DNA [7] .

Figures 6(c)-(e) show different versions of what could be the "mitosis of an artificial neuron" nodes in different frame structure. It can be seen at the bottom of the typical node scheme, a multilevel cross linked structure. Each node needs to transmit, two, three, four signals. Only after acceptance of the phenomenon of artificial neural mitosis, which replaces the neuron could meet the demand of the number of required signals.

If we accept the above approach, then we have met the expectations of the first part of the expansion of the framework of our theory, where the DNA of the neuron mother and her daughter's neurons is: DNA = Common entrance, from the corresponding neuron of the previous layer, as trend vector, same linear transfer function.

If a neuron of the output layer of a network, like a node of a structure had to transmit over a different signal to the external world. We say that is not possible, because, you're right is currently denied. However, in another scenario, as it, the available cells of the epidermis and corresponding to the intestinal epithelium, it would appear, through a biologically inspired by modifying the structure and neuronal functioning, so that the attribute Property neuronal multiplication (mitosis neuronal) is assigned to the artificial neuron. This would be debatable, but viable. Let's concede now, this transcendental to artificial neuron, as a result of biological inspiration derived from some special cells, such as the class of the above named property. Consequently, the Artificial Neural Mitosis (MNA), will become a feature attributable to the artificial neurons, though, this essentially not share with their biological equivalent.

4. GENERAL ARCHITECTURE OF PROPOSED NETWORK

To bring about the first artificial neural network, a neuron is placed on each node of the network structure proposal. Thus, we have a one to one correspondence between nodes and gantry neurons of the various layers of the neural network thus formed [8] . To generate further models neural networks, neuro-structural model primitive neuro be divided into sub-models derived (sub-domains). This is achieved by eliminating successively, the last model obtained, and the layer of the right to illustrate the statement above with Figure 7.

The process of building the sub-neural network models starts with the artificial mitosis, the last layer in the first model (original model). The remaining sub-models of neuro-structural derivatives, network obtained after removal of the respective output layer network model precedent, then, produce mitosis of the output layer of the new sub-model.

All neuro-network models thus obtained operate in a massively parallel and have as output vector, the ends rotational moments associated with the nodes of the output layer. The calculation process is iterative and each network will be implemented through successive approximations; and the same, will be controlled by the learning algorithm. Thus, we can know the outputs of the hidden layers of the original multi-cross-linked network. It is important to note that each network model will shed their own results. Values are retained as data, to operate the following neuro-network model. Finally, the matrix of the results of total rotational moments at the ends of all members of the structure is obtained.

All neuro-derived models have the same input vector.

$\boldsymbol{p} = \left(\mu 11_{vd}\, \mu 12_{vd}\, \mu 13_{vd} \cdots \mu 1N_{vd}\right)^{T}$. An exception, the derivative of neuro-layer model, whose components of the input vector are all factors of rotation of the ends of the rods, which contribute to the nodes of the first layer, that is,

$$\boldsymbol{p} = \left(\mu 11_{vd}\, \mu 11_{ci}\, \mu 12_{cs}\, \mu 12_{vd}\, \mu 12_{ci} \cdots \mu 1(3N-2)_{vd}\, \mu 1(3N-1)_{ci}\right)^{T}$$

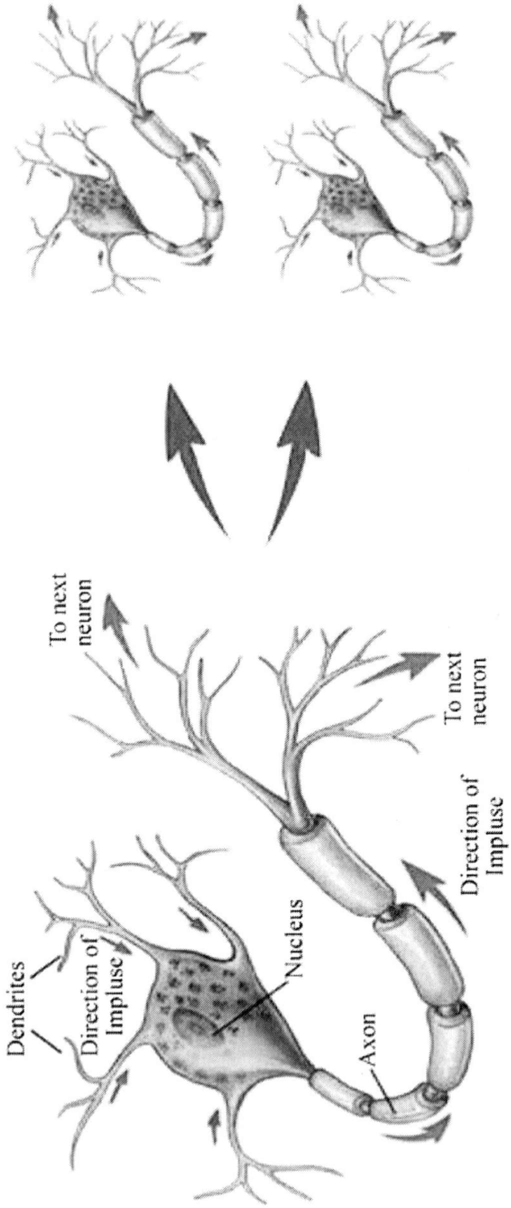

https://www.studyblue.com/notes/note/n/biological-foundations-neuron-communication-/deck/1025438 [8]

Figure 5. Biological cell mitosis.

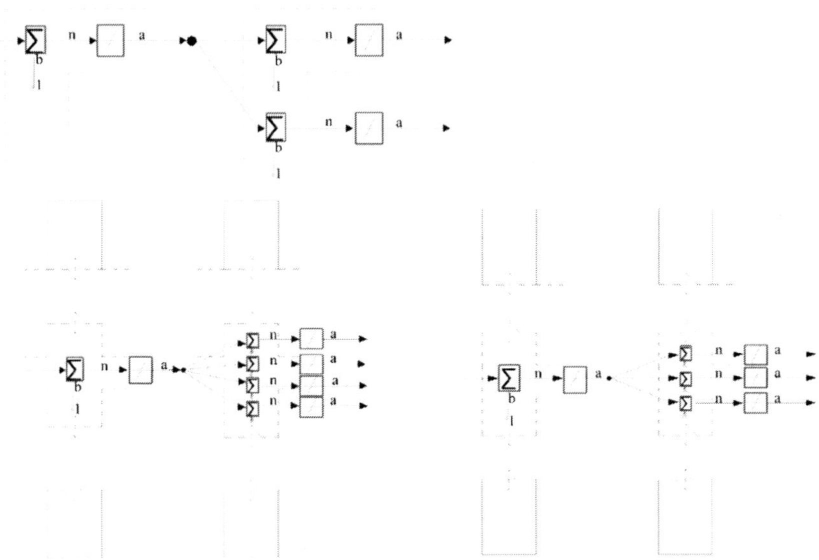

Figure 6. Artificial neural mitosis in the nodes of a structure.

4.1. Division of Neuronal Structure Substructures—Mitosis 1

The "artificial neural network model primitive" or "original model" constitutes the basic network for the development of neural approach "structural analysis with artificial mitosis". This network represents the inspiring model, from which we will aim to get a set of "sub-derived models of neural networks". See the original structural model of Figure 7.

If you look closely, the output layer plane gantry N levels and L layers of Figure 7, we realize that the layer required on each node, as many outlets as members concur in it. Figure 8 shows an overview of artificial neural network model derived mitosis 1, which has the same architecture ANN primitive, but differs from this only in the number of neurons in the output layer. Mitotic neurons are locally attached Model 1, with unidirectional connections except the output layer. The model is a back-forward whose learning algorithm is self-supervised; and therefore will not require a supervisor, and that by itself, it will run internal monitoring to monitor network performance, with a linear transfer function (purelin). This algorithm is also used by the other models derived, with operating in series. It has been proven that the network model is an effective and autonomous system that enables processing of the structural physical parameters, entering the network from an input vector whose components are the distribution factors, which belong to the beams of the first layer. These factors are clearly part of certain characteristics of the original structure [9] [10] .

The outputs of the hidden layer neurons any, are transferred to corresponding processing elements of the back layer, through local connections, after balancing the weight factors. Except for the processing elements of the input layer neurons and output layer, the input of each neuron is the sum of the weighted outputs of the nodes (neurons) in the previous layer plus the corresponding trend.

Each neuron or processing element is activated in accordance with the input into the neuron activation function and the value node trend or threshold of the neuron. L_e input layer consists of an arrangement of $S^1 = N$ neurons, an output layer, containing $S^L = 3N - 1$ processing elements. Between layers of input and output L_e and L_o are (L - 2) hidden layers. Each layer has its own matrix of weights W your own trend vector b, a vector of net input n and a vector output a.

4.2. Sub-Models Derived by Sectioning

For rotational moments at the ends of the members who attend nodes in the hidden layers, the neuro-structural model primitive neuro be divided into sub-models derived (sub-domains). This is achieved by removing successively each time the original model, a layer neuron from right to left. This process begins immediately after, the mitosis 1 generated by the first sub-model derived. The (L - 1) sub-models of neuro-derived remaining structural network are obtained after each respective switching and mitosis of the output layer of the new sub-model

4.3. Sectioning to Mitosis 2

For rotational moments at the ends of the members of the hidden layers of a structure, primitive neuro-structural model must be divided into sub-models derived neuro (sub-domains). This is achieved by removing successively each time the original model, a layer neuron from right to left. The key partitioning process of primitive neuro- structural model sequentially, to give rise to sub domains or neuro-substructures, lies in the fact that the nodes of the new output layer are predominantly high connectivity; and the same, are assigned to processors distributed memory which leads to an iterative application of massively parallel processing, with the support of the neural mitosis [11] . This process is solved the problem of determination sub-end rotational moments of hidden layers of the structure.

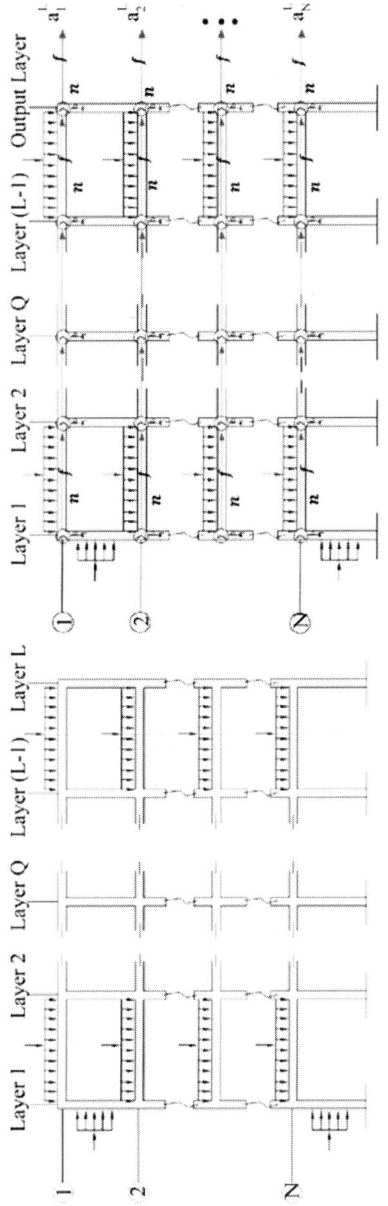

Figure 7. Primitive structure—artificial neural network primitive.

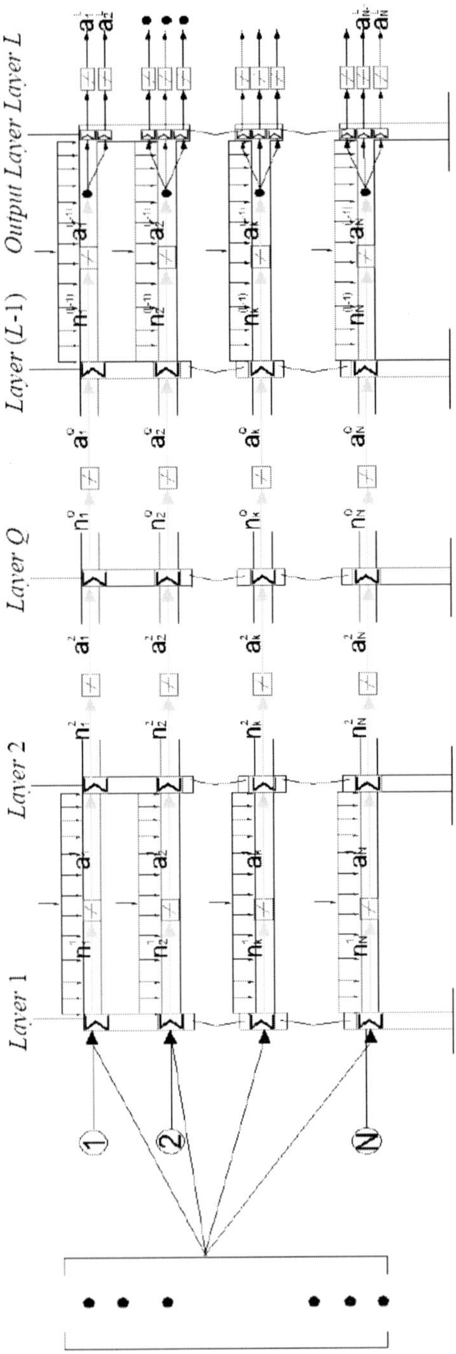

Figure 8. Model mitosis 1.

4.4. Neuronal Architecture Mitosis Model

The model derived mitosis 2 is a network fed forward (back-forward) whose learning algorithm is self-super- vised; and therefore did not require a supervisor, and that by itself, it will run internal monitoring to monitor network performance, with a linear transfer function.

The outputs of the neurons of a layer are transferred to the next layer corresponding through local connections excite or inhibit such exits through weighting factors. Except for the processing elements of the input layer neurons and output layer.

The input of each neuron is the sum of the weighted outputs of the neurons in the previous layer plus the corresponding trend. Each neuron or processing element is activated in accordance with the input into the neuron activation function and knot trend value or threshold of the neuron. InFigure 9 the mitosis neuronal model 2, which is composed of L - 1 layers shown, namely, having a layer unless the original model and the ANN mito sis 1. The input layer L_e is an arrangement of $S^1 = N$ neurons; the output layer L_o contains $S^{(L-1)} = 4N-1$ processing elements. Between layers of input L_e and output L_o are (L - 2) hidden layers. Each layer has its own matrix of weights W, its own trend vector b, a vector of net input n and a vector output a.

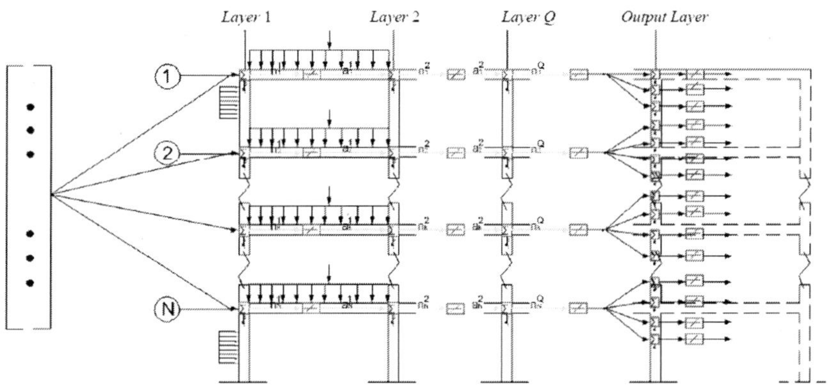

Figure 9. Mitosis artificial 2.

For rotational moments at the ends of the members attending the nodes of the Q-th hidden layer of the structure, primitive neuro-structural model will be severing all layers to the right of the Q-th layer sequentially one to one.

4.5. Neuronal Architecture Model Q-th Artificial Mitosis

Then in Figure 10 an overview, the artificial neural network model built with simple processing elements shown, which have a local memory. Neurons are connected with unidirectional connections, except the output layer. The model is a self-supervised back forward network, the algorithm is similar to the other models that operates in series and therefore will not require a supervisor and that by it will run internal monitoring, to monitor the performance network, with a linear transfer function.

Between layers of input L_e and output L_s^Q, are (Q - 1) hidden layers, which generally own N neurons. Each layer has its own matrix of weights W, its own trend vector b, a vector of net input n and a vector output a.

4.6. Sectioning of the L-th Artificial Mitosis

To bring the last sub-model derived corresponding to the L-th neural mitosis to run the L-th isolation of primitive neural model accompanied the final process of neural mitosis. Precisely, this sub-model is the only one with a different input vector with R = (3N - 1) components.

4.7. Layer Neural Architecture Model: L-th Artificial Mitosis

Figure 11 shows a general scheme, the L-th neural network model derived from a layer with $S^1 = (3N-1)$ neurons which have a small local memory. The input vector to the network has R = (3N - 1) components.

$$\boldsymbol{p} = \left(\mu 1 1_{vd}\, \mu 1 1_{ci}\, \mu 1 2_{cs}\, \mu 1 2_{vd}\, \mu 1 2_{ci} \cdots \mu 1(3N-3)_{cs}\; \mu 1(3N-2)_{vd}\; \mu 1(3N-1)_{ci}\right)^{\mathrm{T}}$$

It has been found that the L-th network model derivative, L-th mitosis is an effective and independent system, which enables processing of the structural physical parameters, entering the network from an input vector whose components distribution are factors that belong to the beams and columns that meet in the first layer of the structure. These factors are clearly part of the original structure certain physical characteristics.

In Figure 11 the neural model of a layer, is composed of neurons $S^1 = 3N - 1$. The layer has its own array of weights W^1, its own vector b^1 trend; net input vector n^1 a component input vector $p = 3N - 1$ component and a vector output a^1.

5. SEQUENCE OF NEURO-STRUCTURAL MODELS

The sequential division of neuro-structural primitive, together model with artificial neural mitosis is a mechanism by which a system of hybrid neural sub-models, capable of giving rise to a registration process, input, processing, and storage is generated and output information from internal and external means of a flat lattice structure of several levels. Figure 12 is shown in simplified form.

The first sub-model derived neuronal differs from sub-neuro-structural model primitive, just in the number of neurons in the output layer. From this differentiation, each model differs from the previous derivative in chronological order; and having a layer of neurons unless this; and possibly the number of neuron of the output layer.

All neural models like its primitive have the same input vector. An exception, the model derived from a layer, which has an input vector with a number of components equal to the number of neurons of a layer model.

5.1 System Networks

The neuro-structural analysis is a new alternative, which is based on two strategies: the division of the porch in sub-structures or sub-domains and artificial mitosis of the output layer, the primitive model.

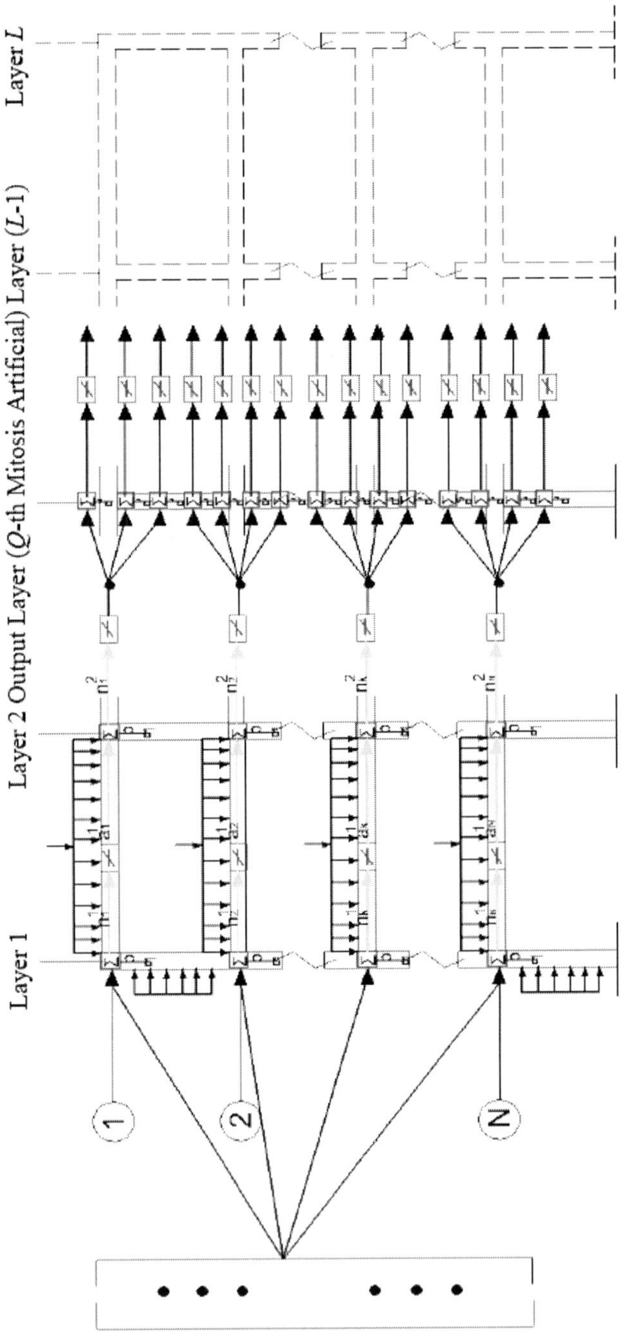

Figure 10. Mitosis artificial Q-th.

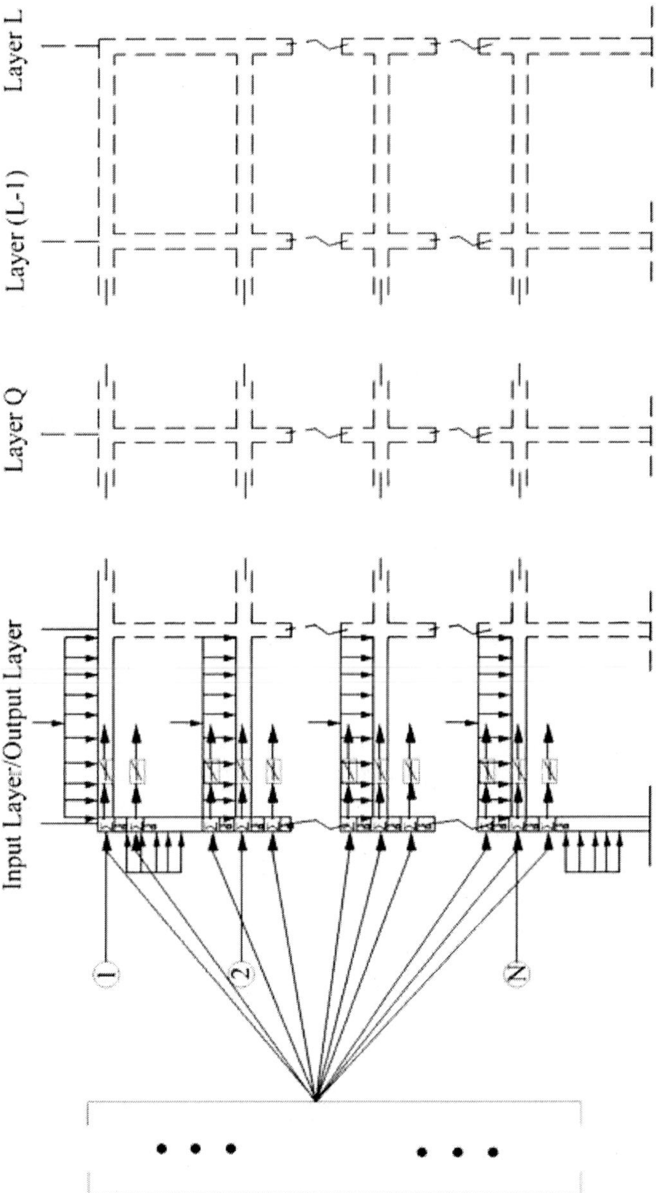

Figure 11. First layer mitosis artificial.

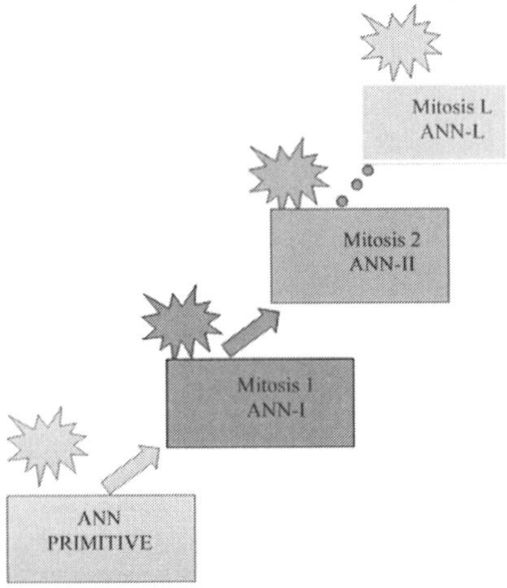

Figure 12. Neural system sub-models series.

The first sub-model derived is called NTSA-I (artificial mitosis 1) shown in Figure 7. NTSA-I calculated rotational moments at the ends of the members which contribute to the final layer of nodes of the output layer structure or network. The sub-model-II ANN (artificial mitosis 2) calculates the rotational moment at the ends, which contribute to the nodes of the penultimate layer, or output layer of the network; and so on, until the sub-model simulation of NTSA-L (sub-neural network model of a single layer), which calculates the rotational moments at the ends of the members who attend at the nodes of the first layer original structure

A macro program called mitosis generates artificial series of calls to all derivatives sub-models; and then the resulting NTSA-I sub-model calls the NTSA-II and so on, until the call of the NTSA-L, to complete the implementation of the series.

6. THE PROCEDURE

The procedure of the NTSA system depends on the solution of three problems for the determination of member constants on fixed end moments, the stiffness at each end of member, and of the over-carry factor distribution factors μ_{ik} for the rotational end moments and distribution factors ϑ_{ik} for the lateral displacement moments at each end for each member of the frame under consideration. The determination of these values is not a part of the presented approach.

The NTSA model has four layer, three inputs and five output values. The 'function network' of the Toolbox of the Matlab creates the net, which generates the first layer weights matrices and biases vectors for the four linear layers required for this problem. These weights and biases can now be trained incrementally using the algorithm. The network must be trained in order to obtain first layer weights and biases. For the second, third and fourth layers, the weights and biases are modified in response to network's inputs and will lead to the correct output vector. There are not target outputs available. The linear network was able to adapt very quickly to the change in the outputs. The fact that it takes only seven iterations for the network to learn the input pattern is quite an impressive accomplishment [1] [2] .

The scheme for entering the calculations systematically is shown in Figure 13. The procedure explained above, is best illustrated by solving the structure in Figure 13, which is loaded in a rather complex fashion. The distribution factors for nodes 1, 2, and 3 constitute the net's input vector.

The fixed end moments for the different loaded members are calculated by using the standard formula available in any structural handbook. Having completed these preliminary calculations, the training can be initiated. The network was set up with the three parameters (distribution factors of the beam) as the input, and the rotational end moments due to rotation as the outputs determined by the first layer [3] .

The calculation starts in the input layer and continues from one layer to the next. Such calculation is carried out quickly. After 6 or 7 iterations have been performed, as explained earlier, it will be noted that there is little or no change in the values of two consecutive sets of calculations. The calculations are now stopped and the values of the last iteration are taken as the correct ones, with the previous values being ignored. For the sake of clarity, these final values have been indicated separately [9] .

The architecture of the ANN will be designed with the "Network Toolbox function of the Artificial Neural Networks" and simulations will be performed using the "Sim tool", both belonging to the Matlab environment. This algorithm, through multiple interactions with the application of the method of the successive approximations, achieves the learning object models "artificial mitosis" [11] .

The first ANN model derived or "artificial mitosis 1", allows us to obtain the rotational moments at all ends of the members who concur in the nodes of the fourth layer and output layer, see Figure 15. The second network model "artificial mitosis 2" gives as results the rotational moments at all ends of the bars that come together in the nodes of the output layer (third layer). The third neural network model of the application is the "artificial mitosis 3", which calculates the rotational moments at the ends of rods in either concurrent nodes of the second layer counted from left to right these.

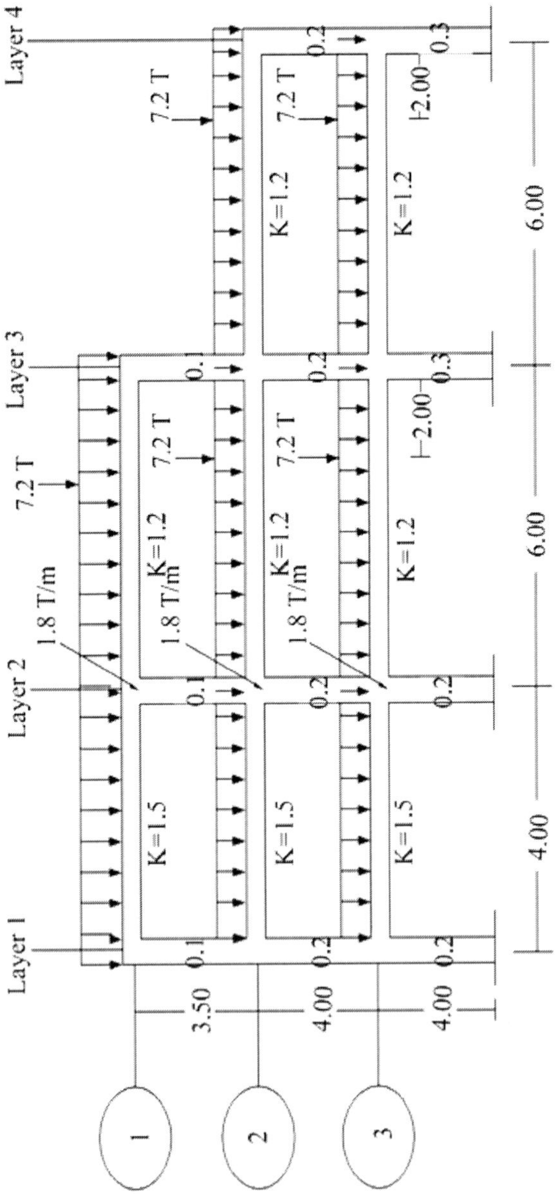

Figure 13. Example of application.

Finally, the fourth model derived or “artificial mitosis 4” consists of a single layer of neurons. This network allows us to calculate the rotational moments, partners at all ends of the members (beams and columns) that access to the nodes of the first layer, the artificial neural network model primitive (Ben Mark) [9] .

The primitive neural network model consists of four layers of neurons, three hidden layers including the input layer and the output layer. We adopt the type of diagram shown in Figure 14, which allows us to write the successive values at the ends of each member. And also it facilitates the systematic income calculations. Show of the numerical calculations will to be performed later, including distribution factors and moments of perfect embedding at the ends of the members.

The model of multi-gantry plane lattice structure will be used as a source of direct inspiration for the design of primitive neural network model. The analogy between models allows the following procedure: Repeat, i.e. double the structural network model, putting in place of each node of a neuron structural model. During the transformation of the structural model in neural model, columns connections between nodes are removed. Thus, the two models differ in the connectivity leading columns. But we should not worry about this, because, associated with each of the layers of primitive neural model and all possible models derived from it, the vectors trends of each layer, taking into account actions that produce connections columns that disappear in the primitive model and its derivatives. In this way, the equivalence between models remains.

The primitive neural network model is shown in Figure 15, this network is back forward. The network consists of simple processing elements, each of which has a local memory. These neurons are connected by unidirectional connections that transfer data from one neuron to the next corresponding layer. The model is a network whose self-supervised algorithm does not require a supervisor, and that by itself, it will run internal monitoring to monitor network performance, using a linear transfer function (purelin). The input vector components are factors corresponding to the beam ends that meet in the input layer to the distribution network [12] .

Here are the original network model inspired by the structural model, keeping one correspondence between nodes in the structural model and the model of primitive neurons network.

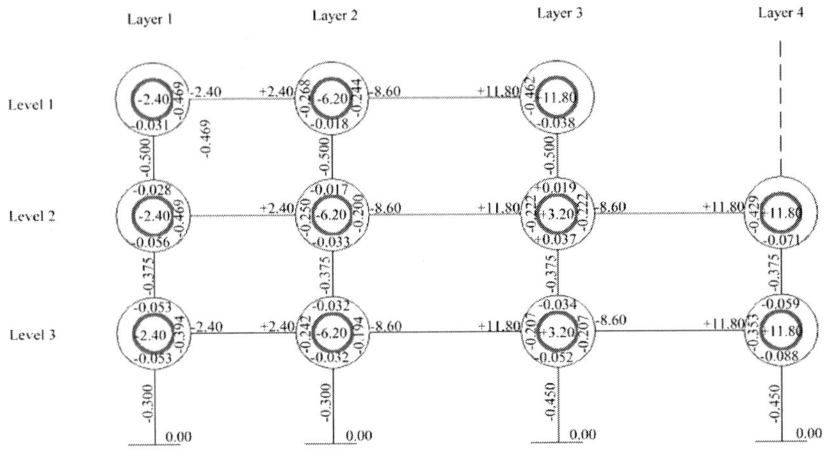

Figure 14. Distribution factors and fixed-end moments.

6.1. Primitive Neural Network

The model consists of four layers of neurons. Each layer has its own matrix of weights W, its own trend vector b, a vector of net inflow n and a vector output a. Following established notation the first, second, third and fourth layer respectively as will matrices $W^1W^2W^3W^4$.

Artificial neural networks are intelligent tools which are extremely useful in situations, for which the rules are not clear enough or are difficult to establish artificial neural networks are robust and fault tolerant. You can calculate a function without a mathematical description of how the output function is operatively associated with the input. She learns to approximate functions even when its form cannot be specified accurately.

6.2. Artificial Mitosis 1 Model

It is an artificial neural network with four layers. The first three layers of "mitosis Model 1" are similar to the original model. The input layer is $S^1 = 3$ neurons. Each neuron is connected locally to its corresponding vector component p input components $R = 3$. The input vector components are key values, called "distribution factors" [2] . The matrix of the weights of the second layer (hidden) W^2 is a diagonal matrix and its elements are the distribution factors of the left ends of the beams are located between the second and third layer of nodes. While the trend vector b^2 of the second layer has as fastening components moments belonging to each node of the layer, the moments due to horizontal displacement of the ends of columns and the rotational moments of the opposite ends of the beams and columns which contribute to the knot or processing unit [1] .

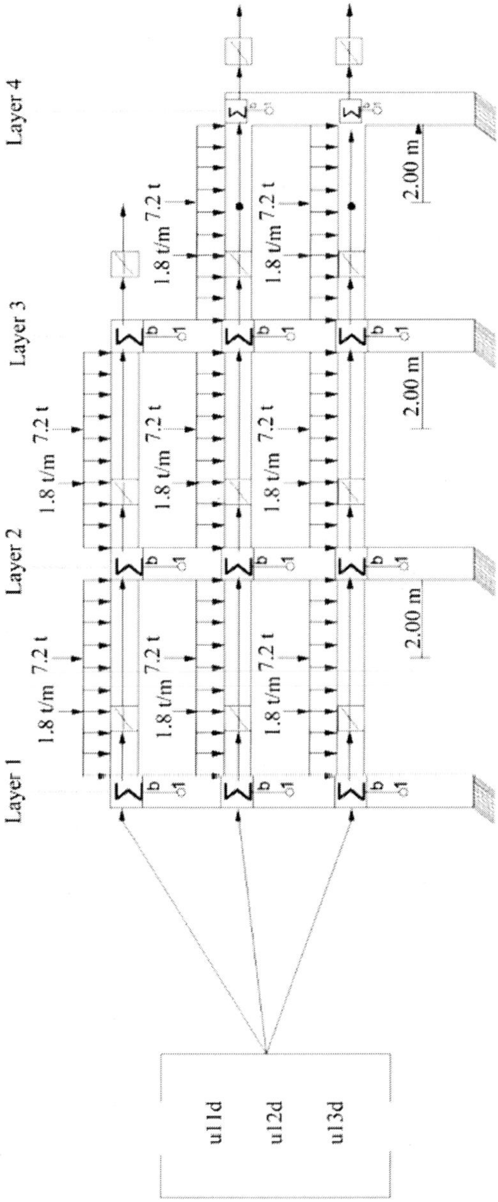

Figure 15. Neuronal model primitive.

The second layer is $S^1 = 3$ neurons. Each neuron is connected locally to the unit of processing of the previous layer. The outputs of the layers one and two are the inputs to the layers two and three. Thus, layer two can be seen as a network of a layer with $R = S^1 = 3$ inputs, $S = S^1 = 3$ neurons and an array of weights W^2 diagonal $S^1 \times S^3$. The input to the layer two is the vector a^1, and a^2 is output vector [10] .

The matrix of the weights of the third layer (hidden) W^3 is a diagonal matrix whose elements are also the (non-adjustable) "distribution factors" left ends of the beams are between the third and fourth layer of neurons. While the bias vector associated b^3 third layer has components similar to the other bias nature. This layer has S^3 neurons. As with the second layer, this is locally connected through the vector a^2 to the next layer. The output layer is also the output of the network; this layer has an array of weights W^4 and $S^4 = 5$ neurons. The matrix of weights has the elements to "factors of distribution" of all the ends of the members attending the nodes of the output layer [1]

Figure 16 shows the network model after suffering the first mitosis the output layer of the original model, taking into account the number of required outputs node double edge and one edge node.

Mitosis Neuronal Architecture Model 1

net =

Neural Network object: numInputs: 1

numLayers: 4 biasConnect: [1; 1; 1; 1]

inputConnect: [1; 0; 0; 0] layerConnect: [4 × 4 boolean]

Figures 17-20 show codes (mitosis 1, mitosis 2, mitosis 3 and mitosis 4) and explained as the neural network of the program Matlab toolbox organises the simulation of networks. MATLAB writes and reads many file during a typical NTSA system analysis.

6.3. Artificial Mitosis 2 Model

It is a neural network model composed of three layers of neurons. The first two layers of Mitosis model 2 are equal to the first two layers of the original model and the two of "mitosis Model 1". Each layer of the model has its own matrix respectively weights: W^1, W^2, W^3 and their delivery trend: b^1, b^2, b^3. The parameters that characterize matrices and vectors weights trends continue thus in the primitive model, as in each of the models subject to mitosis except for its output layer. In Figure 21, the mitosis model 2 wherein the penultimate layer of primitive object model is a mitosis, considering the inherent connections to each of the nodes in the proposed structure is presented.

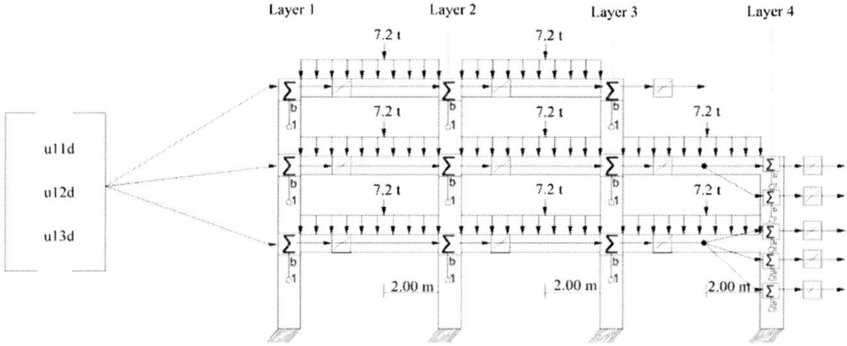

Figure 16. Mitosis model 1—fourth layer output.

```
% FILE: MITOSIS1    ****PARES EXTREMOS****

% *****PORTICO DE VARIOS PISOS CON NODOS DESPLAZABLES*****
%
% CARGAS VERTICALES
%Propiedades Arquitectónicas de la Red
net=network(1,4,[1;1;1;1],[1;0;0;0],[0 0 0 0;1 0 0 0;0 1 0 0;0 0 1 0],[0 0 0
1],[0 0 0 0])

%Rango de las Entradas
net.inputs {1}.range=[-0.5 0;-0.5 0;-0.5 0];
%Neuronas por Capas - Función de Transferencia
%PrimeraCapa
net. layers {1}.size=3;
net. layers {1}.transferFcn='purelin';
net.layers {1}.initFcn='';
%Segunda Capa:
net. layers {2}.size=3;
net. layers {2}.transferFcn='purelin';
net.layers {2}.initFcn='';
%Tercera Capa
net. layers {3}.size=3;
net. layers {3}.transferFcn='purelin';
net.layers {3}.initFcn='';
%Cuarta Capa
net. layers {4}.size=5;
net. layers {4}.transferFcn='purelin';
net.layers {4},initFcn='';

% Datos de la Estructura
```

Figure 17. Mitosis 1 code-1.

```
% FILE: MITOSIS 2    ****PARES EXTREMOS****


% *****PORTICO DE VARIOS PISOS CON NODOS DESPLAZABLES*****
%
% CARGAS VERTICALES
%Propiedades Arquitectónicas de la Red
net=network(1,3,[1;1;1],[1;0;0],[0 0 0;1 0 0;0 1 0],[0 0 1],[0 0
0])
%Rango de las Entradas
net.inputs {1}.range=[-0.5 0;-0.5 0;-0.5 0];
%Neuronas por Capas - Función de Transferencia
%Primera Capa
net.layers {1}.size=3;
net.layers {1}.transferFcn='purelin';
net.layers {1}.initFcn='';
%Segunda Capa:
net.layers {2}.size=3;
net.layers {2}.transferFcn='purelin';
net.layers {2}.initFcn='';
%Tercera Capa
net.layers {3}.size=10;
net.layers {3}.transferFcn='purelin';
net.layers {3}.initFcn='';
% Datos de la Estructura
```

Figure 18. Mitosis 2 code-2.

```
% FILE: MITOSIS3    ****PARES EXTREMOS****


% *****PORTICO DE VARIOS PISOS CON NODOS DESPLAZABLES*****
%
% CARGAS VERTICALES
%Propiedades Arquitectónicas de la Red
net=network(1,2,[1;1],[1;0],[0 0 ;1 0],[0 1],[0 0]);
%Rango de las Entradas
net.inputs {1}.range=[-0.5 0;-0.5 0;-0.5 0];
%Neuronas por Capas - Función de Transferencia
%Primera Capa
net.layers {1}.size=3;
net.layers {1}.transferFcn='purelin';
net.layers {1}.initFcn='';
%Segunda Capa:
net.layers {2}.size=11;
net.layers {2}.transferFcn='purelin';
net.layers {2}.initFcn='';

% Datos de la Estructura
```

Figure 19. Mitosis 3 code-3.

```
% FILE: EXPLOSION4     ****PARES EXTREMOS****

% *****PORTICO DE VARIOS PISOS CON NODOS DESPLAZABLES*****
%
% CARGAS VERTICALES
%Propiedades Arquitectónicas de la Red
net=network(1,1,[1],[1],[0],[1],[0]);

%Rango de las Entradas
net.inputs {1}.range=[-0.5 0;-0.5 0;-0.5 0;-0.5 0;-0.5 0;-0.5 0;-0.5 0;-0.5
0];
%Neuronas por Capas - Función de Transferencia
%Primera Capa
net.layers {1}.size=8;
net.layers {1}.transferFcn='purelin';
net.layers {1}.initFcn='';

% Datos de la Estructura
%Factores de Giro
```

Figure 20. Mitosis 4 code-4.

6.4. Artificial Mitosis 3 Model

Derived network model two layer is sectioning product and artificial mitosis of the second layer of primitive neural network model. Weights matrices and vectors associated trends to the two layers of the neural network model are derived respectively: W^1, W^2 and b^1, b^2. Weights matrix and vector trend corresponding to the first layer of the neural network model remain primitive. See Figure 22:

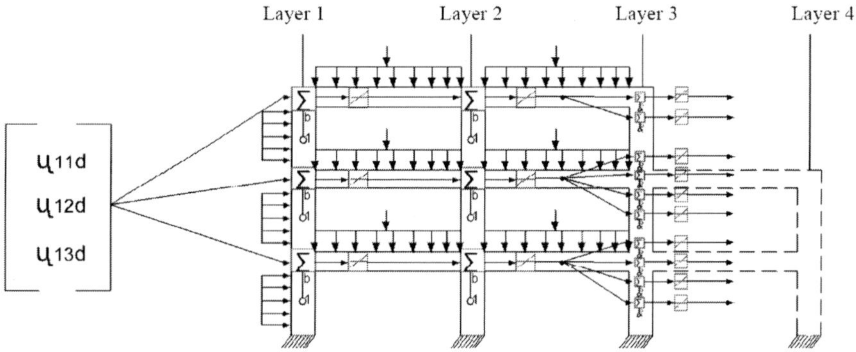

Figure 21. Mitosis model 2—third layer output.

6.5. Artificial Mitosis 4 Model

This is a neural network model of a layer with eight (8) neurons and an input vector p of eight (8) components. In general, the number of components of the input vector to the layer is different from the number of neurons, ie, R ≠ S. But in this particular case, the number of components of the input vector (R = 8) is equal and corresponds to S = 8 number of neurons of the input layer. Each component of the input vector p is connected to the corresponding neuron through the matrix of weights W, which is a diagonal matrix whose elements are the rightmost rotational moments of the beams are between the first and second layer columns or nodes. See Figure 23. Each i-th neuron has a tendency b_i, an adder block, linear transfer function and output a_i. Artificial neurons belonging to artificial neural mitosis 4 model are directly connected with the outside world; and in no case with other neurons [12] . Input vector:

$$p = \left(\mu 11_{vd}\, \mu 11_{ci}\, \mu 12_{cs}\, \mu 12_{vd}\, \mu 12_{ci} \cdots \mu 1(3N2)_{vd}\, \mu 1(3N1)_{ci}\right)^{T}$$

It has components distribution factors of the rotational moments of the members present in the first layer. This model has many components as vector has the layer neurons.

6.6. Results

The results of the simulation are presented in Figure 24.

The results matrix "all" showed above presents the values of the rotational moments at the ends of the various members who access the different layers of flat porch under discussion. The system of neuronal sub-types: mitosis1, mitosis 2, mitosis 3 and mitosis 4 result in a serial process that ends with the mitosis4, during which he lead to the recording, input, processing, storage and output of information from the media internal and external to the portal frame cross-linked multilevel flat. The neuro-structural analysis relies on two strategies: the neural mitosis and sectioning of the portal frame composed of layers.

Finally, a program called mitosis macro generates a series of sequential calls to the respective sub-models derivatives mitosis 3 mitosis 1...mitosis 4 mitosis 2; and thus, the execution of the series is completed. Figure 25 shows the final results of total extreme rotational moments of the portal frame members.

6.6.1. Verification of Results

Refer to example. One way of checking for the validity of our findings is to arbitrarily cut a node and apply the static equilibrium conditions, for node i, as in the Accompanying illustration, Figure 22. In all, the end moment can be written as follow:

$$M_{ik} = ME_{ik} + 2M'_{ik} + M'_{ki} + M''_{ik}$$

$$\text{and } M_{ki} = ME_{ki} + 2M'_{ki} + M'_{ik} + M''_{ik}$$

where,

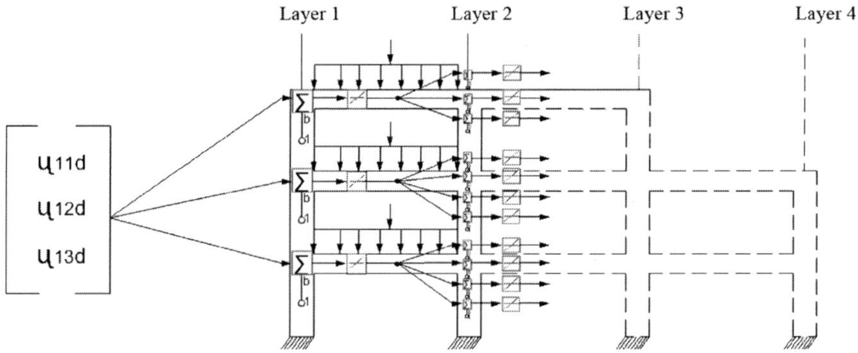

Figure 22. Mitosis model 3—out of the second layer.

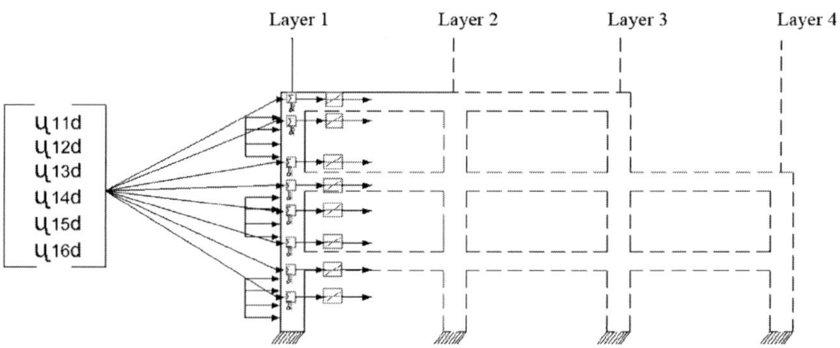

Figure 23. Mitosis model 4—output first layer.

All =

Output Layer 1	Output Layer 2	Output Layer 3	Output Layer 4
0	3.6131	0	0
0	2.8851	-6.8493	0
0	0.2427	-0.5634	0
-0.6354	0.0806	0.0177	0
-0.0420	1.1851	0.2073	0
0.0194	0.9480	0.2073	0
0.2879	0.1564	0.0345	-5.0409
0.0388	0.1736	-0.0349	-0.8343
0.0191	1.3130	-0.2125	-0.6828
0.1423	1.0526	-0.2125	-4.0853
0.0191	0.1736	-0.0534	-1.0184
Mitosis 4	Mitosis 3	Mitosis 2	Mitosis 1

Figure 24. Results—rotational moments at the ends of the members.

M_{ik}, M_{ki}: are the end known as the end moment and exerted by the node i, and k on the corresponding ends of the element, ME_{ik}, ME_{ki}: are fixed end moments induced at the ends, M'_{ik}: is termed as the near end rotational moment, and M'_{ki}: as the far end rotational end moment.

Applying the static equilibrium equations we have: $\sum_{j=1}^{4} M_{ij} = 0$

Considering the node i of the of the structure, Figure 26, of the structure, located between the three layer and level two, shown in Figure 19, where the members: $i-1, i-2, i-3, i-4$:

$$M_{i1} = +11.80 + 2\times 0.2098 + 0.948 = +13.1667t - m$$

$$M_{i2} = 2\times 0.018 - 0.5634 + 0.12\,(\text{is the total displacement moment}) = -0.4074t - m$$

$$M_{i3} = -8.60 + 2\times 0.2098 - 5.0409 = -13.2213t - m$$

$$M_{i4} = 2\times 0.0350 - 0.0347 + 0.42\,(\text{total displacement moment}) = +0.4553t - m$$

$$\sum_{j=1}^{4} M_{ij} = \left(+13.1667 - 0.4074 - 13.2213 + 0.4553\right) \approx 0$$

The validity of the computed the static equilibrium conditions, for node i is verified.

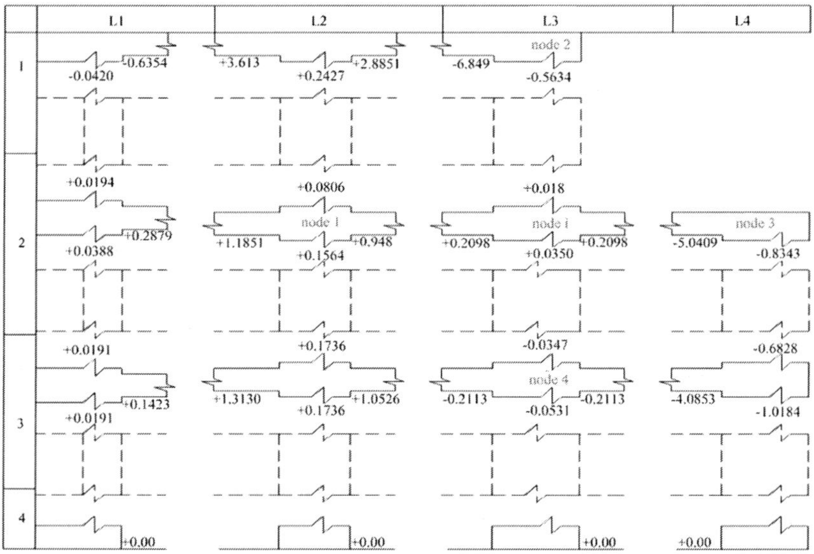

Figure 25. Rotational moments ends in the porch.

6.6.2. Comparison with Kani's Method

A comparison between the presented NTSA model and Kani's Method was performed on the same example, and

it can be shown in Figure 25 and Figure 26. A discrepancy ratio $D_r \frac{C_N}{C_K}$ was used for comparison, where C_N the rotational moment (output network) or total end moment and C_K is Kani's result. The mean value was performed e comparison is shown in Table 1 and Table 2. From looking at the table, one may conclude that the presented model gives a better agreement with Kani's Method. A group of 15 nodes was used for verification. Figure 27 and Figure 28 show the nodes of the structure under study, and also the members who access the aforementioned nodes. In these figures we present the comparison of some of the results: rotational moments at the extremes of members obtained with the application of G. Kani Method (distribution moments: these are the values shown in parentheses), and the results of the rotational moments calculated at the ends on the bars of the structure through the analysis executed by NTSA (whose values are outside the parentheses) that can improve the accuracy and speed of the results (Ben Mark) [9] .

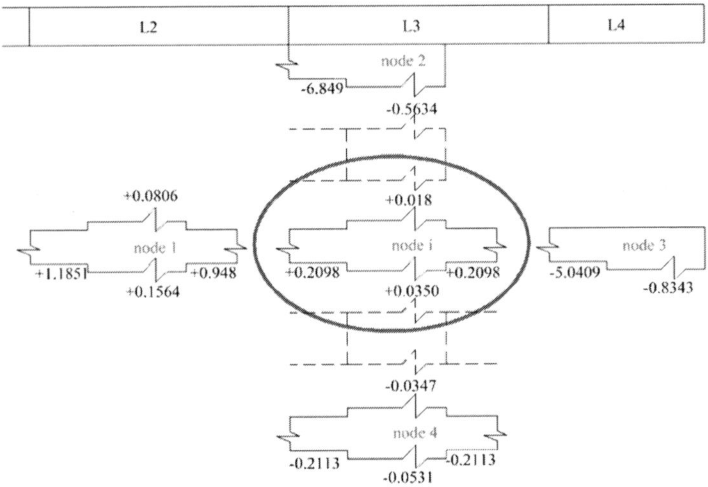

Figure 26. Node i and 1, 2, 3 and 4 equilibrium conditions.

Table 1. Accuracy of formulas for rotational end moments.

Nodes	Layers	Statiscal Mean Value	Statiscal Standard Deviation
Nodes 1s	1	1.0375	0.0178
	2	1.0097	0.0013
	3	1.0067	0.0008
Nodes 2s	1	0.9776	0.0131
	2	0.9995	0.0147
	3	1.0016	0.1107
	4	0.9976	0.0036
Nodes 3s	1	0.9529	0.0036
	2	1.0118	0.0108
	3	1.0578	0.0710
	4	1.0005	0.0032
Nodes 4s	1	1.0000	0.0000
	2	1.0000	0.0000
	3	1.0000	0.0000
	4	1.0000	0.0000

Table 2. Accuracy of formulas for total end moments.

Nodes	Layers	Statiscal Mean Value	Statiscal Standard Deviation
Nodes 1s	1	0.9633	0.0000
	2	1.0289	0.0412
	3	1.0015	0.0054
Nodes 2s	1	1.0023	0.0112
	2	1.0025	0.0064
	3	0.9970	0.0155
	4	0.9964	0.0037
Nodes 3s	1	1.0075	0.0076
	2	1.0061	0.0091
	3	0.9841	0.0186
	4	1.0040	0.0023
Nodes 4s	1	1.0100	0.0000
	2	1.0167	0.0000
	3	0.9783	0.0000
	4	1.0048	0.0000

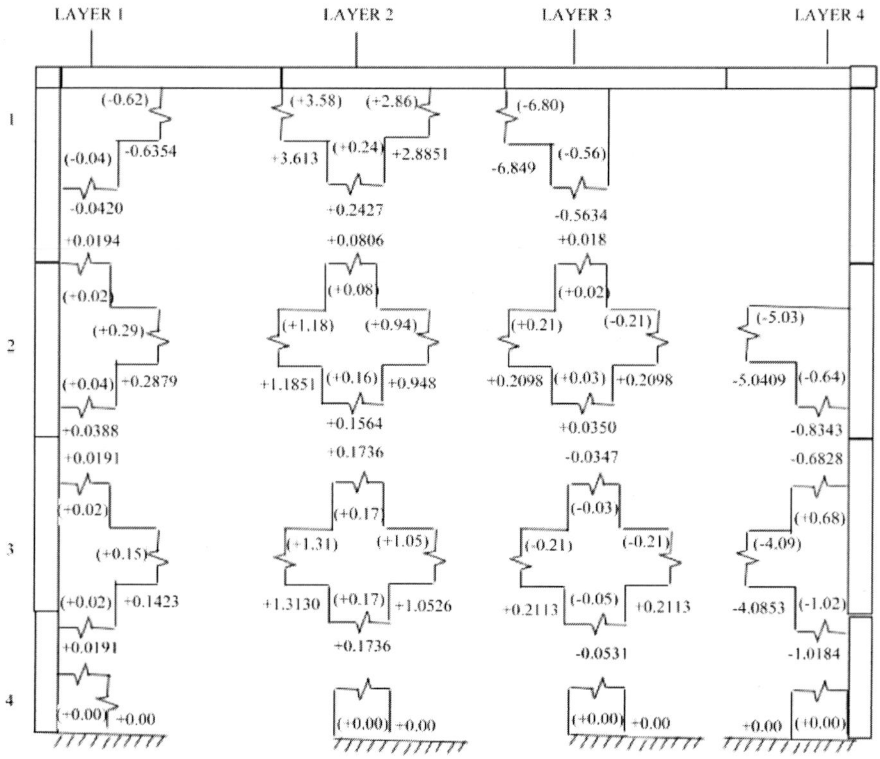

Kani's Method: ( ): NTSA System: --------

Figure 27. Rotational end moments whit horizontal displacement.

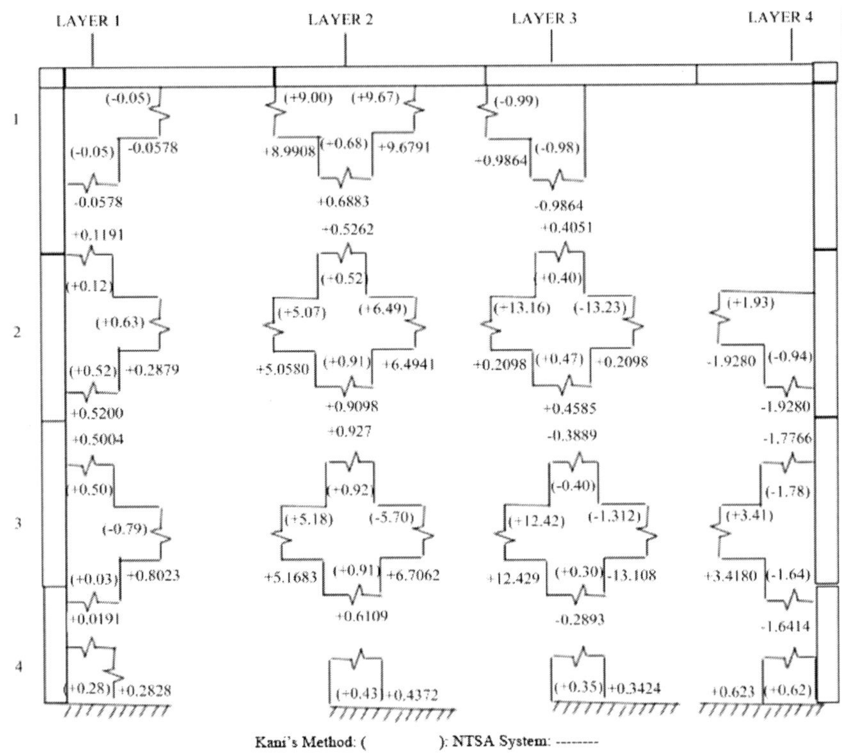

Figure 28. Total end moment's whit horizontal displacement.

7. CONCLUSIONS

1) There is a knack for designing primitive network inspired in a simple lattice model of the proposed structure, since this automatically suggests the neural network topology, connectivity type and even the number of hidden layers and the number of constituent neurons.

2) The design of the various models of artificial neural networks is highly didactic, and from the primitive model the topology design of each NTSA is obtained, including the number of hidden layers, number of neurons per layer, and the way that these are connected.

3) In general, the algorithm is self-supervised learning NTSA common to all models. They differ in the judgments that define the individual architecture.

4) The NTSA models are connected in series allowing interaction between artificial neural network architecture with different connectivity (feed-forward-backward).

5) All NTSA models are linear, which significantly reduces the complexity calculations.

6) In conventional neural networks results hidden layers are not known; through the NSTA system we can know the values or results of the hidden layers of the network.

7) NTSA system introduces a new algorithm based on biology.

ACKNOWLEDGEMENTS

L.R.P.M. thanks to Prof. Alfonzo G. Cerezo for his advice developing this work.

REFERENCES

1. Pinto, L.R. and Zambrano, A.R. (2014) Unsupervised Neural Network Approach to Frame Analysis of Conventional Buildings. Int. J. Communications, Network and Systems Sciences, 7, 203-211.
2. Rivero-Angeles, F.J., Gomez-Ramirez, E., Gomez-Gonzalez, B. and Garrido, R. (2005) Fault Detection in Shear Buildings Subject to Earthquakes Using a Neural Network. Proceeding of the Eighth International Conference on the Application of Artificial Intelligence to Civil, Structural and Environmental Engineering, Edited by B. H. V. Topping, 107, ISBN 1-905088-03-05.
3. Pinto, L. (2008) Tesis Doctoral: AETN Analisis de Estructuras Mediante Topología Neuronal
4. DARPA (1987-1988) Neural Network Study (U.S.). Published by AFCEA International Press, a Division of the Armed Forces Communications and Electronics Association 4406 Fair Lakes Court Fairfax Virginia 22033-3899 USA.
5. Beale, M., Hagan, M.T. and Demuth, H.B. (1995) Neural Network Design. Thomson Learning, Boston.
6. Eaton, L.K. (2001) Hardy Cross and the "Moment Distribution Method". Nexus Network Journal, 3, 15-24.

7. Hoffmann, F. Biological Therapies and Cancer. Produced through an educational grant from La Roche Ltd.

8. Biological Foundations—Neuron Communication. www.studyblue.com/notes/note

9. Kani, G. (1955) Cálculo de Pórticos de Varios Pisos. In: Reverte, S.A., Ed., 1978-1979, Printed in Spain, ISBN-84-291-2051-6, 19-20-21-22.

10. Boso, D., Lefik, M. and Schnefler, B. (2005) Joint Finite Element: Artificial Neural Network Numerical Analysis of Multilevel Composites Artificial Intelligence to Civil, Structural and Environmental Engineering. Edited by B. H. V. Topping, Civil Comp. Ltd., 101, ISBN 1-905088-03-05.

11. Lu, Y., Roychowdhury, V. and Vanderberghe, L. (2007) Distributed Parallel Support Vector Machines in Strongly Connected Networks. Neural Networks. A Publication of the IEEE Computational Intelligence Society.

12. Bebbahani, S. and Nasrabadi, A.M. (2009) Application of Som Neural Network in Clustering.

CHAPTER 6

Modeling regulatory cascades using Artificial Neural Networks: the case of transcriptional regulatory networks shaped during the yeast stress response

Maria E. Manioudaki[1,2] and **Panayiota Poirazi**[1]*

[1]Institute of Molecular Biology and Biotechnology, Foundation for Research and Technology-Hellas, Heraklion, Crete, Greece

[2]Department of Chemistry, University of Crete, Heraklion, Crete, Greece

Over the last decade, numerous computational methods have been developed in order to infer and model biological networks. Transcriptional networks in particular have attracted significant attention due to their critical role in cell survival. The majority of network inference methods use genome-wide experimental data to search for modules of genes with coherent expression profiles and common regulators, often ignoring the multi-layer structure of transcriptional cascades. Modeling methodologies on the other hand assume a given network structure and vary significantly in their algorithmic approach, ranging from over-simplified representations (e.g., Boolean networks) to detailed -but computationally expensive-network simulations (e.g., with differential equations). In this work we use Artificial Neural Networks (ANNs) to model transcriptional regulatory cascades that emerge during the stress response in *Saccharomyces cerevisiae* and extend in three layers. We confine the structure of the ANNs to match the structure of the biological networks as determined by gene expression, DNA-protein interaction and experimental evidence provided in publicly available databases. Trained ANNs are able to predict the expression profile of 11 target genes across multiple experimental conditions with a correlation coefficient >0.7. When time-dependent interactions between upstream transcription factors (TFs) and their indirect targets are also included in the ANNs, accurate predictions are achieved for 30/34 target genes. Moreover, heterodimer formation is taken into account. We show that ANNs can be used to (1) accurately predict the expression of downstream genes in a 3-layer transcriptional cascade based on the expression of their indirect regulators and (2) infer the condition- and time-dependent activity of various TFs as well as during heterodimer formation. We show that a three-layer regulatory cascade whose structure is determined by co-expressed gene modules and their regulators can successfully be modeled using ANNs with a similar configuration.

1. INTRODUCTION

Gene regulation is a fundamental process for the survival and proliferation of any living cell and is tightly controlled during all cellular states. Cells integrate a wide range of environmental signals in order to regulate their gene expression, which is primarily controlled at the level of transcription initiation. Alterations in the external environment affect the activities of specific proteins known as transcription factors (TFs) which can bind to regulatory regions of their target genes and inhibit or enable their transcription. Gene regulation is a multi-layer hierarchical process (Yu and Gerstein, 2006), where proteins regulate genes that produce other proteins. This interacting machinery comprised of TFs and their target genes can be represented as a directed graph thus forming the transcriptional regulatory network. This is often said to represent the master control system of the cell that orchestrates the differential expression of all genes (Yu et al., 2003; Ihmels et al., 2004) and has received significant attention over the last decade (Ihmels et al., 2004; Tanay et al., 2004; Bruggeman et al., 2008; Petricka and Benfey, 2011).

The organization of regulatory networks reflects the organization of the cellular machinery: it consists of functionally coherent groups, the functional modules (Han, 2008), entities that have been recognized as the basic structural unit in any biological system (Hartwell et al., 1999). In the case of transcriptional regulatory networks, the functional modules are groups of genes that are regulated in a coherent manner and thus behave similarly under similar conditions.

The recent increase in the production and accessibility to large-scale experimental data sets led to the development of computational methods that make use of various algorithmic procedures and integrate heterogeneous biological data aiming to infer functional modules that reflect distinct biological entities (Tanay et al., 2004; Fernández, 2007; Zhao et al., 2007; Bruggeman et al., 2008; Han, 2008). The majority of these methods use expression data from microarray experiments to group genes based on their expression profile, assuming that genes which are co-expressed under certain conditions are likely to be co-regulated and/or belong to the same biological pathway (Segal et al., 2003). Module assignment using gene expression data provides information about the organization of gene activities in different intracellular processes. However, it offers little or no information about the type of regulation (e.g., positive or negative) exerted by various regulators onto the members of a given module under different conditions. This limitation was first confronted by Segal et al. (2003) who applied a probabilistic model on gene expression data to identify the modules, their regulators and the conditions under which this regulation takes place. Since then, numerous studies incorporated heterogeneous experimental data such as motif information (Kundaje et al., 2005; Hu et al., 2008; Lee et al., 2008), chromatin immunoprecipitation (ChiP-chip) (Gao et al., 2004; Imoto et al., 2004; Xu et al., 2004; Youn et al., 2010), and protein interaction data (Tornow and Mewes, 2003; Maraziotis et al., 2008) in order to infer gene functional modules and associated networks.

Tools for inferring such networks include MA-networker (Gao et al., 2004), COGRIM (Chen et al., 2007), ReMoDiscovery (Chen et al., 2007), LeTICE (Youn et al., 2010), and GRAM (Bar-Joseph et al., 2003), among others. These approaches identify direct regulatory interactions but are confined to a single level whereas it has been shown that direct regulators form a bottleneck in the hierarchical structure of the multi-layer regulatory network (Yu et al., 2003). Furthermore, existing network-finding approaches provide mostly qualitative (structural) correlations between regulators and target genes. Attempts to model genetic regulations have been reported, although not very successful (Holter et al., 2000;Liebermeister, 2002; Imoto et al., 2003; Yu et al., 2003; Yu and Li, 2005; He and Zeng, 2006; Pournara and Wernisch, 2007; Wang and Li, 2012). Interestingly, quantitative correlations between expression profiles of TFs and their direct targets were possible when TFs were considered to act in a combinatorial manner, each exerting its regulatory function with a different temporal delay (Boone et al., 2007; He et al., 2007). Finally, networks identified based on expression information, such as frequently done in existing approaches, cannot take into account TFs that are constitutively expressed and yet exhibit a condition-specific regulatory action via the formation of heterodimers (Amoutzias et al., 2008).

In this work, we address the abovementioned limitations by constructing three-layer regulatory cascades having as basis not a single component (gene) but a module of genes along with its regulators, as identified by GRAM (Bar-Joseph et al., 2003), one of the most cited module-finding algorithms (Youn et al., 2010). We then use three-layer Artificial Neural Networks (ANNs) to model transcriptional cascades in order to quantitatively predict the expression profile of any gene in a module, given the expression of its indirect upstream regulators. Our models can be extended to take into account TF dimer formation as well as time-dependent combinatorial regulation that may be exerted by multiple regulators.

Transcriptional regulation is particularly important under conditions influencing the organism's survival, such as severe stress. *Saccharomyces cerevisiae* for example, is an organism whose transcriptional profile under stress has been extensively studied (Bammert and Fostel, 2000; Gasch et al., 2000; Rep et al., 2000; Causton et al., 2001; Kwast et al., 2002), primarily due to the simplicity of its genome and its worth as a biotechnological product (Botstein et al., 1997). Thus, we apply our method to a publicly available dataset concerning the response of *S. cerevisiae* cells to various stress conditions.

2. MATERIALS AND METHODS

2.1 Data Acquisition and Pre-Processing

Microarray gene expression data from *S. cerevisiae* cells in response to various environmental stresses (Gasch et al., 2000) were downloaded and used. The dataset comprised of measurements for the expression of 6152 *S. cerevisiae* genes under 19 different conditions over several time points (173 experiments in total) as well as over-expression and knockout experiments. The stress response data included the following conditions: (1) heat shock from 25°C to 37°C, (2) heat shock from various temperatures to 37°C, (3) steady-state temperature growth, (4) temperature shift from 37°C to 25°C, (5) mild heat shock at variable osmolarity, (6) response of mutant cells to heat shock, (7) hydrogen peroxide treatment, (8) response of mutant cells to H_2O_2 exposure, (9) menadione exposure, (10) diamide treatment, (11) DTT exposure, (12) hyper-osmotic shock, (13) hypo-osmotic shock, (14) amino acid starvation, (15) nitrogen source depletion, (16) diauxic shift, (17) stationary phase, (18) steady-state growth on alternative carbon sources, (19) steady-state growth at constant temperatures, (20) over-expression studies, (21) knockout experiments for several time points. The dataset included normalized, background-corrected $\log_2$ Red/Green ratios. Normalization included correction of Cy3 and Cy5 dye biases and background correction to correct for signal intensities outside the spots. The data were $\log_2$ transformed to avoid fractions in signal ratios. Although normalization represents an important step in microarray data analysis and procedure, in this case, since data were previously normalized, no further normalization was performed. Since many—but not all—experiments were performed in duplicates or triplicates, we used the average expression value in our analysis to ensure equivalent contribution of each data-point in the final analysis and also to avoid having replicates of an experimental condition in both the training and test datasets of the ANNs. Furthermore, in an attempt to focus on responses specific to certain stress conditions, we excluded the over-expression and knockout experiments and divided the rest of the experiments in three main functional categories as shown in Table 1.

A dataset containing genome-wide location analysis for the binding of 106 transcriptional regulators to promoter sequences across the *S. cerevisiae* genome (Lee et al., 2002) was also used in the analysis. In the respective study, the authors used a myc epitope tag for each TFs and performed a genome-wide location analysis using microarrays to detect, through hybridization, those promoter regions of the genome that were enriched in epitope tags after chromatin immuno-precipitation experiments. Binding data are represented as confidence values (p-values) for each microarray spot.

2.2 Identification of Gene Modules

In order to identify modules of genes that are co-regulated and co-expressed under a set of conditions, we used the GRAM algorithm (Bar-Joseph et al., 2003) (see Box 1 for a brief description). We used the default settings of the algorithm and the pre-processing option (row-wise normalization). Specifically: (1) a binding cutoff *p*-value of 0.001, (2) rejection of a gene if more that 20% of the expression or binding data are missing, and (3) the minimal size of the initial core set of genes forming a module is equal to 3. This first step resulted in a number of gene modules along with their direct regulators which formed the middle-layer of the ANN model.

2.3 Building the Regulatory Networks

For every GRAM-inferred gene module that is regulated by at least two TFs, we searched for the proteins that regulate these TFs using the YEASTRACT database (www.yeastract.com), to identify the indirect TFs which formed the upper-layer (or input-layer) of the ANN models. Since transcriptional regulation requires the binding of a TF to the promoter region of a gene, in addition to the manually curated bibliographic information, a protein is assumed in this manuscript to transcriptionally regulate the target gene that codes for a specific TF only if it has at least one binding site in the region 1000 base-pairs upstream the promoter of that gene. The region of 1000 base pairs was chosen as this is generally considered the region most likely to contain TF binding sites (Friberg et al., 2005). Binding site data were retrieved from the SGD database (http://www.yeastgenome.org/).

Table 1. Stress conditions organized into three categories.

Category A (heat shock)	Category B (starvation)	Category C (oxidative and osmotic shock)
Heat shock from 25 to 37°C	Amino acid starvation	H_2O_2 treatment
Various temperatures to 37°C	Nitrogen source depletion	Menadione exposure
Steady-state temperature	Diauxic shift	Diamide treatment
Heat shock from 37 to 25°C	Stationary phase	DTT exposure
Heat shock at variable osmolarity	Steady-state on alternative carbon sources	Hyper-osmotic shock
Constant temperature growth		Hypo-osmotic shock

Category A includes experiments related to alterations in temperature; Category B includes experiments related to nutrient depletion; and Category C contains experiments related to oxidative and osmotic shock.

Box 1. The GRAM algorithm. A heuristic algorithm that identifies modules of co-expressed and co-regulated genes based on expression and DNA-binding data. The algorithm first identifies a core set of genes that are bound by the same set of regulators based on a stringent criterion for the *p*-value binding. For this initial set, the method finds those genes that share a similar expression profile and forms a new core set for which the mean expression profile is calculated. Now an extension step is performed. In this step the algorithm searches for genes that are bound by a more relaxed binding criterion and their expression profile is close to the mean expression profile of the core set of genes. For these genes a combined *p*-value is calculated based on the *p*-values of the regulators for every module. A gene is now added to the initial set if its combined *p*-value is less than a more relaxed binding cutoff.

Thus, we built three layer biological cascades (an example is shown in Figure 1) in which the third (output) layer comprises a gene in a module derived by GRAM, the second (middle) layer consists of the TFs that regulate the module and the first (upper) layer includes the TFs that regulate the second layer TFs.

2.4 Modeling and Quantitative Prediction of Gene Expression

Subsequently, we described quantitatively the uni-directional dependencies among the upper-layer and the module output-layer using a well-known machine-learning technique, namely ANNs, which have been previously used to describe interconnections between gene clusters (Huang et al., 2003), identify biomarkers using gene expression data (Pal et al., 2007) and classify gene expression data (Sewak et al., 2009). Unlike previous approaches where the structural characteristics of the ANNs were chosen somehow arbitrarily (Huang et al., 2003), we confined the structure of the ANN to match that of the biological network as this is inferred from expression data, DNA-protein interaction data and bibliographic information (see Figure 2A for an illustration of our methodology).

Each node in the input and middle layers represents one TF and thus the number of nodes in the ANNs equals the number of TFs in the respective biological networks. Examples of such networks are shown in Figures 1, 5, 7, 9 in the Results section. The activation functions in all ANNs are sigmoid-logarithmic in the middle-layer nodes and linear in the output layer. The weights and biases of the network are updated in every training epoch according to the Levenberg–Marquardt optimization (Levenberg, 1944; Marquardt, 1963) implemented in the Neural Network Toolbox in Matlab. For every ANN model the expression profile of the upper-layer TFs is used as input to the model and the expression of the downstream gene is assessed as the output. For each model, training is done using 50% of the experimental conditions and the remaining 50% is used for validation (25%) and testing (25%). Since genes in a module have slightly different expression profiles the procedure of the training/validation/testing is completed independently for every gene. This procedure is repeated 100 times while the training/validation/test data sets are randomly selected for each repetition. The correlation coefficient between the model predictions for the expression of the target gene in the test set and the desired output (the actual expression of the gene) is measured. The model's performance is assessed as the average correlation coefficient taken over the test set for the 100 repetitions and it is considered good if it is higher than 0.70, a threshold which is generally considered stringent in biological simulations (Cohen et al., 2000; Luo et al., 2007).

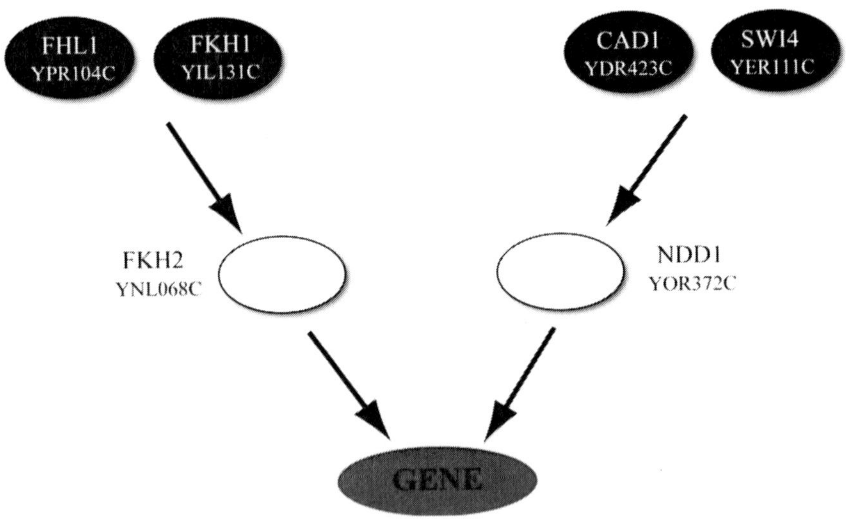

Figure 1. Representative example of a regulatory cascade. The regulatory cascade that is formed based on module 20 in Category A (heat shock). The module, as derived from GRAM, consists of seven genes (YLL006W, YDR043C, YML127W, YFL067W, YLL005C, YLL003W, and YJL225C) which are regulated by the transcription factors FKH2 and NDD1. While the biological cascade is common for all genes, a different ANN is trained/validated/tested based on the expression of every target gene in the module. Most of these genes are implicated in the stress response as evidenced from their MIPS annotation (Mewes et al., 1997).

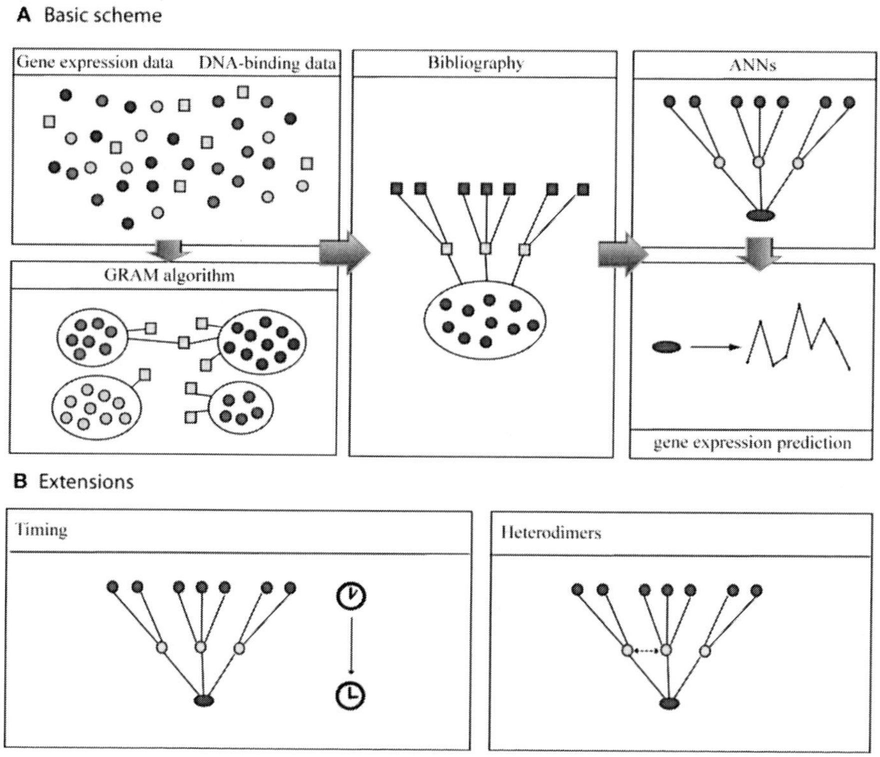

Figure 2. Flow chart of the proposed method. (A) Left panel: Gene expression data from microarray studies and DNA-binding data (ChIP-chip) are used as input to the GRAM algorithm in order to identify gene modules and their regulators. **Middle panel:** Bibliographical information from public databases is used to add another layer of regulators and build the three-layer regulatory cascade. **Right panel:** ANNs with the same structure as the biological cascade are built and trained to predict the expression of the target gene given the expression of the upper-layer regulators. **(B)**The basic scheme was extended to include additional biological aspects such as time-delays among the expression of a transcription factor and the expression of the target gene and formation of heterodimers.

To assess the statistical significance of the models' prediction accuracy, all ANNs are also trained, validated and tested using randomly shuffled data in the expression profile of the output gene.

2.5 Incorporation of Time-Delays in TF Activities

It is well-known that TFs undergo a number of post-transcriptional and post-translational modifications from the moment they are expressed until they become a functional protein (Benayoun and Veitia, 2009). These modifications are likely to require a significant amount of time causing a "delay" between the expression of a TF and its effect on the expression of a targeted gene. This is supported by previous work, where the expression profiles of certain TFs correlated with the expression profiles of their target genes only when activity delays were considered (Yu et al., 2003; He and Zeng, 2006; Boone et al., 2007; He et al., 2007). Based on the above, the issue of "timing" becomes even more important for indirect regulatory interactions such as those exerted by the upper-layer regulators in our model networks. The time difference between the activity of a given TF and its effect on the expression of a target gene is expected to depend on the environmental stress under study. For example while for a heat shock condition gene expression data are measured over a period of minutes or hours, for nitrogen depletion samples these values are measured over a period of days. Thus, we allowed combinatorial interactions among the time-dependent activities of regulatory TFs that laid two layers upstream in the modeled regulatory cascade and their target genes (see Figure 1—Timing for a graphical illustration). For this reason we use the time-series data of the stress-response dataset and shift the profiles of each TF independently. We aim to identify the combination of the shifted TF activities for which the expression of the target gene can be predicted with the optimum accuracy.

Specifically, we first identify all possible combinations of TF activities when their profiles are allowed to be shifted multiple time steps. The only restriction is that a TF profile cannot be shifted so that the TF appears to be activated after its target gene is expressed. Also, the timing of the expression of every TF cannot exceed the available experimental time points. Thus, the expression profile of every TF i in the data space is shifted k_i time steps backwards, where $0 < k_i < n-1$ (n is the total number of time-steps for the condition that has the smallest number of time-steps for this category). As a result, for every set of N TFs we create $n \cdot N$ combinations of possible TF-activity profiles (for an illustrative example see Figure 3 and Box 2). Every such combination is then used as input in the ANNs and the network's performance is assessed as previously discussed.

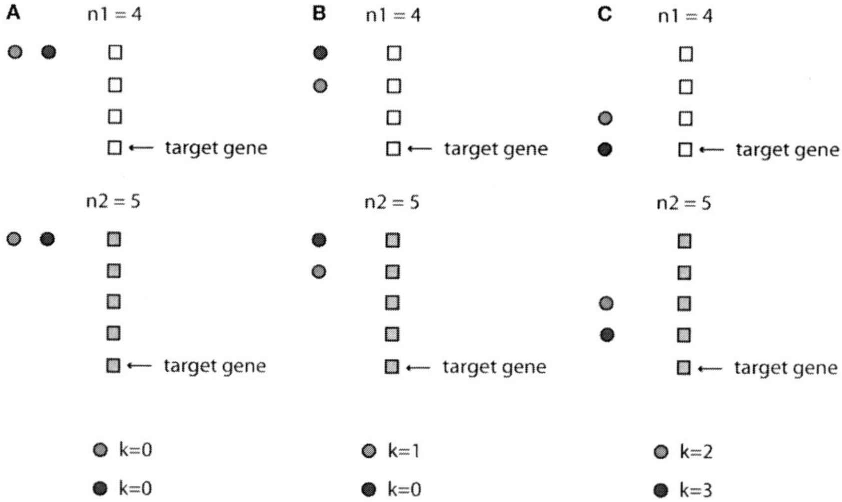

Figure 3. Schematic example of how the TF-delay table is constructed. Assume a category that consists of two conditions, with n1 and n2 time points, respectively (black and white rectangles), and a module that, for simplicity, is regulated by two upper-layer TFs (black and gray circles). The condition with the smallest number of time points is n1 so this will be the "leading" condition. Imagine that the target gene is expressed at time point $n1 = 4$ (This ranges from 1 to 4, since the expression profile of the gene is also shifted in time). Every TF is allowed to be expressed at the time points 1, 2, 3, and 4 (k represents the steps that each TF is shifted). **(A)** Both TFs are expressed at time point 1 ($k = 0$). **(B)** The first (black) TF is expressed at time point 2 ($k = 1$) while the second (gray) at time point 1 ($k = 0$). **(C)** The first (black) TF is expressed at time point 3 ($k = 2$) while the second (gray) at time point 4 ($k = 3$).

2.6 Incorporation of TF Dimers

Over 6000 genes in the *S. cerevisiae* genome appear to be regulated by roughly 200 identified TFs (Harbison et al., 2004). While the set of regulators is relatively small, many of them interact with each other in order to form new transcriptional regulators (e.g., TF dimers). As reviewed by Amoutzias et al. (2008), these heterodimers are unstable complexes able to form or decompose according to the required conditions. Moreover, the genes that correspond to each of the components of a heterodimer may have different patterns of expression. It is possible that one of them is constitutively expressed while its partner's expression is altered in a condition-specific manner. Such complex types of regulatory interactions are generally ignored by most module identification methods which focus on the detection of alterations in gene expression in order to classify the genes and their regulators. To address this limitation (Figure 2B—heterodimers), we used protein-protein interaction data (Maslov and Ispolatov, 2007) and searched for possible interacting pairs of the identified TFs in modules with a single regulator. Of the identified pairs we only took into account those for which DNA binding data were available (Lee et al., 2002). Before considering this protein as an additional regulator we test whether it is constitutively expressed in our dataset. This condition is satisfied if at least 95% of all data points across all stress conditions in a certain category do not exhibit more than a twofold change.

Box 2. A representative example of the TF-delay implementation. Genes in module 20, Category A (**Table 3**) are directly regulated by the TFs FKH2 and NDD1 and indirectly regulated by FHL1, FKH1, CAD1 and SWI4. The "Heat shock" category includes the following set of experiments: "Heat shock from 25 to 37°C," "Various temperatures to 37°C," "Steady-state temperature," "Heat shock from 37 to 25°C," "Heat shock at variable osmolarity," "Constant temperature growth." Of the above, sub-categories "Steady-state temperature" and "Various temperatures to 37°C" included only one time point so they were excluded from time-shift analysis. The remaining included 3–9 time points depending on the sub-category. So, the minimum number of available time points is 3, which means that there are two consecutive available steps (this is mentioned as "MAX" in the table) and one non-consecutive. More specifically, the experiment "29C to 33C" includes 3 time points at which the gene expression was measured, namely at 5 min, at 15 min and at 30 min. The available time steps are therefore 5-to-15 min and 15-to-30 min as well as 5-to-30 min. For simplicity we will consider a time delay step of 1 for all TFs. This would mean that either (1) when the expression of the target gene is measured at 15 min, the expression of the TFs is taken to be the one measured at 5 min or (2) when the expression of the target gene is measured at 30 min, the expression of the TFs is taken to be the one measured at 15 min. The same logic applies for time steps larger than one.

3. RESULTS

3.1 Identification of Transcriptional Modules

Application of the GRAM algorithm on DNA binding data as well as gene expression data from *S. cerevisiae* under stress conditions that fall in three categories (1) heat shock, (2) starvation, and (3) oxidative and osmotic stress resulted in the identification of functional modules and their direct regulators. Genes in these modules were grouped according to their expression profiles and the commonality of their regulators. For each category, we identified 87–100 modules each of which is regulated by 1–3 TFs as detailed in Table 2 and in Table S1. The majority of these modules contained a large number of genes that were regulated by a single TF, suggesting a more general role of these modules that spanned over multiple cellular processes. Since our goal was to identify regulatory networks that were specific to stressful stimuli, we limited our analysis to the 69 modules that were regulated by two or more TFs.

3.2 Inference of Biological Networks

We next searched for the regulators of the TFs that regulate the 69 modules, following the methodology depicted graphically in Figure 2. Based on information from YEASTRACT (www.yeastract.com), for the majority of our modules these regulators were either (1) plentiful (>15), (2) unknown, or (3) did not have a binding site in the promoter region of the gene(s) they regulate (see Methods). As a result, 51/69 modules were excluded from further analysis, leaving a set of 18 modules to be used for building three-layer regulatory cascades. The exclusion of the majority of inferred modules was a result of multiple factors, which, however, did not depend on the 3-layer cascade model itself but on the lack of (unknown TFs or lack of a binding site) or high complexity (too many TFs to find manually) of biological information. If all biological information was available, simulation of all the 3-layer cascades would have been performed. One representative example of such a cascade is shown in Figure 1, which consists of any gene in module 20 (Table S1) whose expression profile is modulated during heat shock (category A), the FKH2 and NDD1 TFs that regulate this module (according to GRAM) and the FHL1, FKH1, CAD1, and SWI4 TFs that regulate FKH2 and NDD1 (according to data from YEASTRACT filtered by our binding criteria). Respective regulatory cascades for all analyzed modules can be found in the Table S2.

Table 2. Information about the modules identified by GRAM.

	Category A	Category B	Category C
Number of modules	99	100	87
Number of modules with one regulator	73	76	68
Number of modules with >2 regulators	26	24	19
Mean number of genes in modules with one regulator	24.6	22.63	15.45
Mean number of genes in modules with >2 regulators	10.5	11.58	8.47

This table gives information about the number of the genes in the modules as well as the number of their regulators.

3.3 Modeling of Biological Networks with Artificial Neural Networks

Each two layer regulatory cascade was subsequently mapped to a structurally constrained ANN in an effort to quantitatively model the effect of upstream TFs on the expression profile of genes that were two levels downstream in the cascade (see Materials and Methods for details). A total of 70 three layer cascades, corresponding to 70 genes contained in 18 modules identified over all stress conditions were trained, validated and tested. Using as input the expression of the upper level regulators, the performance of each ANN model was evaluated by measuring the correlation coefficient r (CC) between the model output and the true expression profile of the target gene over numerous stress conditions. This approach resulted in a relatively small set (11/70) of ANN models with a CC > 0.7 (see Figure 4 and Table S3). The reasons for this could be numerous, ranging from limitations in the model selection to insufficient biological information. In the following sections we address some of these issues in order to improve the performance of our method.

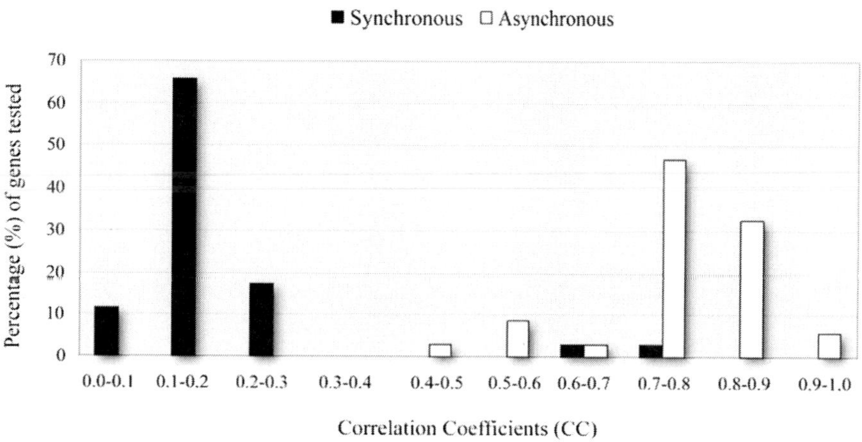

Figure 4. Correlation coefficient (CC) distributions for Synchronous and Asynchronous ANNs. ANN models which consider asynchronous interactions (white) achieve significantly higher performance (larger CC values) than synchronous ANNs (black).

3.4 Asynchronous Networks

One possible limitation of our approach is the assumption that, for a specific condition, the expression of a TF at a given time point has an effect on the expression of its target gene which is evident at the same time point. In other words, the expression profiles of TFs and their targets were assumed to be (relatively) synchronized (Note: Expression values in microarray experiments were measured over distinct time steps in an attempt to capture the gene expression changes during a specific response. However, it is possible that the cause (expression of the TF) and effect (change in the target gene expression) are evident in different sampling times mainly due to the various post-transcriptional modifications that are necessary for the complete activation of the TF). Moreover, in cases where more than one TF regulated a given gene we assume that their effects on gene expression were also synchronous. In reality, this is rarely the case.

To address this problem we next investigated whether a time-shift in the expression profile of each TF, independently of each other, improved the model's performance (see Materials and Methods for details). Specifically, we trained/evaluated and tested the same ANN models using as input the expression profiles of upstream TFs, each of which was shifted one or more time steps backwards in time. Every ANN model implementing a combination of time-shifted TF activities was evaluated using the CC, as previously described. As shown in Figure 4, when a time-shifted (asynchronous) regulation was considered, ANN models could successfully predict the expression profiles of target genes in the majority of the simulated regulatory cascades. It is important to note that usually more than one combinations lead to accurate models (i.e., CC > 0.7). This is rather intuitive since transcriptional regulation is critical for the cell's survival and it is often achieved via alternative pathways (Wagner and Wright, 2007).

A representative example of the improvement in ANN performance via the incorporation of TF delays is shown in Table 3. The table lists the correlation coefficient of the ANN models for three target genes in the network of Figure 1, namely genes YDR043W, YML127W, and YJL225C. "Synchronous" indicates the prediction accuracy of the ANN model when no time-shifts in the TF activities were considered while "Asynchronous" corresponds to the ANN model that incorporates the optimal combination of TF delays, namely the one leading to the highest CC. For gene YDR043W for example, the effects of the TFs FHL1, FKH1, CAD1, and SWI4 were evident in the expression of the target gene after 0, 2, 0, and 1 time steps, respectively. The same TFs appeared to regulate the other two genes in this module with different time delays, indicating that the effect of the same TF under the same conditions on the expression of its target was gene-specific.

Table 3. Maximum correlation coefficients for three cascades modeled by synchronous or asynchronous ANNs.

Module 20	Correlation coefficient		Transcription factor delays				
	Synchronous	Asynchronous	FHL1	FKH1	CAD1	SWI4	MAX
YDR043W	0.65 ± 0.11	0.86 ± 0.13	0	2	0	1	2
YML127W	0.73 ± 0.11	0.85 ± 0.11	0	2	1	1	2
YJL225C	0.48 ± 0.15	0.72 ± 0.14	0	0	0	1	2

In all cases, asynchronous ANNs achieve higher performance. The number of possible time-shifts for each TF depends on the number of experimental time-points available for each condition. This number is indicated with MAX while the number of time points a transcription factor is expressed before the expression of the target gene is referred to as transcription factor delays.

Detailed information regarding the optimal time-shifted combinations of TFs for all simulated cascades can be found in Table S4. It is striking that in the majority (30/34) of simulated regulatory cascades there is a significant improvement in the prediction accuracy of the ANN models, suggesting that the transcriptional regulation of most downstream genes is likely to result fromof a time-dependent combinatorial activation of upstream TFs.

3.5 Incorporation of TF-Dimers

An important limitation of the GRAM algorithm that may have influenced the ANN models is the assumption that TFs regulating a given gene should be differentially expressed under the various stress conditions. This assumption is frequently violated in cases where two TFs form a dimer in order to exert their regulatory action. To address this issue, we also considered constitutively expressed proteins that are found to interact with already identified regulators. Two such proteins were found in our dataset, namely PHO2 and HIR1, which are known to interact with BAS1 and HIR2, respectively.

The TF BAS1 has been implicated in the biosynthesis of histidine, purine and pyrimidine pathways (Tice-Baldwin et al., 1989; Daignan-Fornier and Fink, 1992). According to GRAM, it appeared as a unique regulator for 17 genes in Category A, 21 genes in Category B, and 13 genes in Category C. Moreover, BAS1 is itself regulated by 9 TFs, according to the YEASTRACT database. ANN simulation of the respective regulatory cascade (shown in Figure 5A) showed that in Category A (heat shock) the expression profiles of only 24% of the genes can be predicted with a correlation coefficient above 0.7 while for the rest of the genes in this category as well as genes in the other two categories prediction accuracy was much lower (Figure 6A).

Based on experimental data for protein-protein interactions, PHO2 interacts with BAS1 and is likely to act as a co-regulator as its presence is necessary in order for BAS1 to exert its regulatory action (Maslov and Ispolatov, 2007). In addition, PHO2 was constitutively expressed under both normal and stress conditions in our dataset. Incorporation of both TFs in the ANN model (cascade is illustrated in Figure 5B) significantly improved the model performance, particularly for genes in Categories A and B and less in Category C (see pie charts in Figure 6B). Specifically in Category A, 41% of gene profiles could be predicted with high accuracy, while this number dropped to 32% and 8% for Categories B and C, respectively. Among the ANNs with improved performance, the model corresponding to YOR224C (RPB8) appeared in all three categories. RPB8 is part of the RNA-polymerase III subunit (Archambault and Friesen, 1993;Chédin et al.,

1998; Geiduschek and Kassavetis, 2001) and at least three other genes in the same module encode proteins related to ribosomal components. There is supporting evidence for the involvement of BAS1 and RNA-polymerase and ribosomal proteins in other common regulatory pathways (Som et al., 2005; Kresnowati et al., 2006), an indication that BAS1 is indeed a necessary regulator in various aspects of the cellular machinery.

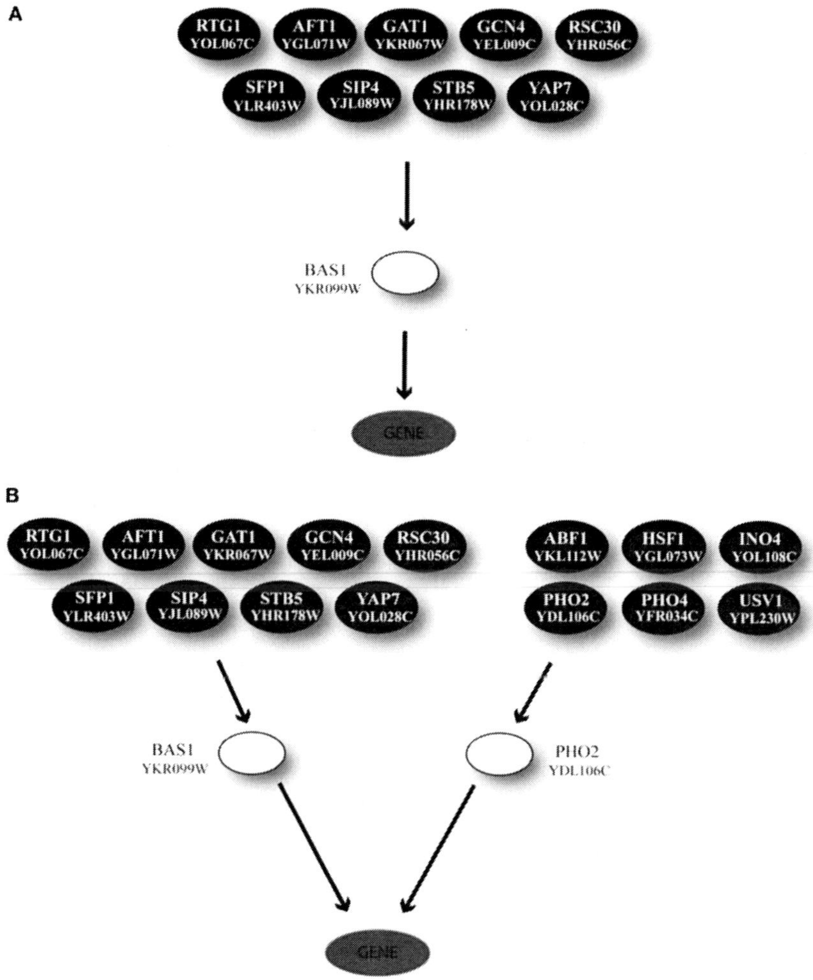

Figure 5. Regulatory cascades containing the TF BAS1 as the GRAM-identified regulator. (A) Regulatory cascade where only BAS1 is considered as a regulator. **(B)** Regulatory cascade where BAS1 as well as PHO2 are considered as regulators of the target gene. These cascades are found in all three stress categories but correspond to different gene modules as identified by the GRAM algorithm.

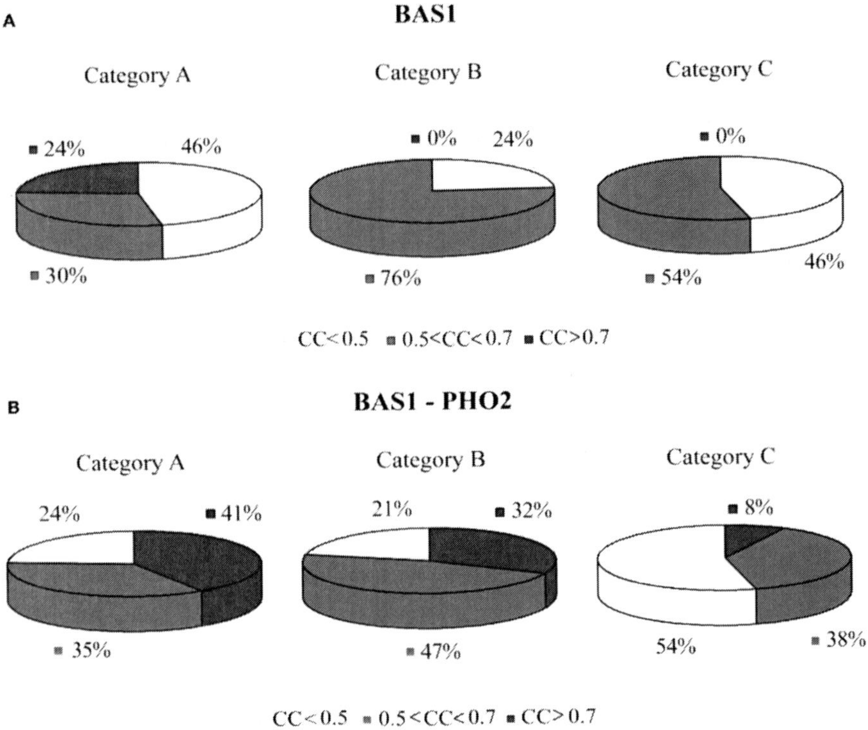

Figure 6. Pie charts showing the grouped correlation coefficients (CC) of the ANN models in the three stress categories. (A) Results correspond to genes in modules regulated by BAS1 alone. Successful ANNs (CC > 0.7) are found only for stress conditions in Category A (heat shock). **(B)** Results correspond to genes in modules regulated by the BAS1-PHO2 dimer. In this case, there is a significant increase in the correlation coefficient of ANNs over all three stress categories.

Another regulator, HIR2, was also identified by GRAM as a unique regulator for 19 genes in Category A, 21 genes in Category B and 5 genes in Category C. HIR2 was itself regulated by 5 TFs as illustrated in Figure7A. The ANN models that simulated this cascade achieved high performance for 78% of the target genes in Category A, but poor performance for the genes in the remaining Categories (pie charts in Figure 8A).

According to experimental evidence, HIR2 acted as a co-regulator of HIR1 for the transcription of histone genes (Sherwood et al., 1993; Desilva et al., 1998). Given that HIR1 was constitutively expressed in over 95% of our stress conditions and interacted with HIR2, a biological cascade that involved both TFs as middle-layer regulators was built (Figure 7B). Simulation of this cascade with ANNs resulted in high performance (CC > 0.7) for 89% (17/19) of the genes in Category A. A large fraction (6/19) of the genes in this category is of unknown function, while 4/19 genes have been associated with the ribosomal machinery and the rest share various other functions. Performance in Category B also improved with 9% of the ANNs achieving a CC above 0.7 while in Category C, although the performance of all ANNs was improved, no gene could be predicted with a CC > 0.7 (Figure 8B). The dramatic difference between ANN model performance in the three categories could suggest that the formation of the heterodimer HIR2-HIR1 is condition-dependent, as is the case for the majority of the known heterodimers (Amoutzias et al., 2008).

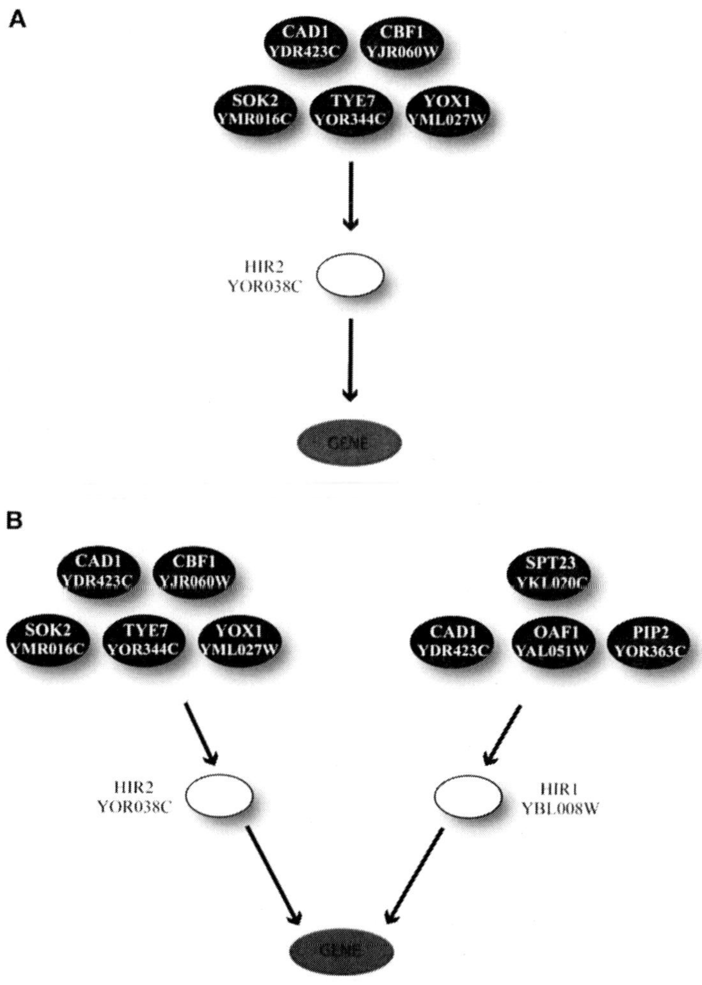

Figure 7. Regulatory cascades containing the TF HIR2 as the GRAM-identified regulator. **(A)** Regulatory cascade where only HIR2 is considered as regulator. **(B)** Regulatory cascade where HIR2 and HIR1 are considered as regulators of the target gene.

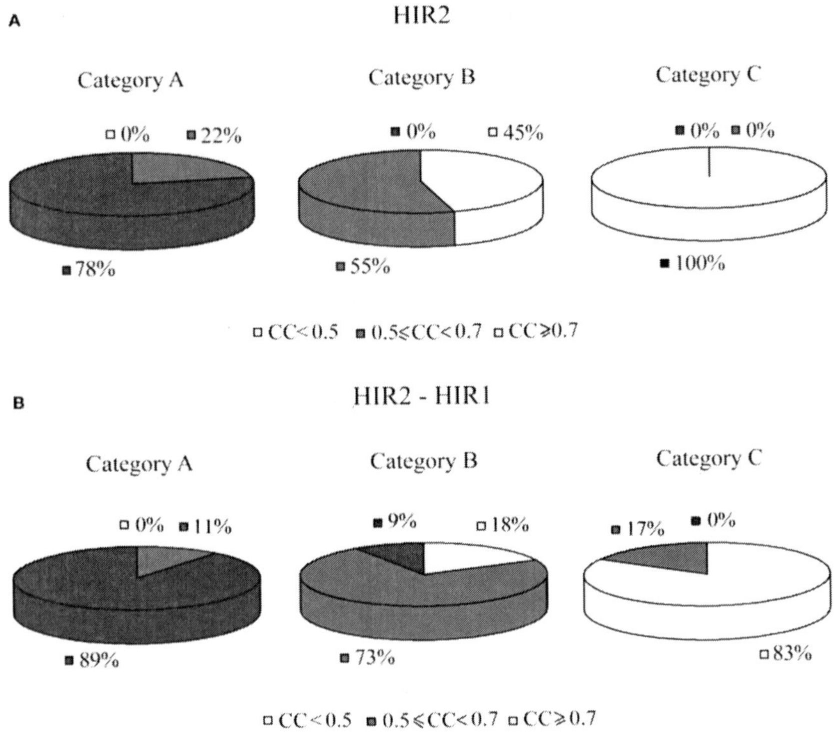

Figure 8. Pie charts that show the grouped correlation coefficients (CC) of the ANN models in the three stress categories. (A) Results correspond to genes in modules regulated by HIR2 alone. ANNs with CC > 0.7 are only seen in Category A (heat shock). For Category C in particular, the correlation coefficient for all genes is very low (CC < 0.5). **(B)** Results correspond to genes in modules regulated by HIR2 and HIR1. In this case the ANN performance is slightly improved in all three categories.

Finally, a special case of transcriptional regulators that are known to have a key role in the stress response in *S. cerevisiae*, MSN2 and MSN4 (Martinez-Pastor et al., 1996; Gasch and Werner-Washburne, 2002), were included in our analysis. In accordance to experimental findings (Haitani and Takagi, 2008), MSN2 was constitutively expressed across all stress conditions in our dataset; subsequently, the GRAM algorithm identified only MSN4 as a transcriptional regulator for 24 genes in Category A, 24 genes in Category B and 16 genes in Category C (the regulatory cascade that includes MSN4 as a middle-layer regulator can be seen in Figure 9A).

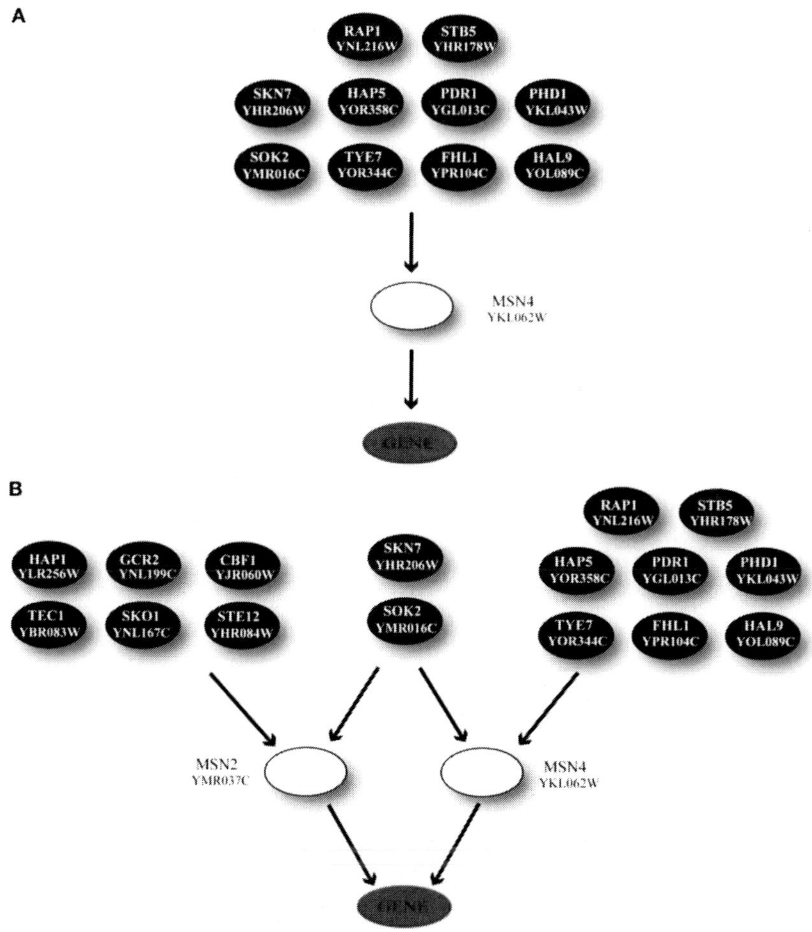

Figure 9. Regulatory cascades containing the TF MSN4 as the GRAM-identified regulator. (A) Regulatory cascade where only MSN4 is considered as a regulator. **(B)** Regulatory cascade where MSN4 and MSN2 are considered as regulators of the target gene.

Considering the importance of both regulators and their ubiquitous presence in stress response we built the regulatory cascade illustrated in Figure 9B which includes both MSN2 and MSN4 as well as their direct regulators. Figure 10 illustrates the performance of the ANNs in the three categories, grouped as models with CC that exceeds the 0.7 threshold and those for which the CC < 0.7 for MSN4 alone (Figure 10A) and MSN4 and MSN2 cascades (Figure 10B). ANN models that contained only MSN4 as a middle-layer regulator had a relatively good performance in Category A and a poor performance in the remaining categories. ANNs that consider both MSN4 and MSN2 had a good performance in all categories. Specifically, in category A (heat shock) more than half of the gene expression profiles could be predicted with high accuracy and it is important to note that the average CC of these ANNs was 0.85. The very high performance under heat shock (in both MSN4 alone and MSN2/4 models) could be related to the unique role of the MSN2/4 TFs to stress responses associated with alterations in temperature. Experimental evidence has shown that despite their overall function as stress response factors, they seem to be especially implicated in pathways that are triggered during heat shock (Ernst et al., 2007).

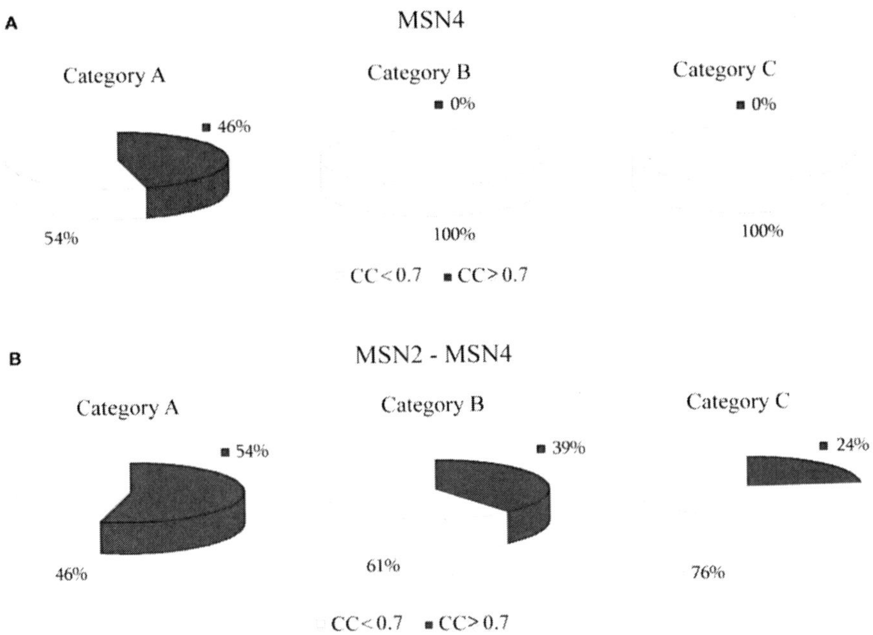

Figure 10. Pie charts that show the grouped correlation coefficients (CC) of the ANN models in the three stress categories. (A) **Results correspond to** genes in modules regulated by MSN4 alone. ANNs with CC > 0.7 are only seen in Category A (heat shock) whereas no good predictions were found in categories B and C. **(B)** Results that correspond to genes in modules regulated by MSN4 and MSN2. In this case the ANN performance is improved in all three categories, with Category A having the majority of the successful ANNs, in agreement with experimental evidence.

To ensure that improved ANN performance was not simply due to the addition of extra free parameters (namely the extra weights between the second middle layer node and its upper layer regulators as well as the weight between the output gene and the added middle layer node), we replaced the PHO2 and HIR1 TFs, respectively, with five (each) randomly selected TFs.

All of the randomly selected replacements were regulated by the same number of upper layer TFs as the replaced factor to ensure the same degree of flexibility (same number of free parameters/weights) in the resulting ANN models. Specifically, the following pairs of second layer TFs were used: (a) BAS1-YBR137W, (b) BAS1-SLY1, (c) BAS1-SAT4, (d) BAS1-CCS1, and (e) BAS1-THR1 instead of the original BAS1-PHO2 dimer and (a) HIR2-YDR249C, (b) HIR2-PMT1, (c) HIR2-YDL026W, (d) HIR2-YJR085C, and (e) HIR2-HIS6 instead of the HIR2-HIR1 dimer. This procedure resulted in ten additional three-layer cascades and a total of 10×86 (10×41 for each BAS1-pair and 10×45 for each HIR2-pair, see previous paragraphs) new ANN models. In all cases the ANN models resulting from the randomly selected TF pairs had poorer performance (fewer ANN models with $CC > 0.7$) than the respective ANNs corresponding to the original dimers. Moreover, in several cases performance was worse than the ANNs incorporating only one of the TFs (as opposed to the dimer), indicating that improved performance does not result from the added flexibility provided by the extra free parameters.

3.6 Statistical Assessment of ANNs' Performance

To assess the accuracy of our ANN models in predicting the expression profile of downstream genes we shuffled the expression values of each target gene and re-trained the models. We found that for all ANNs with $CC > 0.7$, the correlation coefficient for the shuffled data dropped dramatically, indicating that the performance of our ANN models is far from random chance (see Figure 11 for a detailed comparison). For the asynchronous networks in particular, the ANN models corresponding to the various combinations of time-shifted TF activities were also simulated using 100 trials where the expression profile of the target gene was randomly shuffled and the resulting CC was recorded. In all cases, the CC of the optimal set of time-shifted TF activities was found to be significantly higher than the maximum CC obtained by the shuffled ANNs, indicating that the improved performance in asynchronous networks was not a result of overfitting.

4. DISCUSSION

In this work we introduced a semi-dynamic method for modeling the structure of three-layer transcriptional regulatory cascades based on ANNs. The method calculates quantitatively the expression profiles of *S. cerevisiae* genes during various stress conditions, based on the expression of their upstream regulators. *S. cerevisiae* was selected because it is a well-studied organism with an important biotechnological role (Botstein et al., 1997). The availability of a plethora of expression studies concerning the response of *S. cerevisiae* to different environmental changes (Bammert and Fostel, 2000; Gasch et al., 2000; Rep et al., 2000; Causton et al., 2001; Kwast et al., 2002) underlies the importance of this mechanism and provides the large amount of data required for a quantitative approach like the one proposed here.

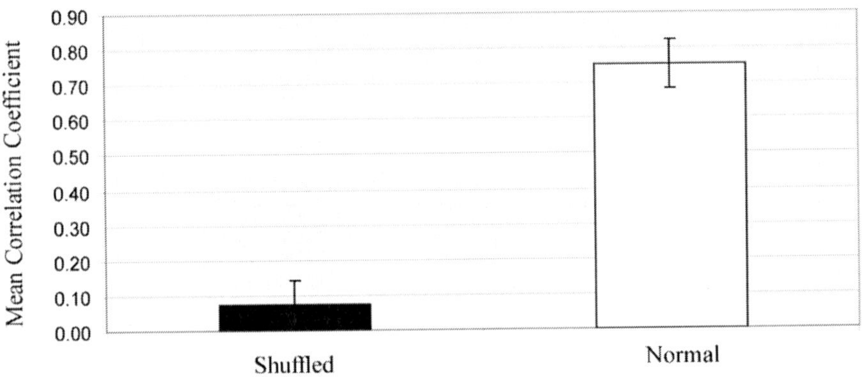

Figure 11. Correlation coefficients of the normal and shuffled data. Bars show the mean and standard deviation of the correlation coefficients achieved by all ANN models when either the normal (white) or randomly shuffled (black) expression profiles for the target gene where used to train the models.

Our approach depends on the use of publicly available software tools (e.g., GRAM or any other module finding tool) for the identification of groups of genes which share common expression profile and a common set of regulators. Using these modules, along with information regarding protein-DNA and protein-protein interaction data, we extend the regulatory modules into three-layer cascades by adding another level of regulation. These cascades are then simulated using ANNs where the expression of the upper level regulators is used to predict the expression of the downstream target gene(s).

Initial use of the ANNs under the assumption that expression profiles corresponding to regulators and regulated genes evolve in synchrony over time and under various conditions, resulted in a relatively poor outcome: only 11 out of 70 models had a high performance (CC > 0.7). Incorporation of time-delays between the profiles of regulators and those of regulated genes, which can vary independently for each regulator, resulted in a massive increase in the percentage of successful ANN models:

Thirty out of thirty four networks tested had a significant improvement in their performance, validating the intuitive assumption that a TF can be expressed several time steps before its effect is evident in the expression of the target gene and that regulators that target the same gene may exhibit their activity in a combinatorial manner. Identification of the optimal combination of TF activities for which the expression of a downstream target gene can be predicted with high accuracy leads to a working hypothesis that describes not only the regulatory components for this gene (network structure) but also their interaction over time (network functionality over time). An important finding is that for every regulatory cascade there is more than one optimal combination of transcriptional activities, in agreement with the observed flexibility of biological systems to overcome perturbations that could hinder their regulatory program.

The method was also found to perform well in three cases where known regulators, which are constitutively expressed under stress conditions, were introduced in the middle layer. Incorporation of protein interaction data as an indication of heterodimer formation to previously identified cascades resulted in a significant improvement in the performance of ANN models. Specifically, we incorporated the interaction of BAS1 with PHO2 and the interaction of HIR2 with HIR1. In the latter case, model performance was condition-dependent with heat shock and starvation categories having the highest performing models, thus pinpointing to the specific stress conditions that favor the formation of the given heterodimer.

Existing network inference approaches range in the field between two extremes. Considering the gene as only existing in two discrete situations that is on and off, as in Boolean approaches (Bornholdt, 2008) or offering an analytical description of the gene state using differential equations (Sakamoto and Iba, 2001). The main limitations of these approaches is the assumption of binary values for the expression of a gene and the requirement for a large amount of parameter values that are not usually available and therefore the restriction of the method to a very small portion of the regulatory network, respectively. Bayesian networks (BN) (Hecker et al., 2009) represent the regulatory interactions by probability in a predefined graph and they are based on identifying the conditional probabilities that best match the expression data on this graph. Their main advantages include the offering of a flexible framework for the combination of different types of data and the ability to avoid over-fitting a model to the training data.

Our method covers these advantages and overcomes the bottleneck of the graph identification by restricting the model structure through the application of clustering methods and bibliography. We demonstrate that these models can capture simple as well as more complicated aspects of biological regulation ranging from semi-dynamic predictions of gene expression profiles over multiple conditions to incorporation of additional types of regulation. Although ANNs have been used in the past to model simple regulatory cascades with moderate accuracy, they did not reproduce the structure of biological networks which enforces several constrains and did not extend to more than one level of regulation (Hart et al., 2006). Our findings show that ANNs can be used to simulate indirect, time-dependent interactions between transcriptional regulators and genes downstream in a cascade as well as their evolution over multiple conditions and time points.

In summary, we propose a quantitative method for modeling biological cascades based on the ANN formalism. Compared to existing methods for network inference and parameter optimization, the proposed scheme uses a combination of a data driven process that identifies clusters of co-expressed and/or co-regulated genes along with a knowledge—driven feature selection approach of previously identified protein-protein interactions in order to reduce the dimensionality of the network components.

Our method offers the algorithmic simplicity of the Boolean network approach using real expression data and although the simulation of these data lacks the analytical description offered by differential equation methods, it can be applied in three layer regulatory networks. Albeit our method was implemented using*S. cerevisiae* data, it is readily applicable to any other organism and/or condition influencing gene transcription. While our method is restricted to time-course gene expression data and requires previous knowledge on gene regulatory interactions, it can be used to predict the expression of downstream regulated genes based on the expression of their regulators and can predict the expression in time points that are not present in the experimental dataset. Moreover, it could be used to identify the time dependencies of regulatory interactions and identify the optimal timing of regulatory combinations. Finally, using our method one could propose candidate co-regulatory pairs of TFs (over a pool of available TFs) based on the increased performance of the ANN for each pair as well as the condition under which this pair is formed. Future work can address the limitations of our approach regarding the identification of upstream regulators, by incorporating an automated search method for inferring these molecules and their interactions from public databases.

AUTHOR CONTRIBUTIONS

Maria E. Manioudaki developed the methods, analyzed, and interpreted the data and wrote the paper. Panayiota Poirazi designed the study, wrote the paper and supervised the project.

CONFLICT OF INTEREST STATEMENT

The authors declare that the research was conducted in the absence of any commercial or financial relationships that could be construed as a potential conflict of interest.

ACKNOWLEDGMENTS

The authors would like to thank members of our lab for useful discussions during the preparation of this manuscript. This work was supported by the Greek General Secretariat for Research and Technology, action 8.3.1 (Reinforcement Program of Human Research Manpower) of the operational program "competitiveness" (PENED 2003-03ED842) and by the "SINERGASIA" program (EDGE 901-13/11/2009).

REFERENCES

1. Amoutzias, G. D., Robertson, D. L., Van De Peer, Y., and Oliver, S. G. (2008). Choose your partners: dimerization in eukaryotic transcription factors. *Trends Biochem. Sci.* 33, 220–229.

2. Archambault, J., and Friesen, J. D. (1993). Genetics of eukaryotic RNA polymerases I, II, and III. *Microbiol. Rev.* 57, 703–724.

3. Bammert, G. F., and Fostel, J. M. (2000). Genome-wide expression patterns in *Saccharomyces cerevisiae*: comparison of drug treatments and genetic alterations affecting biosynthesis of ergosterol. *Antimicrobial Agents Chemother.* 44, 1255–1265.

4. Bar-Joseph, Z., Gerber, G. K., Lee, T. I., Rinaldi, N. J., Yoo, J. Y., Robert, F., et al. (2003). Computational discovery of gene modules and regulatory networks. *Nat. Biotechnol.* 21, 1337–1342.

5. Benayoun, B. A., and Veitia, R. A. (2009). A post-translational modification code for transcription factors: sorting through a sea of signals. *Trends Cell Biol.* 19, 189–197.

6. Boone, C., Bussey, H., and Andrews, B. J. (2007). Exploring genetic interactions and networks with yeast. *Nat. Rev. Genet.* 8, 437–449.

7. Bornholdt, S. (2008). Boolean network models of cellular regulation: prospects and limitations. *J. R. Soc. Interface* 6, S85–S94.

8. Botstein, D., Chervitz, S. A., and Michael, C. (1997). Yeast as a model organism. *Science* 277, 1259–1260.

9. Bruggeman, F. J., Snoep, J. L., and Westerhoff, H. V. (2008). Control, responses and modularity of cellular regulatory networks: a control analysis perspective. *IET Syst. Biol.* 2, 397–410.

10. Causton, H. C., Ren, B., Koh, S. S., Harbison, C. T., Kanin, E., Jennings, E. G., et al. (2001). Remodeling of yeast genome expression in response to environmental changes. *Mol. Biol. Cell* 12, 323–337.

11. Chédin, S., Ferri, M. L., Peyroche, G., Andrau, J. C., Jourdain, S., Lefebvre, O., et al. (1998). The yeast RNA polymerase III transcription machinery: a paradigm for eukaryotic gene activation. *Cold Spring Harb. Symp. Quant. Biol.* 63, 381–389.

12. Chen, G., Jensen, S. T., and Stoeckert, C. J. (2007). Clustering of genes into regulons using integrated modeling-COGRIM. *Genome Biol.* 8:R4.

13. Cohen, B. A., Mitra, R. D., Hughes, J. D., and Church, G. M. (2000). A computational analysis of whole-genome expression data reveals chromosomal domains of gene expression. *Nat. Genet.* 26, 183–186.

14. Daignan-Fornier, B., and Fink, G. R. (1992). Coregulation of purine and histidine biosynthesis by the transcriptional activators BAS1 and BAS2. *Proc. Natl. Acad. Sci. U.S.A.* 89, 6746–6750.

15. Desilva, H., Lee, K., and Osley, M. A. (1998). Functional dissection of yeast Hir1p, a WD repeat-containing transcriptional corepressor. *Genetics* 148, 657–667.

16. Ernst, J., Vainas, O., Harbison, C. T., Simon, I., and Bar-Joseph, Z. (2007). Reconstructing dynamic regulatory maps. *Mol. Syst. Biol.* 3, 74.

17. Fernández, A. (2007). Molecular basis for evolving modularity in the yeast protein interaction network. *PLoS Comput. Biol.* 3:e226.

18. Friberg, M., Von Rohr, P., and Gonnet, G. (2005). Scoring functions for transcription factor binding site prediction. *BMC Bioinformatics*6:84.

19. Gao, F., Foat, B. C., and Bussemaker, H. J. (2004). Defining transcriptional networks through integrative modeling of mRNA expression and transcription factor binding data. *BMC Bioinformatics* 5:31.

20. Gasch, A. P., Spellman, P. T., Kao, C. M., Carmel-Harel, O., Eisen, M. B., Storz, G., et al. (2000). Genomic expression programs in the response of yeast cells to environmental changes. *Mol. Biol. Cell* 11, 4241–4257.

21. Gasch, A. P., and Werner-Washburne, M. (2002). The genomics of yeast responses to environmental stress and starvation. *Funct. Integr. Genomics* 2, 181–192.

22. Geiduschek, E. P., and Kassavetis, G. A. (2001). The RNA polymerase III transcription apparatus. *J. Mol. Biol.* 29, 1–26.

23. Haitani, Y., and Takagi, H. (2008). Rsp5 is required for the nuclear export of mRNA of HSF1 and MSN2/4 under stress conditions in *Saccharomyces cerevisiae*. *Genes Cells* 13, 105–116.

24. Han, J. (2008). Understanding biological functions through molecular networks. *Cell Res.* 18, 224–237.

25. Harbison, C. T., Gordon, D. B., Lee, T. I., Rinaldi, N. J., Macisaac, K. D., Danford, T. W., et al. (2004). Transcriptional regulatory code of a eukaryotic genome. *Nature* 431, 99–104.

26. Hart, C. E., Mjolsness, E., and Wold, B. J. (2006). Connectivity in the yeast cell cycle transcription network: inferences from neural networks. *PLoS Comput. Biol.* 2:e169.

27. Hartwell, L. H., Hopfield, J. J., Leibler, S., and Murray, A. W. (1999). From molecular to modular cell biology. *Nature* 402, C47–C52.

28. He, F., Buer, J., Zeng, A. P., and Balling, R. (2007). Dynamic cumulative activity of transcription factors as a mechanism of quantitative gene regulation. *Genome Biol.* 8:R181.

29. He, F., and Zeng, A. P. (2006). In search of functional association from time-series microarray data based on the change trend and level of gene expression. *BMC Bioinformatics* 7:69.

30. Hecker, M., Lambeck, S., Toepfer, S., Van Someren, E., and Guthke, R. (2009). Gene regulatory network inference: data integration in dynamic models-a review. *BioSystems* 96, 86–103.

31. Holter, N. S., Mitra, M., Maritan, A., Cieplak, M., Banavar, J. R., and Fedoroff, N. V. (2000). Fundamental patterns underlying gene expression profiles: simplicity from complexity. *Proc. Natl. Acad. Sci. U.S.A.* 97, 8409–8414.

32. Hu, J., Hu, H., and Li, X. (2008). MOPAT: a graph-based method to predict recurrent cis-regulatory modules from known motifs. *Nucleic Acids Res*. 36, 4488–4497.

33. Huang, J., Shimizu, H., and Shioya, S. (2003). Clustering gene expression pattern and extracting relationship in gene network based on artificial neural networks. *J. Biosci. Bioeng*. 96, 421–428.

34. Ihmels, J., Levy, R., and Barkai, N. (2004). Principles of transcriptional control in the metabolic network of *Saccharomyces cerevisiae.Nat. Biotechnol*. 22, 86–92.

35. Imoto, S., Higuchi, T., Goto, T., Tashiro, K., Kuhara, S., and Miyano, S. (2004). Combining microarrays and biological knowledge for estimating gene networks via bayesian networks. *J. Bioinform. Comput. Biol*. 2, 77–98.

36. Imoto, S., Kim, S., Goto, T., Miyano, S., Aburatani, S., Tashiro, K., et al. (2003). Bayesian network and nonparametric heteroscedastic regression for nonlinear modeling of genetic network. *J. Bioinform. Comput. Biol*. 1, 231–252.

37. Petricka, J. J., and Benfey, P. N. (2011). Reconstructing regulatory network transitions. *Trends Cell Biol*. 21, 442–451.

38. Kresnowati, M. T., Van Winden, W. A., Almering, M. J., Ten Pierick, A., Ras, C., Knijnenburg, T. A., et al. (2006). When transcriptome meets metabolome: fast cellular responses of yeast to sudden relief of glucose limitation. *Mol. Syst. Biol*. 2, 49.

39. Kundaje, A., Middendorf, M., Gao, F., Wiggins, C., and Leslie, C. (2005). Combining sequence and time series expression data to learn transcriptional modules. *IEEE/ACM Trans. Comput. Biol. Bioinform*. 2, 194–202. doi: 10.1109/TCBB.2005.34

40. Kwast, K. E., Lai, L. C., Menda, N., James, D. T. 3rd., Aref, S., and Burke, P. V. (2002). Genomic analyses of anaerobically induced genes in *Saccharomyces cerevisiae*: functional roles of Rox1 and other factors in mediating the anoxic response. *J. Bacteriol*. 184, 250–265.

41. Lee, H.-G., Lee, H.-S., Jeon, S.-H., Chung, T.-H., Young-Sung, L., and Won-Ki, H. (2008). High-resolution analysis of condition-specific regulatory modules in *Saccharomyces cerevisiae*. *Genome Biol*. 9.

42. Lee, T. I., Rinaldi, N. J., Robert, F., Odom, D. T., Bar-Joseph, Z., Gerber, G. K., et al. (2002). Transcriptional regulatory networks in*Saccharomyces cerevisiae*. *Science* 298, 799–804.

43. Levenberg, K. (1944). A method for the solution of certain non-linear problems in least squares. *Q. Appl. Math.* 2, 164–168.

44. Liebermeister, W. (2002). Linear modes of gene expression determined by independent component analysis. *Bioinformatics* 18, 51–60.

45. Luo, F., Yang, Y., Zhong, J., Gao, H., Khan, L., Thompson, D. K., et al. (2007). Constructing gene co-expression networks and predicting functions of unknown genes by random matrix theory. *BMC Bioinformatics* 8:299.

46. Maraziotis, I., Dimitrakopoulou, K., and Bezerianos, A. (2008). An *in silico* method for detecting overlapping functional modules from composite biological networks. *BMC. Syst. Biol.* 2:93.

47. Marquardt, D. W. (1963). An algorithm for least-squares estimation of nonlinear parameters. *J. Soc. Ind. Appl. Math.* 11, 431–441.

48. Martinez-Pastor, M. T., Marchler, G., Schuller, C., Marchler-Bauer, A., Ruis, H., and Estruch, F. (1996). The *Saccharomyces cerevisiae*zinc finger proteins Msn2p and Msn4p are required for transcriptional induction through the stress response element (STRE). *EMBO J.* 15, 2227–2235.

49. Maslov, S., and Ispolatov, I. (2007). Propagation of large concentration changes in reversible protein-binding networks. *Proc. Natl. Acad. Sci. U.S.A.* 104, 13655–13660.

50. Mewes, H. W., Albermann, K., Heumann, K., Liebl, S., and Pfeiffer, F. (1997). MIPS: a database for protein sequences, homology data and yeast genome information. *Nucleic Acids Res.* 25, 28–30.

51. Pal, N. R., Aguan, K., Sharma, A., and Amari, S. (2007). Discovering biomarkers from gene expression data for predicting cancer subgroups using neural networks and relational fuzzy clustering. *BMC Bioinformatics* 8:5.

52. Pournara, I., and Wernisch, L. (2007). Factor analysis for gene regulatory networks and transcription factor activity profiles. *BMC Bioinformatics* 8:61.

53. Rep, M., Krantz, M., Thevelein, J. M., and Hohmann, S. (2000). The transcriptional response of *Saccharomyces cerevisiae* to osmotic shock. Hot1p and Msn2p/Msn4p are required for the induction of subsets of high osmolarity glycerol pathway-dependent genes. *J. Biol. Chem.* 275, 8290–8300.

54. Sakamoto, E., and Iba, H. (2001). "Inferring a system of differential equations for a gene regulatory network by using genetic programming" in *Evolutionary Computation 2001. Proceedings of the 2001 Congress on*, Vol. 1 (Seoul), 720–726.

55. Segal, E., Shapira, M., Regev, A., Pe'er, D., Botstein, D., Koller, D., et al. (2003). Module networks: identifying regulatory modules and their condition-specific regulators from gene expression data. *Nat. Genet*. 34, 166–176.

56. Sewak, M. S., Narender, P. R., and Duan, Z. H. (2009). Gene expression based leukemia sub-classification using committee neural networks. *Bioinform. Biol. Insights* 3, 89–98.

57. Sherwood, P. W., Tsang, S. V., and Osley, M. A. (1993). Characterization of HIR1 and HIR2, two genes required for regulation of histone gene transcription in *Saccharomyces cerevisiae*. *Mol. Cell. Biol*. 13, 28–38.

58. Som, I., Mitsch, R. N., Urbanowski, J. L., and Rolfes, R. J. (2005). DNA-bound Bas1 recruits Pho2 to activate ADE genes in*Saccharomyces cerevisiae*. *Eukaryotic Cell* 4, 1725–1735.

59. Tanay, A., Sharan, R., Kupiec, M., and Shamir, R. (2004). Revealing modularity and organization in the yeast molecular network by integrated analysis of highly heterogeneous genomewide data. *Proc. Natl. Acad. Sci. U.S.A*. 101, 2981–2986.

60. Tice-Baldwin, K., Fink, G. R., and Arndt, K. T. (1989). BAS1 has a Myb motif and activates HIS4 transcription only in combination with BAS2. *Science* 246, 931–935.

61. Tornow, S., and Mewes, H. W. (2003). Functional modules by relating protein interaction networks and gene expression. *Nucleic Acids Res*. 31, 6283–6289.

62. Wagner, A., and Wright, J. (2007). Alternative routes and mutational robustness in complex regulatory networks. *BioSystems* 88, 163–172.

63. Wang, S. Q., and Li, H. X. (2012). Quantitative modeling of transcriptional regulatory networks by integrating multiple source of knowledge. *Bioprocess Biosyst. Eng*. 35, 1555–1565.

64. Xu, X., Wang, L., and Ding, D. (2004). Learning module networks from genome-wide location and expression data. *FEBS Lett*. 578, 297–304.

65. Youn, A., Reiss, D. J., and Stuetzle, W. (2010). Learning transcriptional networks from the integration of ChIP-chip and expression data in a non-parametric model. *Bioinformatics* 26, 1879–1886.

66. Yu, H., and Gerstein, M. (2006). Genomic analysis of the hierarchical structure of regulatory networks. *Proc. Natl. Acad. Sci. U.S.A.* 103, 14724–14731.

67. Yu, H., Luscombe, N. M., Qian, J., and Gerstein, M. (2003). Genomic analysis of gene expression relationships in transcriptional regulatory networks. *Trends Genet.* 19, 422–427.

68. Yu, T., and Li, K. C. (2005). Inference of transcriptional regulatory network by two-stage constrained space factor analysis.*Bioinformatics* 21, 4033–4038.

69. Zhao, J., Ding, G.-H., Tao, L., Yu, H., Yu, Z.-H., Luo, J.-H., et al. (2007). Modular co-evolution of metabolic networks. *BMC Bioinformatics* 8:311.

CHAPTER 7

Forecasting of Runoff and Sediment Yield Using Artificial Neural Networks

Avinash Agarwal[1], R. K. Rai[2], Alka Upadhyay[2]

[1]National Institute of Hydrology, Roorkee, Uttarakhand, India

[2]DHI (India) Water & Environment Pvt. Ltd., New Delhi, India

ABSTRACT

Runoff and sediment yield from an Indian watershed during the monsoon period were forecasted for different time periods (daily and weekly) using the back propagation artificial neural network (BPANN) modeling technique. The results were compared with those of singleand multi-input linear transfer function models. In BPANN, the maximum value of variable was considered for normalization of input, and a pattern learning algorithm was developed. Input variables in the model were obtained by comparing the response with their respective standard error. The network parsimony was achieved by pruning the network using error sensitivity - weight criterion, and model generalization by cross validation. The performance was evaluated using correlation coefficient (CC), coefficient of efficiency (CE), and root mean square error (RMSE). The single input linear transfer function (SI-LTF) runoff and sediment yield forecasting models were more efficacious than the multi input linear transfer function (MI-LTF) and ANN models.

KEYWORDS

Artificial Neural Network, forecasting, runoff, sediment Yield

1. INTRODUCTION

Rainfall-runoff-sediment yield is the most complex hydrological phenomenon to comprehend due to tremendous spatial variability of watershed characteristics and precipitation patterns, making the physical modeling quite complex and involved. The quantity of runoff and sediment yield resulting from a given rainfall depends mainly on rainfall intensity, duration, and distribution besides others, such as initial soil moisture, land use, slope, etc. The runoff is critical to many water resources activities, for example, design of flood protection works, protection of agricultural lands, planning of water storage and release, etc. The erosion in the watershed may be occurred due to rainfall and runoff, and degrades its land. The sediment transport caused the reduction of storage capacity of rivers and reservoirs. Also, sediments can carry pollutants such as radioactive materials, pesticides and nutrients, and their transportation is generally avoided.

A number of linear and non-linear hydrological models have been developed since 1930's for describing the processes of rainfall-runoff, runoff-sediment yield, and rainfall-runoff-sediment yield in a watershed fluvial system, and these are also useful in forecasting. These models can be broadly classified as lumped, conceptual, hydrological, and hydraulic models. The physically based classical models requires wide range of parameters related to land uses, soil characteristics, soil horizon, watershed treatment, man made activities, conservation practices, soil moisture variation, topographic data, surface roughness, etc. These parameters vary significantly over a space and time, and very difficult to monitor. Under these circumstances, classical models require assumption of the parameters. Most common assumption to be made for describing the rainfall-runoff-sediment yield process is: sediment produced in the catchments is uniformly distributed over the catchments, uniform soil group, approximation of slopes, etc. These assumptions create lumped-ness in the physically based distributed classical model. On the other hand, ANN based approach yields satisfactory results without going into the details of catchments characteristics. The artificial neural network (ANN) approach comprises both linear and nonlinear concepts in model building, and can be operated with the dynamic or memory less input-output system. It has the following major advantages [1]. 1) An ANN model does not require a prior knowledge of the system and, therefore, can be applied to solve the problems not clearly defined, 2) The model has more tolerance to noise and incomplete data, and thus, requires less data for model development, and 3) The results are the outcome of the collective behavior of data, and thereby, the effect of outlier is minimized. In ANN, the gradient descent search optimization embedded with back propagation algorithm [2] is quite popular in ANN for exploring diverse areas such as bio-medical, engineering, image processing, water resources, and others [3,4].

The hydrologic applications of ANN include modeling of daily rainfall-runoff-sediment yield process, snowrainfall process; assessment of stream ecological and hydrological responses to climate change; rainfall-runoff forecasting, sediment transport and groundwater quality forecasting, and groundwater remediation. A good overview of ANN application to rainfall-runoff simulation and forecasting is available elsewhere [5−16]. Imrie improved the generalization by adding a guidance system to the cascade correlation learning architecture and extrapolation properties using an activation function. Wilby interpreted the internal behavior of an ANN-based rainfall-runoff model by deleting all the nodes other than the hidden nodes and comparing the state variables and internal fluxes. Danh and Elshorbagy proposed back propagation artificial neural network (BPANN) models for runoff forecasting using fixed stopping criterion and independent variables, respectively, and compared them for performance with the available conceptual models.

It is apparent from above that the generalization of ANN models for varying data sets and their application to sediment modeling do not appear to have been studied extensively or even attempted. Thus, the objective of the study is to develop memory-based linear transfer function, ANN-based runoff and sediment yield forecasting models for daily and weekly time periods; and to evaluate the model performance for their forecasting abilities using the data of the Vamsadhara River watershed of India.

2. MODEL DEVELOPMENT

The linear transfer function (LTF) models require minimum input and computation, and yield the results of desired accuracy [17]. In an LTF, two or more time-dependent observations are linked as:

$$S_t = \sum_{j=1}^{p} a_j S_{t-j} + \sum_{j=1}^{q} b_j R_{t-j+1} + \sum_{j=1}^{r} c_j Q_{t-j+1} \tag{1}$$

where, S_t is the dependent observation; R_t and Q_t are independent observations; p, q, r are the response or time memory; and a_j, b_j, c_j are the parameters associated with the jth variable. Least square method can be used to solve the set of t linear equations for parameters. Representing $[a_1, a_2,\ldots, a_p, b_1, b_2,\ldots,b_q, c_1, c_2,\ldots,c_r]^T$ as $\hat{H}$, the variance of the response {var ($\hat{H}$)} is given as [17]: Var ($\hat{H}$) = $[A^T A]^{-1} s^2$, where A is the input matrix and s^2 is the variance of error term (e_t) expressed as:

$$\sigma^2 = (t-p-q-r)^{-1}\sum_{j=1}^{t} e_j^{\ 2} \tag{2}$$

The Var ($\hat{H}$) helps in parsimonious selection of response or time memory by comparing the respective standard error (SE) associated with each response ($\hat{H}$ 3 $\sqrt{\mathrm{Var}(\hat{H})}$) as follows. The initially selected response value is increased one by one and if the resulting $\hat{H}$ is less than the respective standard error {$\sqrt{\mathrm{Var}(\hat{H})}$}, otherwise the response value is decreased one by one. The obtained response value represents the number of successive past events of the variable affecting output [17].

In a feed forward BPANN scheme, nodes of the input layer receive the normalized data set (input). The weighted sum corresponding to each node of next layer is calculated and passed to next layer usually through a sigmoid activation function. For updating the weights of interconnection, the error (E) is calculated at the output as;

$$E = \frac{1}{2}\sum_{1}^{k}[d_{(k)} - O_{(k)}]^2 \tag{3}$$

This error is propagated back to hidden layer and finally to input layer. Here, $d_{(k)}$ is the observed output at the kth node of the output layer and $O_{(k)}$ is the estimated output at the kth node of the output layer. In all iterations, weights are updated using $W_{(ij)\,n+1} = W_{(ij)\,n} + DW_{(ij)\,n}$.

The speed of convergence in gradient descent is normally increased by introducing a momentum term b ($0 < b < 1$) considering the effect of previous weight change as:

$$W_{(ij)\,n+1} = W_{(ij)n} + \Delta W_{(ij)n} + \beta[W_{(ij)n} - W_{(ij)n\text{-}1}] \tag{4}$$

The change in weights (DW) in the direction of negative gradient is given by $DW_{(ij)} = -\ a^{\partial E/\partial W_{(ij)}}$, where a is learning rate such that $0 < a < 1$. The learning rate governs the rate of change of weights.

The network parsimony (minimum network structure) can be achieved through A information criterion (AIC) [18], B information criterion (BIC) [19], or by pruning of network [20]. AIC and BIC utilize SE statistic for deciding the number of free parameters [21]. In pruning, [20] suggested the use of error sensitivity $Se_{(ij)}$ with respect to weight $W_{(ij)}$ for elimination of the respective weight without involvement of excessive calculations. $Se_{(ij)}$ is defined as:

$$Se_{(ij)} = -\frac{\partial E}{\partial W_{(ij)}} \int_{W^{f}_{(ij)}}^{W^{i}_{(ij)}=0} W^{f}_{(ij)} \qquad (5)$$

Finally, $Se_{(ij)}$ with respect to $W_{(ij)}$ reduces to:

$$Se_{(ij)} = \sum_{1}^{it} \frac{\Delta W_{(ij)}^{2}}{\alpha} \cdot \frac{W^{f}_{(ij)}}{W^{f}_{(ij)} - W^{i}_{(ij)}} \qquad (6)$$

where, i and f indicate the initial and final values of weights. Since the weight update is available during learning, only summation of squares of change in weight, i.e. $\Sigma \Delta W_{(ij)}^{2}$, is determined for estimation of the sensitivity of error. A comparatively low sensitivity suggests the pruning of the corresponding node.

The model can be generalized by cross-validation, but without ascribing a level of accuracy in the beginning of model formulation. The performance of the developed model is checked in all iterations. In this method, the network is trained on training data set and its statistical performance is evaluated on training and cross-validation data sets. The method monitors the generalized performance, and the process of learning is stopped when there is no more improvement in training along with the performance in cross validation period. For performance, the following statistical and hydrological criteria were used in the present analysis.

2.1. Correlation Coefficient (CC)

The correlation between the observed and estimated values is accounted by the correlation coefficient expressed as follows.

Table 1. Summary of annual hydrological data of Vamsadhara River basin.

Year	Weighted Average Rainfall (mm)	Runoff (m^3/s)	Runoff (mm)	Sediment yield (kg/s)	Runoff-Rainfall Ratio	Sediment Yield-runoff ratio/ Sediment Concentration (kg/m^3))
1984	965.2	13245.2	146.3	28917.2	0.15	2.18
1985	1079.6	17195.9	190	29802.3	0.18	1.73
1986	971.7	20519.6	226.7	36233	0.23	1.77
1987	715.8	6506.6	71.9	12230.4	0.1	1.88
1988	1053.7	25454.9	281.2	42742.8	0.27	1.68
1989	1074.2	19789.9	218.7	30699.6	0.2	1.55
1992	1028.7	40017.2	442.1	62668.2	0.43	1.57
1993	773.3	15005.5	165.8	20177.2	0.21	1.34
1994	1142.3	31896.4	352.4	33675.8	0.31	1.06
1995	1010.6	34237.4	378.3	30291.5	0.37	0.88

$$CC = \frac{\sum_{j=1}^{n}\{(Y_j-\bar{Y})(\hat{Y}_j-\bar{\hat{Y}})\}}{\{\sum_{j=1}^{n}(Y_j-\bar{Y})^2\sum_{j=1}^{n}(\hat{Y}_j-\bar{\hat{Y}})^2\}^{1/2}} \times 100 \quad (7)$$

where, $\bar{Y}$ and $\bar{\hat{Y}}$ are mean of observed and estimated values, Y and $\hat{Y}$ are the observed and estimated values respectively and n is the number of observation.

2.2. Coefficient of Efficiency (CE)

Nash and Sutcliffe proposed this non-dimensional criterion on the basis of standardization of the residual variance with initial variance. It is estimated as follows:

$$CE = \left\{1 - \frac{\text{Residual Variance}}{\text{Initial Variance}}\right\} x100 = \left\{1 - \frac{\sum_{j=1}^{n}(Y_j - \hat{Y}_j)^2}{\sum_{j=1}^{n}(Y_j - \bar{Y})^2}\right\} x100 \quad (8)$$

A perfect agreement between the observed and estimated values yields the coefficient of efficiency as 100 percent. For zero agreement, all the estimated values must be equal to the observed mean. A negative efficiency shows that the estimated values are less than the observed mean. It can be used to compare the model's relative performance on different watersheds.

2.3. Root Mean Square Error (RMSE)

An alternate criterion of residual error is the root mean square error [22] which is expressed as the measure of mean of the residual variance summed over the period, that is:

$$RMSE = \sqrt{\frac{\text{residual variance}}{n}} = \left(\frac{\sum_{j=1}^{n}(Y_j - \hat{Y}_j)^2}{n}\right)^{1/2} \quad (9)$$

3. STUDY AREA

The Vamsadhara river basin is located between $18^{0}15'$ to $19^{0}55'$ north latitudes and $83^{0}20'$ to $84^{0}20'$ east longitudes (**Figure 1**) situated between Mahanadi and Godavari river basins in South India. The river originates from Kalahandi and travels through a distance of about 254 km before it joins the Bay of Bengal. Its principal tributaries are Chauldua, Phaphalia, Ganguda, Sanna Nadi, and Mahendrathanaya. The watershed (7820 km^2) has six raingauge stations: Kutraguda, Mohana, Gudari, Mohandragarh, Gunur, and Kashinagar. Weighted rainfall for study was estimated by considering the Thiessen polygons as shown in **Figure 1**. The calculated weights for each rainguage, starting from rainfall stations one to six were 0.2640, 0.1835, 0.2696, 0.1096 and 0.0224 respectively. The yearly weighted rainfall, runoff and sediment yield are shown in **Table 1**. The mean annual rainfall and temperature varies from 170 to 1280 mm, and 10 to 43 $^{\circ}$C, respectively and humidity is above 95% during the monsoon. The soils of area can be classified into mixed red, black soils, red sandy soils, yellow soils, coastal sands, and forest soils, and its surface is mostly covered with Kankar and Murum. The hydro-meteorological data are being collected by India Meterological Department (IMD) and Central Water Commission (CWC), Godavari Mahanadi Circle Division, South Eastern Region, Bhubaneswar, Orissa. The measurements include rainfall, stage and sediment concentration in a one litre of sample. Rainfall is measured by India Meteorological Department (IMD) in the units of mm/day using a self recording raingauge. The runoff is recorded by CWC by area velocity method and observed once in a day starting at 8.00 h. The velocities are measured by current metre and using sounding rod. The runoff recorded by CWC is reported in the units of m^3/s. The sediment yield, that is the suspended is measured by taking 11 water sample and is reported in the units of kg/m^3.

In this study, the daily rainfall, runoff, and sediment yield data of monsoon period (June 1–October 31) for 1984-1987 were used for model calibration, and the data of 1988-89 and 1992-95 for cross validation and verification.

4. ANALYSIS OF THE RESULTS

In the present study, the acceptable significance limits only for CC and CE for the model were considered as > 75% and > 60% respectively. The criterion RMSE is basically used to compare the models and therefore no limits for these criteria have been fixed for evaluating the performance of a particular model. The descriptive statistic of data used in model development, cross-validation, and verification is shown in **Table 1**. It is seen that runoff-rainfall ratios for 1992, 1994, and 1995 are higher than those in other years, implying that the verification with former years data may underestimate the runoff. However, sediment concentration values in 1993, 1994, and 1995 are lower than those in the other years, and therefore, the developed models may overestimate the sediment yield in those verification years.

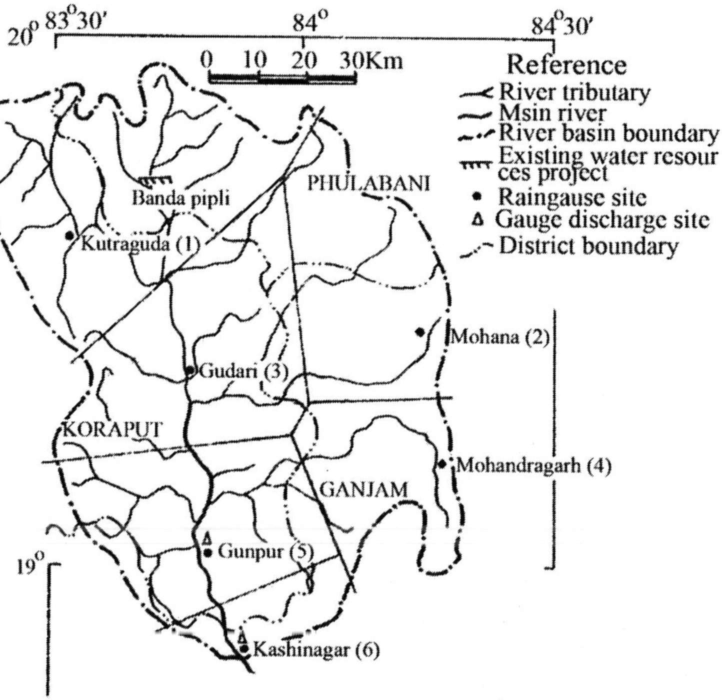

Figure 1. Index map of Vamsadhara River basin showing catchment details.

4.1. Runoff Forecasting

4.1.1. Model development

1) SI-LTF model:

Following the above-described procedure, the single-input linear transfer function (SI-LTF) rainfall (R)- runoff (Q) models were developed for daily and weekly time steps, and the results are shown in**Table 2**. In this table, the value of time memory for input rainfall is a maximum, and for runoff, it is a maximum of three. The time memories of runoff are 3 and 2 respectively for daily and weekly models, indicating that the lumping of time periods from daily to weekly reduced the time memory of the runoff. It can be inferred from the values of CC and CE that the daily runoff forecasting model performed better than the weekly model (**Table 2**). Daily observed and predicted runoff by SI-LTF model during calibration period is shown in **Figure 2**.

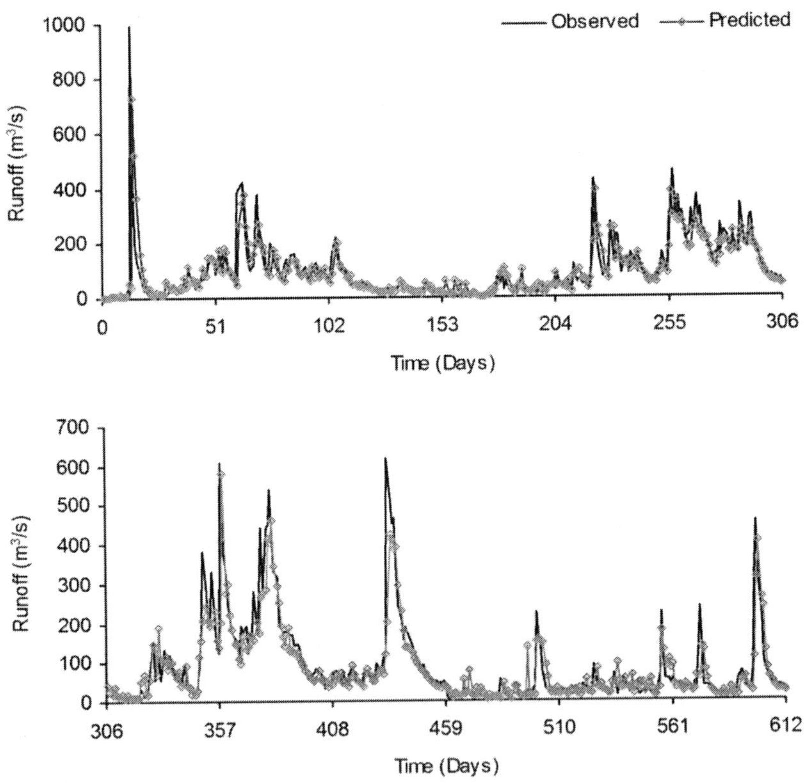

Figure 2. Daily observed and predicted runoff from SI-LTF model during calibration period (1984-87).

2) MI-LTF model:

The multi-input linear transfer function (MI-LTF) models account for the spatial variation of rainfall over the watershed. Therefore, rainfall values of all raingauge stations, viz., R1 to R6 along with runoff (Q) were utilized in models of the considered time steps, and the results are shown in **Table 2**. It is observed that the rainfall at Mohana (R2) and Gudari (R3) does not influence the output of any time scale model though these areas are located near the centroid of the watershed. It may be attributed to the presence of a dense forested area in the middle of the watershed acting as a sink. The time memory of the runoff in multi-input daily and weekly is 2. Again the results infer that the daily runoff forecasting model performed better than weekly model.

3) ANN model:

Artificial neural network (ANN) models developed for daily and weekly bases with sigmoid as activation function by pattern learning were subjected to maximum 5000 iterations. The highest value of respective variable in series was considered for normalization of input and output variables. The learning rate (a) and momentum term (b) were assumed constant (0.5) for error convergence. All interconnecting weights were updated using the error of input-output pairs. The pattern learning is governed by the error of each data set, and the interconnecting weights simultaneously adjusted. The processing is however slow as it continuously improved the weights for each data set.

The three-layer network structure was selected so that the number of nodes in the input layer equaled the number of input variables. Due to non-availability of guidelines [1], the number of hidden nodes in the hidden layer(s) was initially taken twice the input nodes [23] and it was increased by one at a time considering the improved generalization and the above pruning criterion. Corresponding to one output, only one node was taken in the output layer. Thus, a three-layer network structure with varying numbers of hidden nodes in the hidden layer was tried, and the performance, in both cross-validation and validation, of the finally selected ANN models for all above mentioned time steps for pattern learning is shown in **Table 3**. In this table, the generalized case exhibits least errors in both calibration and crossvalidation. Here, the data of 1984 to 1987 were used in calibration, and the data of 1988 to 1989 and 1992 to 1995 in cross-validation, referred as the first and second verification periods in the text, respectively. Performance of daily models both in calibration and cross-validation is better than the weekly model suggesting that the daily model is better than weekly model.

Table 2. Model order and corresponding parameters for different time periods of runoff forecasting in SI-LTF and MI-LTF models for Vamsadhara River basin.

Model and variable		Model Parameter Associated with Input time Memory			Performance of Developed Model		
		t-1	t-2	t-3	CC (%)	CE (%)	RMSE
SI-LTF (Daily)	R	3.390	--	--	--	--	--
	Q	0.524	0.127	0.181	80.0	64.0	65.0
SI-LTF (Weekly)	R	6.192	--	--	--	--	--
	Q	0.332	0.202	--	62.7	37.9	490.0
	R1	0.728	0.479	--	--	--	--
MI-LTF (Daily)	R5	0.296	--	--	--	--	--
	R6	2.166	--	--	--	--	--
	Q	0.448	0.249	--	80.0	63.3	66.0
	R4	0.894	--	--	--	--	--
MI-LTF (Weekly)	R6	5.173	--	--	--	--	--
	Q	0.333	0.203	--	62.9	38.1	490.0

4.2. Sediment Yield Forecasting

4.2.1. Model Development

1) SI-LTF model:

These models consisted of rainfall (R), runoff (Q), and sediment yield (Sy), and the results are given in **Table 5**. Apparently, the results of daily model show the effect of one day preceding sediment yield on today's computed sediment yield. Daily observed and predicted sediment yield by SI-LTF model during calibration period is shown in **Figure 5**. On the other hand, the weekly model has no preceding sediment yield term(s) in the model, implying that the preceding weekly or higher time step sediment yield have no effect on the current output. The model performance reduces with increased lumping of data to larger time periods. Based on the evaluation criteria, though no model passes the criteria, these could be used cautiously for the study area.

2) MI-LTF model:

Similar to the above, the rainfall values of six rainfall gauging sites were considered in the models developed for various time periods, and the results are given in **Table 5**. Similar to the single-input model, the daily multi-input model indicated the effect of successive past sediment yield on the output whereas weekly did not. In addition, the rainfall of Kutraguda (R1), Mohan (R2) and Gudari (R3) did not influence the output of any of the models despite their being in the center and upper part of the watershed, perhaps due to the presence of forest and better soil conservation practices in the Koraput district, which covers the polygon area of Kutraguda, Mohan and Gudari rain gauge sites (**Figure 1**). As seen and inferred from **Table 5**, the lumping of time periods improved the model performance in calibration.

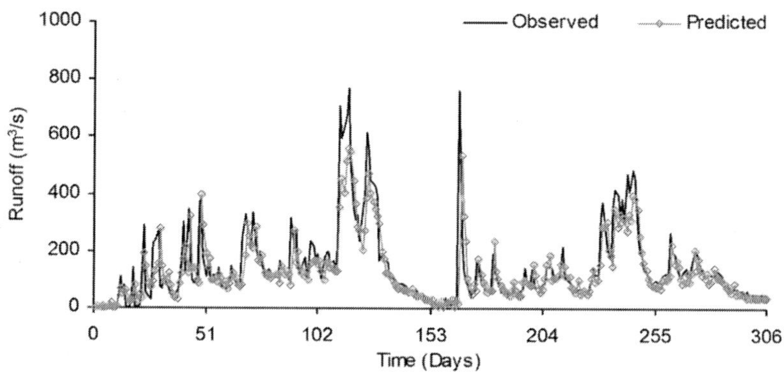

Figure 3. Daily observed and predicted runoff from SI-LTF model during verification period (1988-89).

Table 3. Pattern learned artificial neural network runoff forecasting models for different time periods for Vamsadhara River basin.

			Performance in model development, %					
			Calibration (1984-1987)			Cross validation (1988-1989)		
Model	Structure	Iteration	CC (%)	CE (%)	RMSE	CC (%)	CE (%)	RMSE
Daily	4.8.1	2490	83.8	70.1	60.0	79.5	62.5	82.0
Weekly	3.6.1	53	65.7	43.2	470.0	68.1	37.7	600.0

3) ANN model:

Similar to above, the ANN models for various time steps were developed for sediment yield. The models were calibrated on 1984 to 1987 data set, cross-validated for maximum generalization with the data of first verification period (1988-1989). The ANN structure was selected as explained earlier, and the number of hidden nodes were selected on the bases of improved generalization and pruning criteria. The values of CC and CE indicated no model to perform well.

4.2.2. Model Verification

Similar to the models performance in calibration, **Table 7** shows non-workability of all the models in verification probably due to high variation of sediment concentration (**Table 1**) and nonhomogeneous data. However, the daily model worked better than the weekly model. Daily observed and predicted sediment yield by SI-LTF model during both the verification periods is shown in **Figure 6** and 7.

5. CONCLUSIONS

The following conclusions can be drawn from the study:

1) A three-layered ANN structure with the number of nodes in hidden layer as twice of the input nodes was the best generalized model obtained in less than 5000 iterations.

2) The daily SI-LTF, MI-LTF, ANN models worked well in runoff forecasting, whereas others failed in calibration as well as in both the verification periods. The daily SI-LTF model was superior to daily MI-LTF model in calibration as well as in both the verification periods. However, the daily ANN model showed an improvement over the former in calibration, but was slightly poor in cross-validation as well as in verification.

3) Any of the proposed SI-LTF, MI-LTF, ANN models can be useful for sediment yield forecasting with little variation in accuracy.

Table 4. Performance evaluation of SI-LTF, MI-LTF and ANN runoff forecasting models during verification stages.

Model	Performance evaluation criteria					
	Verification (1988 – 1989)			Verification (1992 - 1995)		
	CC, %	CE, %	RMSE	CC, %	CE, %	RMSE
SI-LTF (Daily)	80.7	63.5	80	86.7	72.4	103
SI-LTF (Weekly)	68.9	37.8	590	70.7	40.4	956
MI-LTF (Daily)	80.6	62.8	82	85.6	60.7	108
MI-LTF (Weekly)	68.2	39.5	592	70.6	39.9	961
ANN (Daily)	--	--	--	84.5	63.9	--
ANN (Weekly)	--	--	--	71.4	26.3	--

Table 5. Model order and corresponding parameters for different time periods of sediment yield forecasting in SI-LTF and MI-LTF models for Vamsadhara River basin.

Model and variable		Model parameters associated with input time			Performance evaluation of developed model		
		t-1	t-2	t-3	CC, %	CE, %	RMSE
SI-LTF (Daily)	R	15.98	--	--	--	--	--
	Q	0.31	0.23	--	--	--	--
	Sy	0.15	--	--	64.4	41.4	284.0
SI-LTF (Weekly)	R	17.60	3.21	--	--	--	--
	Q	0.25	--	--	--	--	--
	Sy	--	--	--	35.0	8.7	1551.0
MI-LTF (Daily)	R5	3.54	--	--	--	--	--
	R6	13.56	0.58	--	--	--	--
	Q	0.61	--	--	--	--	--
	Sy	0.07	--	--	65.0	42.2	282.0
MI-LTF (Weekly)	R4	10.73	4.79	--	--	--	--
	R6	4.95	--	--	--	--	--
	Q	0.17	--	--	--	--	--
	Sy	--	--	--	40.7	13.8	1507.0

Table 6. Pattern learned artificial neural network sediment yield forecasting models for different time periods for Vamsadhara River basin.

Model	Structure	Iteration	Performance in model development					
			Calibration (1984-1987)			Cross validation (1988-1989)		
			CC	CE	RMSE	CC	CE (%)	RMSE
Daily	4,8,1	65	63.4	40.0	290	57.3	31.0	280.0
Weekly	3,6,1	793	46.8	21.7	1400	44.0	15.7	1300.0

Table 7. Performance evaluation of SI-LTF, MI-LTF and ANN sediment yield forecasting models during verification stages.

Model	Performance evaluation criteria					
	Verification period (1988 – 1989)			Verification period (1992 - 1995)		
	CC	CE	RMSE	CC	CE	RMSE
SI-LTF (Daily)	60.3	35.3	268	67.9	45.0	347
SI-LTF (Weekly)	39.1	1.4	1367	45.5	18.6	2143
MI-LTF (Daily)	60.4	33.4	276	67.6	45.5	349
MI-LTF (Weekly)	40.4	3.1	1355	48.1	17.4	2158
ANN (Daily)	--	--	--	67.2	44.8	351
ANN (Weekly)	--	--	--	25.7	5.1	2300

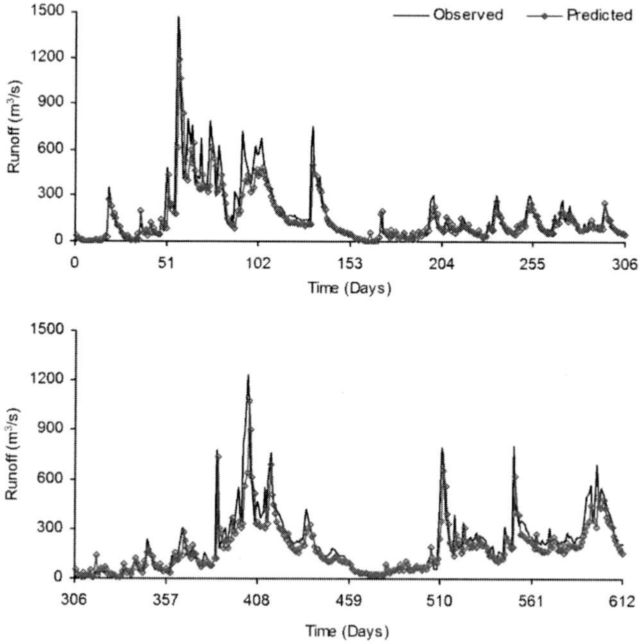

Figure 4. Daily observed and predicted runoff from SI-LTF model during verification period (1992-95).

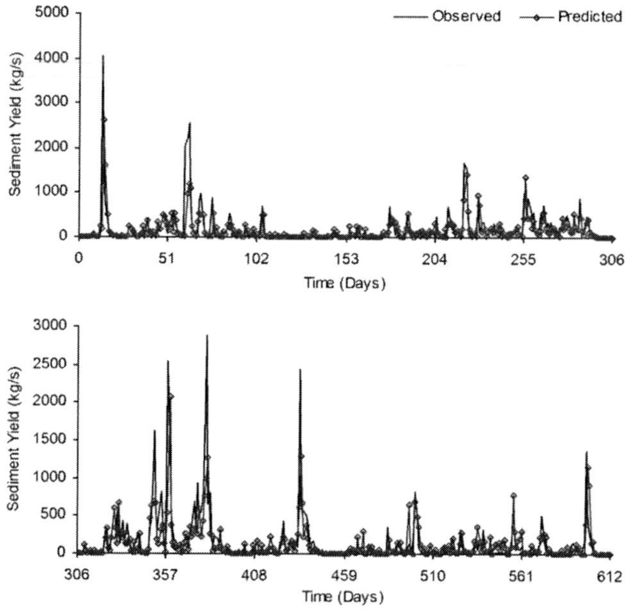

Figure 5. Daily observed and predicted sediment yield from SI-LTF model during calibration period (1984-87).

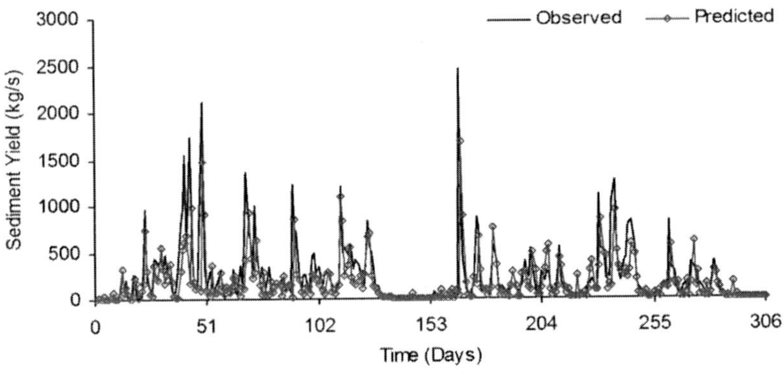

Figure 6. Daily observed and predicted sediment yield from SI-LTF model during verification period (1988-89).

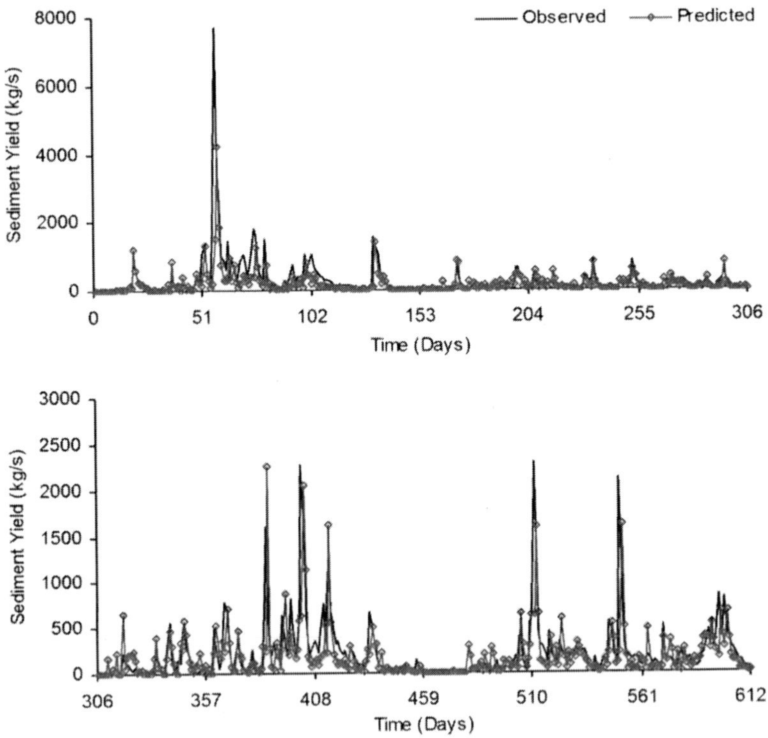

Figure 7. Daily observed and predicted sediment yield from SI-LTF model during verification period (1992-95).

REFERENCES

1. S. Tokar and M. Markus, "Precipitation-runoff modelling using artificial neural networks and conceptual models," Journal of Hydrologic Engineering, Vol. 5, No. 2, pp. 156–161, 2000.

2. D. E. Rumelhart, G. E. Hinton, and R. J. Williams, "Learning internal representations by error propagation," Parallel Distributed Processing, MIT Press, Cambridge, Vol. I, pp. 318–362, 1986.

3. D. E. Rumelhart, B. Widrow, and M. A. Letr, "The basic ideas in neural networks" Communications of the ACM, Vol. 37, No. 3, pp. 87–92, 1994.

4. Y. Shamseldin, K. M. O'Connor, and G. C. Liang, "Methods for combining the outputs of different rainfall-runoff models," Journal of Hydrology, Vol. 197, pp. 203–229, 1997.

5. "ASCE (2000a) Task Committee on application of artificial neural networks in hydrology," Artificial Neural Networks in Hydrology, I: Preliminary Concepts, Journal of Hydrologic Engineering, ASCE, Vol. 5, No. 2, pp. 124–137.

6. "ASCE (2000b) Task Committee on application of artificial neural networks in hydrology" Artificial Neural Networks in Hydrology, II: Hydrologic Application, Journal of Hydrologic Engineering, ASCE, Vol. 5, No. 2, pp. 115–123.

7. Jagadeesh, B. Zhang, and R. S. Govindaraju, "Comparison of ANNs and empirical approaches for predicting watershed runoff," Journal of Water Resources Planning and Management, Vol. 126, No. 3, pp. 156–166, 2000.

8. S. Tokar and M. Markus, "Precipitation-runoff modelling using artificial neural networks and conceptual models," Journal of Hydrologic Engineering, Vol. 5, No. 2, pp. 156–161, 2000.

9. H. M. Nagy, K. Watanabe, and M. Hirano, "Prediction of sediment load concentration in rivers using artificial neural network model," J. Hydraulic Engrg., Vol. 128, No.6, pp. 588–595, 2002.

10. M. P. Rajurkar, U. C. Kothyari, and U. C. Chaube, "Modeling of the daily rainfall-runoff relationship with artificial neural networks," Journal of Hydrology, Vol. 285, pp. 96–113, 2004.

11. K. P. Sudheer and S. K. Jain, "Radial basis function neural network for modeling rating curves." J. Hydrologic Engrg., Vol. 8, No. 3, pp. 161–164, 2003.

12. T. W. Kim and J. B. Valdes, "Nonlinear model for drought forecasting based on a conjunction of wavelet transforms and neural networks," J. Hydrologic Engrg., Vol. 8, No. 6, pp. 319–328, 2003.

13. H. Moradkhani, K. L. Hsu, H. V. Gupta, and S. Sorooshian, "Improved streamflow forecasting using selforganizing radial basis function artificial neural networks," Journal of Hydrology, Vol. 295, pp. 246–262, 2004.

14. J. Olsson, C. B. Uvo, K. Jinno, A. Kawamura, K. Nishiyama, N. Koreeda, T. Nakashima, and O. Morita, "Neural networks for rainfall forecasting by atmospheric downscaling." J. Hydrologic Engrg., Vol. 9, No. 1, pp. 1–12, 2004.

15. Keskin, M. Erol, and T. Özlem, "Artificial neural network models of daily pan evaporation" Journal of Hydrologic Engrg., Vol. 11, No. 1, pp. 65–70. 2006

16. N. S. Raghuwanshi, R. Singh, and L. S. Reddy, "Runoff and sediment yield modeling using artificial neural networks: Upper siwane river," J. Hydrologic Engrg., India, Vol. 11, No. 1, pp. 71–79, 2006.

17. J. Johnston, Econometric Methods, McGraw Hill, Tokyo, 1972.

18. H. Akaike, "A new look at the statistical model identification," IEEE Transactions on Automatic Control, Vol. AC-19, pp. 716–723, 1974.

19. Rissanenj, "Modeling by short data description," Automation, Vol. 14, pp. 465–471, 1978.

20. E. D. Karnin, "A simple procedure for pruning back propagation trained neural networks," Institute of Electrical and Electronics Engineers Transactions on Neural Networks, Vol. 1, No. 20, pp. 239–242, 1990.

21. Z. X. Xu and J. Y. Lijy, "Short term inflow forecasting using an artificial neural network model," Hydrological Process, Vol. 16, pp. 2423–2439, 2002.

22. P. S. Yu, C. L. Liu, and T. Y. Lee, "Application of a transfer function model to a storage runoff process," Stochastic and Statistical Methods in Hydrology and Environmental Engineering, Vol. 3, pp. 87–97, 1994.

23. K. W. Hipel, A. Ian, U. S. Panum, and V. P. Singh, "Stochastic and statistical methods in hydrology and environmental engineering," The series analysis in hydrology and environmental engineering, Kluwer Academic Publishers, The Netherlands, Vol. 3, 1994.

CHAPTER 8

Artificial Neural Network classification of operator workload with an assessment of time variation and noise-enhancement to increase performance

Alexander J. Casson[1,2*]

[1]Sensing, Imaging and Signal Processing Group, School of Electrical and Electronic Engineering, The University of Manchester, Manchester, UK

[2]This Work was Carried Out While the Author was with the Optical and Semiconductor Devices Group, Department of Electrical and Electronic Engineering, Imperial College London, London, UK

Workload classification—the determination of whether a human operator is in a high or low workload state to allow their working environment to be optimized—is an emerging application of passive Brain-Computer Interface (BCI) systems. Practical systems must not only accurately detect the current workload state, but also have good temporal performance: requiring little time to set up and train the classifier, and ensuring that the reported performance level is consistent and predictable over time. This paper investigates the temporal performance of an Artificial Neural Network based classification system. For networks trained on little EEG data good classification accuracies (86%) are achieved over very short time frames, but substantial decreases in accuracy are found as the time gap between the network training and the actual use is increased. Noise-enhanced processing, where artificially generated noise is deliberately added to the testing signals, is investigated as a potential technique to mitigate this degradation without requiring the network to be re-trained using more data. Small stochastic resonance effects are demonstrated whereby the classification process gets better in the presence of more noise. The effect is small and does not eliminate the need for re-training, but it is consistent, and this is the first demonstration of such effects for non-evoked/free-running EEG signals suitable for passive BCI.

1. INTRODUCTION

Augmented Cognition is a recent research concept focusing on creating the next generation of Human-Computer Interaction devices (Schmorrow and Stanney, 2008). Closed-loop Brain Computer Interfaces (BCIs) are a classic example of such systems. In these, a human uses a computer whilst simultaneously the computer monitors the human and changes its operation based upon the results. Changes might be made to the outputs presented, to the input streams which are used, or to the levels of automation and assistance that are provided, amongst others. For example, workload monitoring systems aim to detect when an operator is in a high or a low workload state to potentially change the speed at which information is presented. As such the work flow and operating environment can be optimized in a real-time and time-varying manner (Wilson and Russell, 2007). Alternatively, workload monitoring could be used to enhance human training: the mental load of a new task can be objectively measured and training times increased or decreased to terminate the training only when the task involves a low level of effort (Ayaz et al., 2012).

Emerging passive BCIs use the spontaneously produced *free-running* EEG (electroencephalogram) signals that are naturally present on the scalp due to the normal functioning of the brain without any specific stimuli present or the user consciously controlling their brain activity (Zander and Kothe, 2011). Workload classification is quickly emerging as a *killer app* for passive BCI as it is particularly suited to being based upon free-running EEG as opposed to evoked responses. At the extreme fatigue end: sleep onset is characterized in free-running EEG by a reduction in alpha (7–12 Hz) activity, with this being replaced by theta (4–7 Hz) activity (Rechtschaffen and Kales, 1968). This has long been used clinically for sleep staging and there are numerous recent papers (see for example Christensen et al., 2012; Dijksterhuis et al., 2013) demonstrating that frequency band changes can be used to identify less extreme changes in vigilance level.

As with many BCI systems the challenge for workload monitors now is in moving out-of-the-lab and into uncontrolled environments which are significantly corrupted by noise, and in creating systems that are reliable, robust and re-usable. For machine learning based systems an essential parameter in this is the training required to set up the classification process. During training, EEG data which is known to arise during a particular workload state is presented to the classifier, which it then uses to *learn* classification boundaries which can be applied to new EEG data where the workload state is not known. Ideally this classifier generation and training process:

1 Leads to the best classification performance when applied to new EEG data.

2 Requires little training data.

3 Maximizes the time before re-training is required.

Traditionally much focus has been attached to maximizing condition 1, ensuring that the classification system has good generalisability and can be correctly applied to previously unseen data. However, for practical systems this is not the only objective, and conditions 2 and 3 which determine the *temporal performance* of the system are of critical importance. Recent work (Estepp et al., 2011; Christensen et al., 2012) has suggested that the performance of workload monitors is not constant, instead degrading as the time gap from the training session is increased. (Christensen et al., 2012) showed that some of this performance loss could be recovered by increasing the amount of training data used and including data from multiple EEG sessions: each new EEG session would start with a short new training period to supplement the existing training data that was collected in previous sessions, potentially some time ago.

However, each time the classifier has to be re-trained the user must be placed into a known high or low workload state and new EEG data collected. This requires both a considerable amount of effort, and decreases the time during which the system can be practically used to perform useful classifications. It is thus essential to devise and investigate new techniques that can potentially be used to improve temporal performance without requiring more training sessions to be carried out, and this paper begins this investigation.

This paper presents an Artificial Neural Network based workload classifier for determining operator state from the EEG (Section 2) with its performance evaluated in two novel ways. Firstly, focus is given to the temporal performance of the classification process, quantifying how long a system developed using very little training data can be used before re-training is required (as opposed to maximizing the general performance, condition 1 above). Whilst very good short term performance is obtained, it is verified that the performance does degrade with time.

Secondly, *noise-enhanced processing* is investigated as a technique to mitigate the change in performance. In this, artificially generated corrupting noise is deliberately added to the otherwise raw EEG input signal to evoke *stochastic resonance* from the classification process. This is an effect in non-linear processes whereby performance actually gets better in the presence of small amounts of noise (Kay, 2000; Chen et al., 2007). The method is particularly interesting to explore as noise is widespread in ultra-portable out-of-the-lab EEG recordings, and creating algorithms that are firstly robust to the presence of noise and then even enhanced by the presence of noise can completely change how ultra-portable EEG systems are designed.

As an illustration, nearly all modern dry EEG electrodes are based upon having *fingers*, rather than discs, for easier penetration through the hair, as illustrated in Figure 1. However, electrode contact noise is a function of the electrode area (Huigen et al., 2002), and electrode fingering decreases this area. In-depth measurements of dry electrode performance have been presented (Chi et al., 2010; Gandhi et al., 2011;Slater et al., 2012) but most studies report only a correlation coefficient between EEG recorded at nearby locations with wet and dry electrodes. Typical values reported are: >0.93 (Xu et al., 2011); 0.89 (Matthews et al., 2007); 0.83 (Gargiulo et al., 2008); 0.81–0.98 (IMEC, 2012); 0.68–0.90 (Patki et al., 2012); 0.39–0.85 (Estepp et al., 2009). New results in this paper show that even with up to 15 μVrms of artificial noise added to the raw EEG traces, correlations in-line with those reported for dry electrodes are found.

Figure 1. Two state-of-the-art commercially available dry EEG electrodes show the common *fingered* configuration. **Photographs by the author.**

Further, in general electronics there is often a trade-off between power consumption and the effective noise present (Harrison and Charles,

2003; Xu et al., 2011; Cuadras et al., 2012). A number of high quality, highly miniaturized wearable EEG units have become commercially in recent years, but with battery lives typically in the 8 h range (Casson, 2013). This power performance still falls far short of *pick up and use*devices and substantial improvements in system power consumptions are still required to realize units that can be trusted to be re-usable session after session. If BCI applications can be enhanced by noise the design of ultra-portable EEG systems is completely changed: more noise is now desirable and this potentially allows the use of even smaller electrodes and lower power consumption processing electronics.

A preliminary investigation of using noise-enhancement in workload monitoring was presented in Casson (2013). However, this was restricted to the investigation of two subjects and the use of a single Artificial Neural Network based algorithm. The results here have been extended to eight subjects and include the use of an array of neural networks to enable collective decision making.

2. MATERIALS AND METHODS

2.1. Workload Task and Data

In this work scalp EEG data is used to classify an operator as being in either a high or low workload state whilst they are performing a flight simulator task (Comstock and Arnegard, 1992; Miller Jr., 2010). This task was designed to represent aircraft operations, and particularly those of remote piloting. The data was recorded as part of the 2011 Cognitive State Assessment Competition (Estepp et al., 2011; Christensen et al., 2012) and is publicly available from the competition organizers.[1] It consisted of two EOG channels (vertical and horizontal) and 19 EEG channels in standard 10–20 locations (Fp1, Fp2, F7, F3, Fz, F4, F8, T3, T5, C3, Cz, C4, T4, T6, P3, Pz, P4, O1, O2). All channels used a mastoid reference and ground and a 256 Hz sampling rate.

A total of 118 EEG recordings were performed using eight subjects (15 tests in six subjects, 13 tests in two) on five separate days spread over a month. Each recording day consisted of three 15 min EEG sessions, allowing the temporal performance of the workload monitor to be evaluated on a number of scales. Firstly, within each each 15 min session the EEG data can be divided into training and testing periods which are separated in time by seconds or minutes. Recordings on the same day are separated by minutes or hours, while recordings on different days are separated by days and weeks, and up to a month.

The simulator task was set up such that the difficulty varied dynamically in order to induce known high and low workload states in the operator. No measure of task error was present in the publicly available data and instead the user workload is inferred from the set simulator difficulty. A total of 5 min was spent in each state, with at least a 1 min transition present between task segments classed as high and low workload. Here, only the high and low workload monitoring data segments are analyzed, with the transition segments being discarded. All subjects were trained in the operation of the simulator before the workload monitoring experiment was carried out, eliminating learning effects in the operators themselves.

2.2. Classification Engine

The workload classifier is a feedforward backpropogation Artificial Neural Network (ANN) with five hidden layers (Duda et al., 2001). The network is trained on EEG frequency domain information from seven frequency bands: 0–4 Hz, 4–7 Hz, 7–12 Hz, 12–30 Hz, 30–42 Hz, 42–84 Hz, 84–128 Hz; calculated using a 1024 point FFT. This gives a total of 147 input features, and these are generated in 30 s epochs with 25 s overlap such that the assessment of operator state is updated every 5 s. All features are zero mean and unit standard deviation normalized before being passed to the network for classification. The network training used the scaled conjugate gradient backpropagation method and incorporated an early stopping training/validation/testing data split to avoid overfitting. Only 50% of the training vectors in each block of training data were used directly for network training. During network development 10 different networks with random starting weights/biases were used with different random selections of the feature vectors placed in the training sets.

After the classification has been performed, the output of the ANN is a binary state placing the subject into one of the workload categories, with this assigned as the class with the maximum output from the ANN. The classification performance is reported as the percentage of epochs in each testing set which are correctly matched with the known high or low workload in that period in the simulation.

2.3. Average and Temporal Performance Assessment Process

Multiple trained ANNs are generated and evaluated here to assess the performance of the classification process in three ways: the time independent average performance, the temporal performance, and the temporal performance when noise enhancement techniques are applied.

The time independent average performance is found by using a leave-one-out cross validation procedure where all but one of the EEG recordings are used to train the ANN, and the remaining EEG record is used to test the ANN. This process is done on a per-subject basis, and is repeated using all of the different EEG records as the test data. The output is multiple out of sample test performances which are averaged to obtain an overall figure. This is a standard ANN development procedure (Duda et al., 2001), but it uses large amounts of training data (approximately 150 min per subject) and it does not maintain the temporal ordering of the EEG recordings. In many instances the training data will come from time points after the test data (the procedure is non-causal) and the overall average figure does not reflect changes in ANN performance over time.

To assess the temporal performance a modified procedure is used, which maintains causality and uses much less training data. Firstly, a unique ANN is generated for each of the 118 EEG sessions using training data from the start of an EEG session. This network is used to monitor the operator workload in the remainder of this session and demonstrates the *same session* performance when the training and test data segments are very close together in time. The procedure for sub-dividing each EEG session into training/validation data and prospective testing data is illustrated in Figure 2. Here, the first 20 epochs (125 s duration) from each of the high and low workload periods are used for generating the network. The remainder of the data is used for testing. For the performance evaluation 11 epochs (80 s) of test data are assessed at a time: this vector is then stepped through all of the available test data epochs to show the achieved classification performance as a function of time. These values can then be averaged to produce an overall figure which does reflect changes in ANN performance over time.

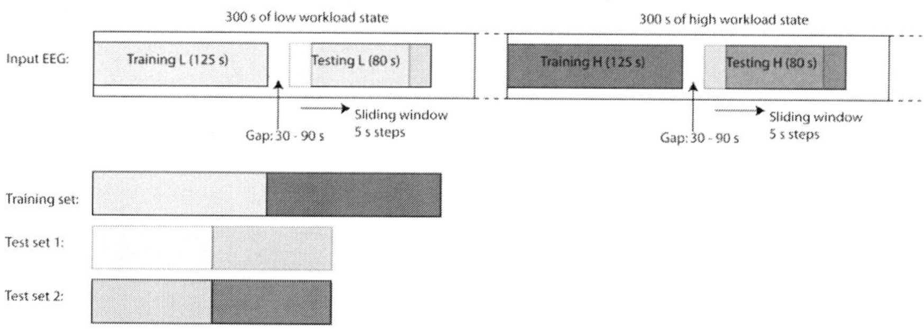

Figure 2. Training/validation and test data sets are generated from a sliding window allowing the time distance from the training data to be varied and performance evaluated as a function of this distance.

The *cross session* temporal performance is then evaluated by re-using the ANNs trained as above (from a part of one session only) and testing them using EEG data collected in different recording sessions. In this case the test EEG is not split into training/test segments, and is instead all used as test data. Multiple time scales are investigated using different configurations of the available data:

- Time-scale of minutes: the ANN trained on the first recording session on a day is tested using data from the second recording session on that day. This is repeated for all 5 days of recording.
- Time-scale of hours: the ANN trained on the first recording session on a day is tested using data from the last recording session on that day. This is repeated for all 5 days of recording.
- Time-scale of days: the ANN trained on the very first recording session is tested using data from all of the other EEG recording sessions no matter when they were performed, some up to a month later.

All of these networks are kept subject specific, and it can be seen that the testing is purely prospective: data from different sessions is not mixed, testing data can only be from the future compared to training data, and the objective is to use little training data to accurately classify operator states which are significantly removed in time from the training data.

2.4. Noise-Enhanced Performance Assessment Process

To investigate noise-enhanced processing two analyses are carried out here. Firstly, to determine suitable noise levels to inject, a raw recorded EEG trace is compared to the same EEG trace after it has had artificial white Gaussian noise deliberately added to it. The correlation coefficient is then found, allowing a comparison with the correlations found in typical wet vs. dry EEG electrode studies (see discussion in Section 1). The white Gaussian noise is generated in MATLAB using the wgn function with independent noise streams added to each of the EEG channels. For this analysis a single complete 12.5 h EEG recording is used; the previously publicly available data from (De Clercq et al., 2006; Vergult et al., 2007). This long EEG record is split into multiple shorter duration EEG sections, and the correlation in each section plotted against the duration of the section. This allows the maximum, minimum and median correlation coefficients over time to be found.

Secondly, to assess the noise-enhancement, there are numerous different ways in which noise corrupted EEG data can be passed to the ANN in order to optimally evoke stochastic resonance from the classification procedure, and only one approach is evaluated here. In this, illustrated in Figure 3, 10 identical ANNs are used in parallel each driven by EEG traces which have been corrupted by independent noise sources. The classification process is thus repeated 10 times with the output class being decided by a majority vote (in the case of a tie the output is put into the high workload state). The use of artificial noise thus allows multiple attempts of the classification for each individual EEG epoch, which would normally only be possible once.

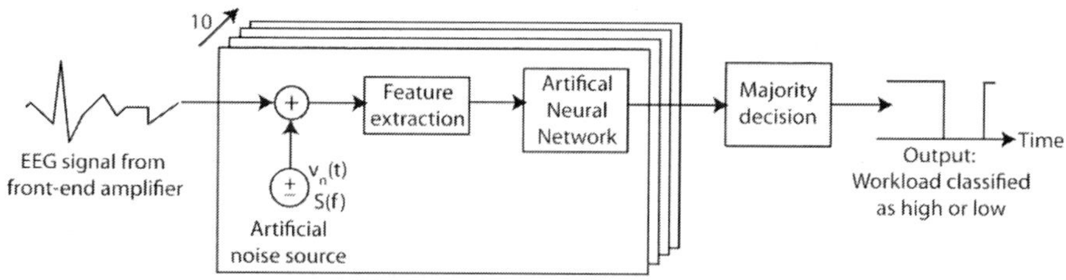

Figure 3. Input referred additive noise is used to artificially corrupt an EEG recording before it is passed to an array of workload classification Artificial Neural Networks. By varying the Root-Mean-Squared (RMS) amplitude of the noise introduced the effect on the algorithm detection performance can be investigated. Here the artificial noise source is white, has an instantaneous value $v_n(t)$, power spectral density $S(f)$ and bandwidth of 128 Hz.

This novel *testing with noise* is employed here to assess the cross session performance. *Training with noise* is a common technique used to increase the accuracy of Artificial Neural Networks by adding small levels of noise to the training data before training the network (Duda et al., 2001). (The aim is to do this multiple times and make the available training data more variable and more representative of future unknown data.) However, this is not employed here. The ANNs are created and trained as detailed above for assessing temporal performance, using the raw recorded EEG signals to generate the input features. Artificial noise is only added in during the testing process. To demonstrate that the results are repeatable, multiple-runs of this 10 ANN configuration using independently generated noise cases have been carried out.

3. RESULTS

3.1. Average and Temporal Performance

Across all subjects the average performance from the leave-one-out cross validation procedure is 73%. The per subject performances are given in Table 1. This is a satisfactory level showing that the networks can be used for determining the operator state, and this information potentially fed back in order to optimize operating procedures.

Table 1. Classification performance when using the leave-one-out cross validation to assess average performance.

Subject	Performance(%)
A	68.4
B	73.5
C	72.6
D	62.8
E	79.8
F	75.5
G	68.4
H	79.6

The *same session* temporal performance is shown in Figure 4. For illustrating the spread of results, Figure4 breaks down the performance per subject, and plots the mean performance as the time distance between the training and test data is increased. Vertical lines illustrate the distribution of results across the 15 EEG records from each subject (13 in subjects C and D). Combining all of these results together the overall average performance, where all the performances in Figure 4 are averaged to produce a single value for each person and then averaged again across the 8 subjects, is 86%.

This is higher than the 73% from the cross-validation results because the ANNs have been trained on only a small amount of data and the ANNs generalize well to new data close in time to this, at the cost of worse performance as the time gap increases. From Figure 4 it is apparent that even over the time span of seconds to minutes the performance is not constant, with noticeable variations present. In subject B the average classification performance gets better over time, but for all of the other subjects there is a decrease in the average performance as the training data becomes increasingly distant in time, with a mean correlation of −0.77.

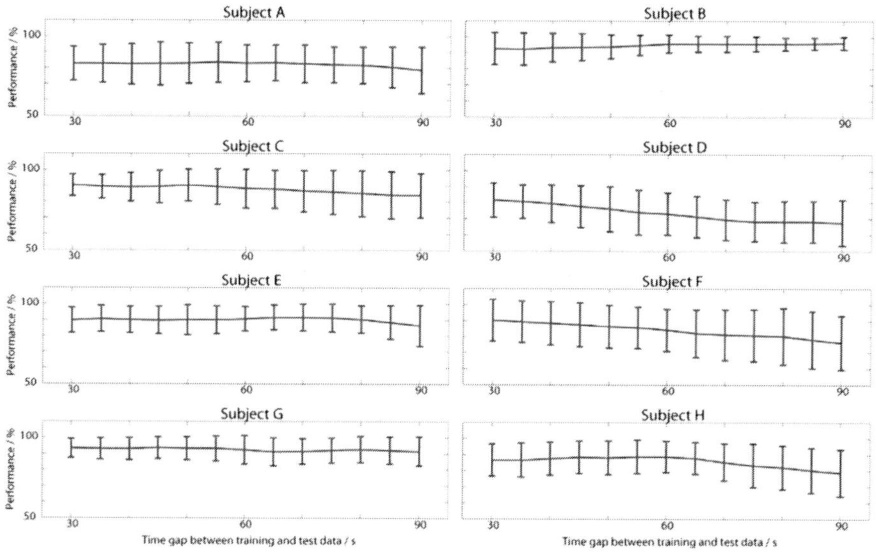

Figure 4. Classifier performance when test data is from the same EEG session as the training data. Plotted lines show the mean performance across all 15 sessions in each subject (13 for C and D) and the mean absolute deviation from this.

The impact of extending the time gap to days is shown in the *cross session* results in Figure 5 which illustrates how well the ANN generated in the very first EEG session can be directly re-used over a long time span. The overall average performance is 57%. To compare this to chance a re-sampling approach has been used where the state classification from the ANN is replaced with that from a random number generator. Random values are drawn from a uniform distribution with it being equally likely to mark an EEG epoch as high or low workload. This artificially generated classification output is then analyzed in exactly the same way as the true ANN output. Simulations over 1000 runs show that the mean performance of this random classifier is 50.0%, and the best chance performance is 52.1%. This indicates that the 57% performance of the ANN is above the chance level, although it is unlikely to be of practical use.

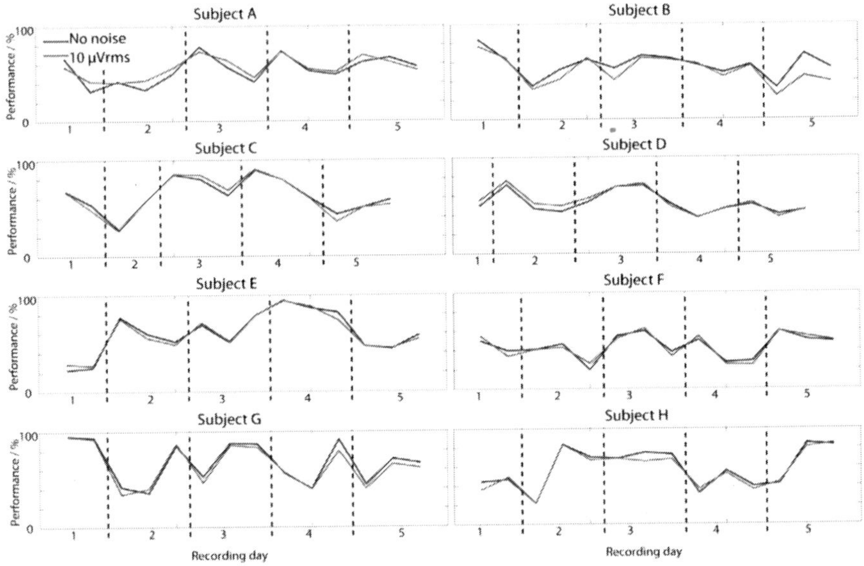

Figure 5. Workload classification performance in 14 EEG sessions (12 in subjects C and D) when the ANN is trained on 250 s of data from the first recording session and tested on all of the other EEG records, up to a month apart from the training session. Results are shown for both the no noise classifier and the ANN array with 10 μVrms of added noise.

Nevertheless, within this 57% wide per-session performance variances are seen. Many of the records achieve very good classification performances above 70%, even many days after the training session. Similarly, however, many perform at the chance level, and indeed some perform substantially worse than chance such that better classification would be actually obtained by inverting the output of the classification process. From this, it is clear that in some cases it is possible to directly re-use the workload classifier across multiple days, but this must be coupled with a method for assessing whether good classification performance is likely to be obtained.

3.2. Noise-Enhanced Performance

Figure 5 also shows the *cross session* performance of the ANN array when used with 10 μVrms of artificially generated noise. This result is noise-robust: in the 10 μVrms case the average performance is decreased only to 56%, while with 5 μVrms of added noise it remained at 57%.

To put this noise level in context, Figure 6 shows the correlation coefficients between the 12.5 h raw recorded EEG trace and noise corrupted versions of it, as the coefficient is calculated over different time spans. Coefficients in excess of 0.9 are readily achieved, even in the presence of up to 15 μVrms noise. Partly this is because the underlying correlation present is not accurately estimated when very short sections of data are analyzed. There is a consistent tendency for the median correlation to be underestimated at the cost of much larger variances.

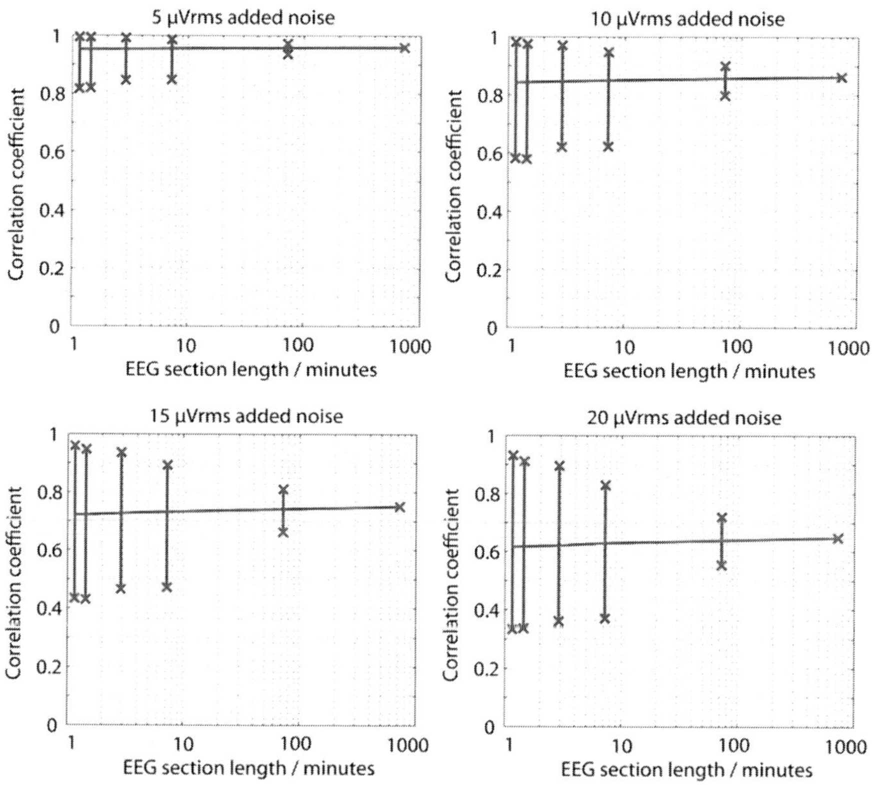

Figure 6. Correlation coefficients between a raw EEG trace and a noise corrupted copy of the same EEG trace as the EEG section length used for calculation is changed. Vertical lines show the maximum, minimum and median correlation values found over a complete 12.5 h EEG recording.

Given this, the array of ANNs was tested using noise levels of 0, 5, 10, and 20 μVrms. The resulting temporal performance is summarized in Figure 7 which demonstrates the average classification performance of the *cross session* ANNs when all of the possible training/test configurations are used to investigate each time gap. In the no noise case this finds a very similar performance degradation to that in (Christensen et al., 2012, Figure 6). The achieved performance stabilizes over the time frame of hours to 56%, substantially below the starting performance level (86%), with the temporally independent cross validation results (73%) being between the two.

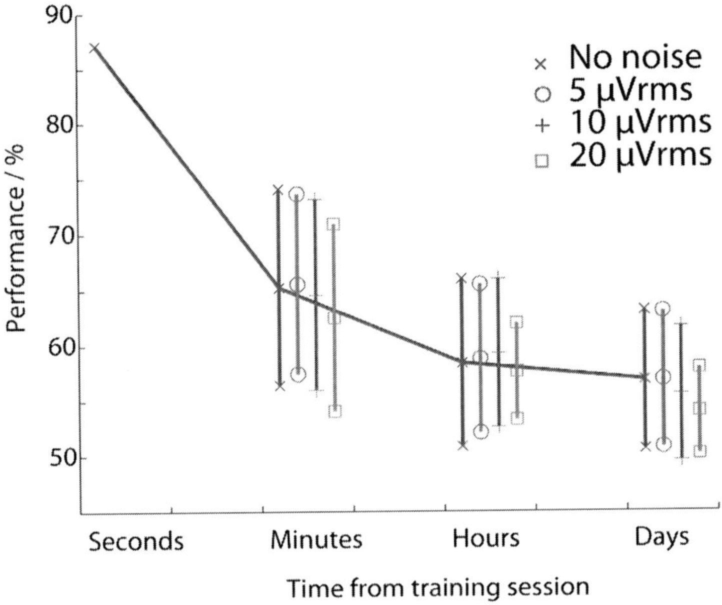

Figure 7. Workload classification performance as a function of the time between the training and testing data set. This is evaluated using the array of Artificial Neural Networks and injecting different levels of artificially generated noise into the input EEG signals.

Christensen et al. (2012) demonstrated that some of this performance loss could be recovered by re-training the network with a small amount of known workload state data from each new EEG session, but this is undesirable due to the time and effort required. To avoid this process Figure 7 also tests the hypothesis of using noise-enhanced processing as a potential mitigation approach which does not require network re-training. A single noise case run of the parallel ANN configuration is presented in Figure 7. Apparent is that the 5 and 10 µVrms cases both show small improvements in classification performance over the time span of minutes to hours. Further, all of the noise levels show a reduction in the mean deviation of the performances, showing that the performance is less variable if additional noise is introduced. However, neither of these effects is substantial: the performance improvement in Figure 7 is approximately 0.3%, and the improvement in the results spread is less than 1% in most cases.

A repeated measures ANOVA, with null hypothesis that there is no difference between the mean performances in the noise and no noise cases, rejects the null hypothesis ($p = 0.13$ for the 5 μVrms, minutes case). This indicates that the mean performance change is not significant. Nevertheless, to demonstrate that the noise-enhancement in Figure 7 (which used a single run of the parallel ANN) is repeatable, 10 independent runs of the parallel ANN configuration have been performed with different input noise signals generated for each run. The performance values in the 5 μVrms of added noise case, when the training and test EEG data sections are minutes apart, are given in Table 2. This shows that no performance decreases are present. Instead there is a clear and consistent effect of improved performance with the small amount of noise added. Modeling the probability of the noise-enhanced algorithm being better than the no noise one as a Binomial distribution $B(10, 0.5)$ (that is, there is a 50% chance of the noise added case outperforming the no noise case) the probability of all 10 noise added runs having better performance than the no noise one is less than 0.001. Assuming the noise added case is always better,$B(10, 1)$, for the 10 runs performed the 95% confidence intervals go down to $p = 0.69$ suggesting as a lower bound that the performance is enhanced in approximately 70% of noise cases.

Table 2. Classification performance when training and test data are taken from EEG sessions performed on the same day, minutes apart. Results are for 10 independent runs with 5 μVrms of added white Gaussian noise.

No noise	**65.2%**
Noise run 1	65.6
Noise run 2	65.5
Noise run 3	65.6
Noise run 4	65.4
Noise run 5	65.7
Noise run 6	65.4
Noise run 7	65.5
Noise run 8	65.2
Noise run 9	65.3
Noise run 10	65.3

4. DISCUSSION

This paper has presented a passive BCI based workload classification system investigating two of the most important factors for practical out-of-the-lab systems: the performance variation over time; and the noise robustness of the classification process. In-line with previous work, very good classification performance is achieved using little training data when the time gap between the training data and testing data is small.

4.1. Average and Temporal Performance

The cross validation ANNs, which used all of the possible training data regardless of when it was recorded compared to the test data, achieved an average performance level of 73%. This procedure is a standard approach for obtaining the best generalization performance when applied to new EEG data (maximizing criteria 1 from the introduction). However, it uses large amounts of test data from many different recordings to ensure the training data is as representative as possible. This is because when the training features are all temporally close together there is the potential for specific features, such as eye blinks or changes in muscle tone, to be present in the training sections which are not present in subsequent EEG recording sessions. This affects how Independent, Identically Distributed (IID) the feature vectors are, and places a lot of weight on a single training session to be representative in terms of equipment set up, user familiarity and user neurophysiology. Changes in performance in subsequent test sessions could then be due to a change in the mental status, or due to changes in the feature IID distributions.

It is for this reason that the *same session* tests, using ANNs trained using just 250 s of data (maximizing condition 2 from the introduction), achieved better performance (86%), but only over the short time frame where the EEG, feature IID distribution, and user mental state in the test period are very similar to that in the training period. As the time gap between the training data and the testing data increased there was a consistent decrease in the classification performance (Figure 4) as the used training EEG becomes less representative of the current testing EEG. This decrease in performance is apparent even over the relatively small time scale of minutes. Over larger time scales, substantial decreases in classification performance are observed with, as would be expected, the performance becoming worse than the 73% achieved when using more temporally spread training data. The average performance leveled off at approximately 57% for prolonged multi-day testing (Figure 7) with this degradation occurring over the time frame of hours. This performance level is above chance, but is unlikely to be meaningful for practical use. Moreover, a substantial variance in day-to-day performance was observed (Figure 5). On some days the existing neural network could be directly re-applied without further training of the network and very acceptable performances obtained. However, on many days the network achieved poor classification and would not be re-usable.

Only one point in the trade-off between the amount/spread of training data used and the amount/speed of the performance drop off has been investigated in this paper. Classically this trade-off is altered by using more training data: either in advance as in the cross validation approach; or by periodically re-training the network (periodically putting the user in a known workload state to generate new training data).Christensen et al. (2012) estimated that adding in 2.5 min of known workload state data per class was sufficient for re-training. However, generating this data requires time and effort, which places an emphasis on maximizing the time before re-training is required (maximizing condition 3 from the introduction).

4.2. Noise-Enhanced Processing

Noise-enhanced signal processing is proposed here as an algorithmic method for recovering some of the ANN performance when there is a large time gap between the training and test data sections, and so increasing the time before re-training is required. It has the benefits of not requiring any additional training data to be collected and being transparent to the end user.

Stochastic resonance is a well known but counter-intuitive effect where the performance of a non-linear system actually increases as more noise is present in the system (McDonnell and Abbott, 2009; McDonnell and Ward, 2011). Noise-resonance is commonly seen in biological systems, including neurons and the brain itself (Wiesenfeld and Moss, 1995; Gluckman, 1996; McDonnell and Abbott, 2009) and can be used in signal detection (Kay, 2000; Chen et al., 2007). For example the technique has been used for improving the performance of algorithms for detecting micro-calcifications (a key early sign of cancer) in breast mammograms (Peng et al., 2009), in radar target classification (Jouny, 2010), and is widely used in chromatography (Zhang et al., 2014). It has recently been applied to detecting transient signals in the EEG, both evoked and natural (Casson and Rodriguez-Villegas, 2011; Sampanna and Mitaim, 2013). Given the natural associations between noise and the EEG it is a very relevant technique to attempt to utilize, although it has not previously been applied to free-running EEG, suitable for passive BCI applications.

By using an array of ANNs to perform *testing with noise*, as a complement to the more common *training with noise*, Figure 5 shows that the workload classification process over the time span of days is robust to 5 and 10 μVrms of noise being added into the EEG traces. Figure 7 and Table 2 then demonstrate small stochastic resonance effects whereby the classification performance over the time span of minutes was consistently enhanced by the presence of small amounts of noise. To this end, the application of noise-enhancement has been successful, and this paper is the first demonstration of these effects in free-running EEG. However, while consistent, the achieved improvements are less than 1% and far too small to meaningfully eliminate any re-training required to get good multi-day use of the workload classification training with 250 s of data. It is clear that for free-running EEG (unlike evoked and transient EEG, Casson and Rodriguez-Villegas, 2011; Sampanna and Mitaim, 2013) the potential stochastic resonance effects are either very small or are not yet being exploited optimally.

Only one arrangement for introducing noise has been investigated here, but the process introduces many new degrees of freedom in terms of how much noise is added, its spectral composition and the parallel processing and multiple-runs options that it enables. Noise-enhancement and stochastic resonance effects are only just starting to be exploited in BCI applications, and if further established these new degrees of freedom are potentially highly interesting for re-visiting in other BCI problems where they can be exploited to improve performance.

5. CONCLUSIONS

This paper has investigated the temporal performance of Artificial Neural Networks for operator workload classification. Networks trained using 150 min of EEG data in a leave-one-out cross validation procedure obtained an average classification performance of 73% (Table 1). In contrast networks trained on only 250 s of EEG data, and so being much quicker to set up, achieved 86% average performance (Figure 4) over a short time frame when the test EEG/user state was very similar to the training data. However, these networks do not generalize well (due to changes in the IID distribution of the features, changes in user state, and similar) which leads to a drop off in classification performance as the time gap between the training and testing data increases (Figure 7). This shows that short training periods can be used with the ANN classifiers, but the classifiers will only work for a short amount of time.

To overcome this, noise-enhanced processing was explored as an algorithmic technique for recovering some of the lost performance, increasing the amount of time that the classifier can work for. Noise-enhanced processing is explored because it is potentially transparent to the end user and has interesting correlates with the design of EEG units/electrodes. The results (Table 2) show that noise enhanced processing has an effect: consistently better performances are obtained; although the current improvements are very small. Nevertheless, the consistent improvement is an interesting result and the first evidence that stochastic resonance effects could be exploited in free-running EEG and passive BCI applications.

CONFLICT OF INTEREST STATEMENT

The author declares that the research was conducted in the absence of any commercial or financial relationships that could be construed as a potential conflict of interest.

FOOTNOTES

1. Competition contact for data: Justin Estepp, Air Force Research Laboratory,cognitive.state@gmail.com

REFERENCES

1. Ayaz, H., Shewokis, P. A., Bunce, S., Izzetoglu, K., Willems, B., and Onaral, B. (2012). Optical brain monitoring for operator training and mental workload assessment. *Neuroergonomics* 59, 36–47.

2. Casson, A. J. (2013). "Towards noise-enhanced augmented cognition," in *Foundations of Augmented Cognition, Vol. 8027, of Lecture Notes in Computer Science*, eds D. D. Schmorrow and C. M. Fidopiastis (Berlin; Heidelberg: Springer), 259–268.

3. Casson, A. J., and Rodriguez-Villegas, E. (2011). Utilising noise to improve an interictal spike detector. *J. Neurosci. Methods* 201, 262–268.

4. Chen, H., Varshney, P. K., Kay, S. M., and Michels, J. H. (2007). Theory of the stochastic resonance effect in signal detection: part I-fixed detectors. *IEEE Trans. Signal Process.* 55, 3172–3184.

5. Chi, Y., Jung, T.-P., and Cauwenberghs, G. (2010). Dry-contact and noncontact biopotential electrodes: methodological review. *IEEE Rev. Biomed. Eng.* 3, 106–119

6. Christensen, J. C., Estepp, J. R., Wilson, G. F., and Russell, C. A. (2012). The effects of day-to-day variability of physiological data on operator functional state classification. *Neuroimage* 59, 57–63.

7. Comstock, J. R., and Arnegard, R. J. (1992). *The Multi-Attribute Task Battery for Human Operator Workload and Strategic Behavior Research.* Technical report, TM-104174, National Aeronautics and Space Administration Langley Research Center, Hampton, VA.

8. Cuadras, A., Casas, O., and Pallas-Areny, R. (2012). "Power-noise trade-off in signal amplifiers," in *IEEE I2MTC* (Montevideo).

9. De Clercq, W., Vergult, A., Vanrumste, B., Van Paesschen, W., and Van Huffel, S. (2006). Canonical correlation analysis applied to remove muscle artifacts from the electroencephalogram. *IEEE Trans. Biomed. Eng.* 53, 2583–2587.

10. Dijksterhuis, C., de Waard, D., Brookhuis, K., Mulder, B., and de Jong, R. (2013). Classifying visuomotor workload in a driving simulator using subject specific spatial brain patterns. *Front. Neurosci.* 7:149.

11. Duda, R. O., Hart, P. E., and Stork, D. G. (2001). *Pattern Classification.* New York, NY: Wiley.

12. Estepp, J. R., Christensen, J. C., Monnin, J. W., Davis, I. M., and Wilson, G. F. (2009). "Validation of a dry electrode system for EEG," in*Proceedings HFES* (San Antonio, TX).

13. Estepp, J. R., Klosterman, S. L., and Christensen, J. C. (2011). "An assessment of non-stationarity in physiological cognitive state assessment using artificial neural networks," in *IEEE EMBC* (Boston, MA).

14. Gandhi, N., Khe, C., Chung, D., Chi, Y. M., and Cauwenberghs, G. (2011). "Properties of dry and non-contact electrodes for wearable physiological sensors," in *International Conference BSN* (Dallas, TX).

15. Gargiulo, G., Bifulco, P., Calvo, R. A., Cesarelli, M., Jin, C., and van Schaik, A. (2008). "A mobile EEG system with dry electrodes," in *IEEE BioCAS* (Baltimore, MD).

16. Gluckman, B. J. (1996). Stochastic resonance in a neuronal network from mammalian brain. *Phys. Rev. Lett.* 77, 4098–4101.

17. Harrison, R. R., and Charles, C. (2003). A low-power low-noise CMOS amplifier for neural recording applications. *IEEE J. Solid State Circuits.* 38, 958–965.

18. Huigen, E., Peper, A., and Grimbergen, C. A. (2002). Investigation into the origin of the noise of surface electrodes. *Med. Biol. Eng. Comput.* 40, 332–338.

19. IMEC (2012). *Holst Centre and Panasonic Present Wireless Low-Power Active-Electrode EEG Headset.* Available online at:http://www.imec.be/

20. Jouny, I. (2010). Stochastic resonance and suboptimal radar target classification. *Proc. SPIE* 7696, 729612.

21. Kay, S. (2000). Can detectability be improved by adding noise? *IEEE Signal Process. Lett.* 7, 8–10.

22. Matthews, R., McDonald, N. J., Hervieux, P., Turner, P. J., and Steindorf, M. A. (2007). "A wearable physiological sensor suite for unobtrusive monitoring of physiological and cognitive state," in *IEEE EMBC* (Lyon).

23. McDonnell, M. D., and Abbott, D. (2009). What is stochastic resonance? definitions, misconceptions, debates, and its relevance to biology. *PLoS Comput. Biol.* 5:e1000348.

24. McDonnell, M. D., and Ward, L. M. (2011). The benefits of noise in neural systems: bridging theory and experiment. *Nat. Rev. Neurosci.* 12, 415–426.

25. Miller, W. D. Jr. (2010). *The U.S. Air Force-Developed Adaptation of the Multi-Attribute Task Battery for the Assessment of Human Operator Workload and Strategic Behavior*, Technical report, AFRL-RH-WP-TR-2010-0133, U.S. Air Force Research Laboratory, Wright-Patterson airforce base.

26. Patki, S., Grundlehner, B., Verwegen, A., Mitra, S., Xu, J., Matsumoto, A., et al. (2012). "Wireless EEG system with real time impedance monitoring and active electrodes," in *IEEE BioCAS* (Hsinchu).

27. Peng, R., Chen, H., and Varshney, P. (2009). Noise-enhanced detection of micro-calcifications in digital mammograms. *IEEE J. Sel. Areas Signal Process.* 3, 62–73.

28. Rechtschaffen, A., and Kales, A. (eds.). (1968). *A Manual of Standardized Terminology, Techniques and Scoring System for Sleep*

Stages of Human Subjects. Washington, DC: Public Health Service, U.S. Government Printing Office.

29. Sampanna, R., and Mitaim, S. (2013). "Noise enhanced array signal detection in P300 speller paradigm using ICA-based subspace projections," in *IEEE EMBC* (Osaka).

30. Schmorrow, D. D., and Stanney, K. M. (eds.). (2008). *Augmented Cognition: A Practitioner's Guide*. Santa Monica, CA: Human Factors and Ergonomics Society.

31. Slater, J. D., Kalamangalam, G. P., and Hope, O. (2012). Quality assessment of electroencephalography obtained from a "dry electrode" system. *J. Neurosci. Methods* 208, 134–137.

32. Vergult, A., De Clercq, Q., Palmini, A., Vanrumste, B., Dupont, P., Van Huffel, S., et al. (2007). Improving the interpretation of ictal scalp EEG: BSS-CCA algorithm for muscle artifact removal. *Epilepsia* 45, 950–958.

33. Wiesenfeld, K., and Moss, F. (1995). Stochastic resonance and the benefits of noise: from ice ages to crayfish and SQUIDs. *Nature* 373, 33–36.

34. Wilson, G. F., and Russell, C. A. (2007). Performance enhancement in a UAV task using psychophysiological determined adaptive aiding.*Hum. Factors* 49, 1005–1019.

35. Xu, J., Yazicioglu, R. F., Grundlehner, B., Harpe, P., Makinwa, K. A. A., and Van Hoof, C. (2011). A 160 μW 8-channel active electrode system for EEG monitoring. *IEEE Trans. Biomed. Circuits Syst*. 5, 555–567.

36. Zander, T. O., and Kothe, C. (2011). Towards passive braincomputer interfaces: applying braincomputer interface technology to humanmachine systems in general. *J. Neural Eng*. 8:025005.

37. Zhang, W., Guo, J., Xiang, B., Fan, H., and Xu, F. (2014). Improving the detection sensitivity of chromatography by stochastic resonance. *Analyst* 139, 2099–2107.

CHAPTER 9

Assessment of Surface Soil Moisture Using High-Resolution Multi-Spectral Imagery and Artificial Neural Networks

Leila Hassan-Esfahani *, Alfonso Torres-Rua, Austin Jensen and Mac McKee

Utah Water Research Laboratory, Utah State University, 8200 Old Main Hill, Logan, UT 84341, USA

ABSTRACT

Many crop production management decisions can be informed using data from high-resolution aerial images that provide information about crop health as influenced by soil fertility and moisture. Surface soil moisture is a key component of soil water balance, which addresses water and energy exchanges at the surface/atmosphere interface; however, high-resolution remotely sensed data is rarely used to acquire soil moisture values. In this study, an artificial neural network (ANN) model was developed to quantify the effectiveness of using spectral images to estimate surface soil moisture. The model produces acceptable estimations of surface soil moisture (root mean square error (RMSE) = 2.0, mean absolute error (MAE) = 1.8, coefficient of correlation (r) = 0.88, coefficient of performance (e) = 0.75 and coefficient of determination (R^2) = 0.77) by combining field measurements with inexpensive and readily available remotely sensed inputs. The spatial data (visual spectrum, near infrared, infrared/thermal) are produced by the AggieAir™ platform, which includes an unmanned aerial vehicle (UAV) that enables users to gather aerial imagery at a low price and high spatial and temporal resolutions. This study reports the development of an ANN model that translates AggieAir™ imagery into estimates of surface soil moisture for a large field irrigated by a center pivot sprinkler system.

KEYWORDS

remote sensing; high-resolution; multi-spectral; soil moisture; Artificial Neural Network; UAV

1. INTRODUCTION

Soil moisture content (SMC) is an important factor in managing irrigated farms. SMC includes two main components: surface soil moisture (SSM) (held in the upper 10 cm of soil) and root zone soil moisture (held in the upper 200 cm of soil). Surface soil moisture is a key component for addressing energy and water exchanges at the land surface/atmosphere interface and can be estimated using different techniques, such as *in situ* measurements, physically based models, remote sensing, *etc.* Grayson and Western addressed the estimation of soil moisture by applying: (1) field (or *in situ*) measurements; (2) remote sensing techniques; and (3) soil water balance simulation models [1,2]. Soil moisture constitutes a very small volume in terms of the total global water balance, but it plays a significant role in water resources planning and management [2]. Many current crop production management decisions that are made by growers, production managers, and crop advisors in precision agriculture are already based on observation from remotely sensed data such as satellite imagery. The objective of this research is to generate surface soil moisture (SSM) estimates using high-resolution, remotely sensed data, collected at 15 cm pixel resolution, as inputs to a learning machine algorithm (Artificial neural networks (ANNs))developed under supervised learning procedures. ANNs are used to build the SSM estimation model. To our knowledge, this is the first study to document estimation of surface soil moisture using remotely sensed data at such a fine spatial resolution and readily available in the sense of temporal resolution. The results will contribute not only to efficient and reliable high-resolution multi-spectral remote sensing validation, but also to better utilization of remotely sensed soil moisture products for enhanced irrigation modeling and scheduling.

Various techniques for retrieving soil moisture content have been the subject of research for almost four decades. Gravimetric measurements of soil moisture are very reliable but are laborious. Measuring SMC with imbedded sensors, such as time and frequency domain reflectometers (TDRs and FDRs), does not require a huge investment of time or facilities; however, most of these methods suffer from some of these same disadvantages. *In situ* measurements can be exhaustive and expensive if large areas are involved, as these measurements are mainly "local," with a particular footprint representing moisture conditions in only a fraction of a cubic meter of soil [3]. Because of the spatial heterogeneity of soil moisture due to different soil conditions, vegetation, topography, or impact of human activities, local measurements when are carried out on a larger scale such as fields or watersheds, might result in inaccuracies [4]. Remote-sensing techniques might provide a useful tool to address these data acquisition difficulties.

Some of the early works in estimating SMC using remote sensing [5,6,7,8,9,10] established that thermal remote sensing, in concert with *in situ* measurements, can be used to measure, or at least quantitatively infer, soil moisture content. The possibility of estimating SSM (0–7.6 cm) from visible and near-infrared (NIR) reflectance data has also been demonstrated [11]. Optical and thermal remote-sensing techniques or passive and active microwave sensors offer large-scale monitoring of SSM [11,12,13]. Some meteorological satellites, such as the Advanced Microwave Scanning Radiometer (AMSR-E), the European Remote Sensing (ERS) satellite scatterometer or the Meteorological Satellite (METEOSAT), offer the possibility of monitoring operational SSM [3]. However, the coarse spatial resolutions (ERS-Scat: 50 km, AMSR-E: 56 km and METEOSAT: visible and infrared (IR) 5 km) of the instruments are often not consistent with the scale of hydrologic processes of interest [14,15]. A number of studies on soil moisture estimations introduced the error sources that have degraded the accuracy of satellite remotely sensed soil moisture content such that it is critical to calibrate soil moisture estimation algorithms and to validate derived products using ground-truth data. The error sources comprise radio-frequency interference (RFI) [16], vegetation water content [13,17], surface roughness [16], and land surface heterogeneity [18]. It has been stated in the literature that a space-borne sensor designed to interpret SMC on the basis of soil microwave emission, and therefore the relationship between soil dielectric constant and water content, will show considerable systematic uncertainty of around 4% with maximum figures at relatively low water content in SMC retrieval [19].

Remote-sensing measurements in the thermal IR band has given rise to the thermal inertia (TI) approach for SMC retrieval. The TI approach relates SMC to the magnitudes of the differences between daily maximum and minimum soil and crop canopy temperatures [6]. This approach retrieves SMC from models that describe TI as a function of water content [20,21].

The implementation of the TI approach is simple because knowledge of soil physical properties and climate can produce representative SMC profiles up to a depth of 1 m. The limitation of the approach, however, is its sensitivity to the uncertainty of soil physical properties, which are complex to determine spatially and are typically obtained with point measurements [22]. The TI method provides large-scale spatial coverage, but the functions are empirical and have the drawback of being site- and time-specific, such that none of them are general enough to be applied extensively [21]. Monitoring soil moisture by remote sensing includes another set of approaches that permits SSM retrieval from the information contained in satellite-derived surface temperature (Ts) and vegetation index (VI). However, one of the major drawbacks of the Ts-VI method is that, in order to have enough points in a remote-sensing image to use in the determination of the boundaries of extreme conditions, a sufficiently large number of pixels must be sampled. This limitation is a handicap when dealing with smaller scale imagery on the order of the size of a typical farm field [11].

The difficulties associated with the above introduced approaches have led researchers to look for data-driven modeling tools, such as artificial neural networks (ANNs), support vector machines (SVMs), and relevance vector machines (RVMs), to estimate soil moisture [2,3,23,24,25,26,27]. For example, Landsat data has been used for soil moisture estimation using relevance vector and support vector machines [3]. One of the major advantages of the machine learning approach to SMC estimation is that it can provide estimates having resolutions commensurate with remotely sensed data [3].

2. MATERIALS AND METHODS

2.1. Artificial Neural Networks (ANNs)

This section presents a brief description of ANNs relevant to this study. A three-layered feed-forward neural network (FFNN) model was developed that includes "I" input neurons, "h" hidden neurons, and "o" output neurons, which can be shown symbolically as ANNs (i,h,o) [28]. Connection weights and bias connect these neurons. Input is multiplied by the connection weights. These products are simply summed, fed through a transfer function to generate a result, and then output. The hidden layer neurons usually use a sigmoidal activation function, while the output layer neurons utilize a linear activation function. The activation functions are used to transform inputs to targeted outputs with a nonlinear regression procedure. Each ANN model requires training and testing operations. In the training operation, by minimizing the cost function (Mean Squared Error (MSE) in this study), the connection weights and bias values are optimized. Once trained, an independent set of data that was not used for training is applied to test the neural network model [26]. The issue that threatens the application of ANN-based models is the randomness of predicted output, which is fixed in this study [29]. This was carried out by applying seed generation function. Since weights are initialized randomly, seed generation function was reset to overcome the randomness of the results by fixing the weights initialization and make the results reproducible. Also the models were run for a wide range of seed values. The training operation of ANNs was performed by a back-propagation algorithm, which is the most commonly used supervised training algorithm in the multilayer feed-forward networks. The network weights are simultaneously modified by the back-propagation algorithm which seeks to minimize the difference between the targets and the computed outputs. In this kind of algorithm the processing operation is performed in a forward direction, from inputs to hidden layers and eventually to an output layer [30]. A back-propagation method uses a least-mean-square-error method and generalized-delta rule to optimize the network weights. The derivative chain rule and the gradient-descent method are utilized to adjust the network weights [31]. Forward pass and reverse pass are two main phases of the training operation. In the first phase, the input data are multiplied by the initial weights, forming weighted inputs that then are added to yield the net to each neuron. This net generates the output of the neuron after passing through an activation or transfer function.

In back-propagation networks, a derivative of the activation function modifies the network weights. Therefore, continuous-transfer functions are targeted. The Log-sigmoid transfer function and Hyperbolic tangent sigmoid transfer function are the most common continuous-transfer functions in back-propagation networks [32]. The Log-sigmoid transfer function was used in this study. The output of the neuron is transmitted to the next layer as an input, and this procedure is repeated until it reaches the output layer. The error between the network outputs and the target outputs is computed at the end of each forward pass and it is checked with a specific value. If the error passed this value, the procedure continues with a reverse pass; otherwise, training is stopped [33]. In the reverse pass, the weights in the network are modified using the error value. The modification of weights in the output layer is different from the modification of weights in the hidden layers. In the output layer, the target outputs are provided, whereas in the intermediate layers, target values do not exist [31]. Therefore, back propagation uses the derivatives of the objective function regarding the weights in the entire network to distribute the error to neurons in each layer in the entire network [33].

2.2. Selection of Possible Input Variables

One of the critical issues in training learning machine algorithms such as ANNs is to select the appropriate input variables. The idea is to choose the combination of variables that are highly correlated with soil moisture. Previous studies have shown good correlation between soil water content and infrared (IR) skin temperature and normalized difference vegetation index (NDVI), and between IR heating rate and thermal images [34,35]. Optical and microwave remotely sensed data have been used for surface soil energy balance modeling [6,11,12,36]. After collecting these variables from independent datasets, the correlation and dependency among these variables were evaluated in the study reported here. Some Vegetation Indices (VIs) are considered as input variables with some contributions in soil moisture estimations [4,37,38,39].

2.3. Study Area, Instrumentation, Techniques and Data

2.3.1. Study Area

The study area is a farm in Scipio, Utah (39°14′N, 112°6′W), equipped with a center pivot irrigation system covering an area of approximately 84 acres. The main crops are alfalfa and oats, grown from April to October. Figure 1 shows the location of the farm in Utah, and provides information about the heterogeneity within the farm due to different crop types and the presence of an access road. Generally the center pivot lateral rotates clockwise and supplies irrigation water to the field at a constant rate from an upstream reservoir. In the current study a full rotation of the center pivot takes three days and six hours to irrigate the field fully to field capacity. This study was carried out for the crop growing cycle starting 16 May 2013 and ending 17 June 2013 (4 days).

2.3.2. Instrumentation: AggieAir Minion (Remote Sensing Platform)

AggieAir is the remote sensing platform applied in the current study. This platform is comprised of an autonomous unmanned aerial vehicle (UAV) that carries a multispectral sensor payload. The UAV navigates over the area of interest based on a pre-programmed flight plan and captures images using the on-board sensor payload system. The UAV is a small aircraft (8 feet wing span, 14-pound take-off weight) that can fly for an hour at a speed of

30 miles per hour. In this study, the UAV was equipped with visual, near-infrared, and thermal cameras and flew over the study area on four dates in 2013 (16 May, 1 June, 9 June, and 17 June), acquiring imagery with the optical cameras at 0.15 m resolution and with the thermal camera at about 60 cm. The wavelength range peaks around 420, 500, 600 and 800 nm, respectively, for blue, green, red and NIR sensors. Detailed information about the operation of the AggieAir system has been previously published by Jensen [40].

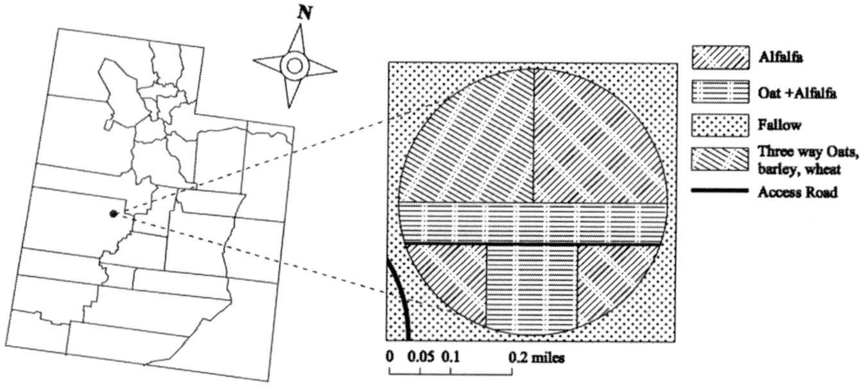

Figure 1. The location of the study area in Utah (schematic Utah counties map (on the left) and cropping pattern for 2013 irrigation season (on the right)), (39°14′N, 112°6′W).

After the AggieAir UAV completes a flight mission, the aircraft may have acquired 300–400 images from each camera: visual, near-infrared, and thermal (Figure 2a). The images can be georeferenced directly using the position and orientation of the UAV when the image was exposed (Figure 2b) [40]. EnsoMOSAIC is used to orthorectify the AggieAir imagery with high accuracy [40,41]. EnsoMOSAIC generates hundreds of tie-points between overlapping images and uses photogrammetry and block adjustment to refine the position and orientation information for each image, thereby accurately georeferencing each image (Horizontal Accuracy: 1–2 pixels Vertical Accuracy: 1.5–2 pixels (when all error sources are controlled)). EnsoMOSAIC also generates an internal digital elevation model (DEM) to compensate for distortions in the imagery caused by changing elevations. The resulting product is an orthorectified mosaic (Figure 2c) that is in 8-bit digital format. AggieAir uses a modified "reflectance mode" method to convert the digital numbers of the mosaic to reflectance values [40]. This radiometric normalization is the ratio of the digital number from the mosaic to the digital number from a spectralon white reflectance panel with known reflectance coefficients, multiplied by the reflectance factor which accounts for the zenith angle of the sun at the time, date, and location of the photos. The product of this method is an orthorectified mosaic in reflectance values. The reflectance values (for all four flights) range from 0.11 to 0.36, 0.20 to 0.49, 0.15 to 0.51 and 0.51 to 0.61 for blue, green, red and NIR, respectively. Thermal values range from 10.2 to 43.3 degrees Celsius.

a.

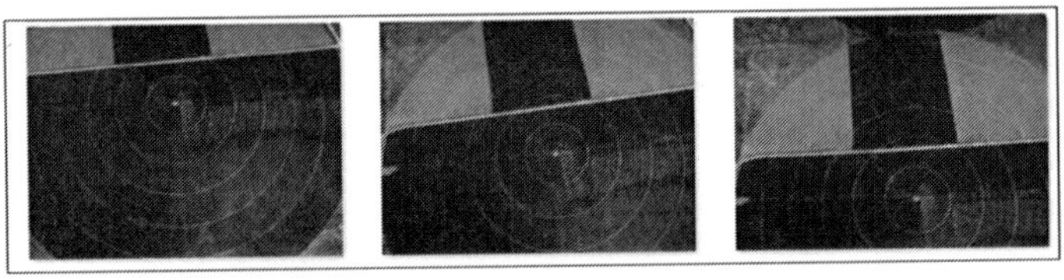

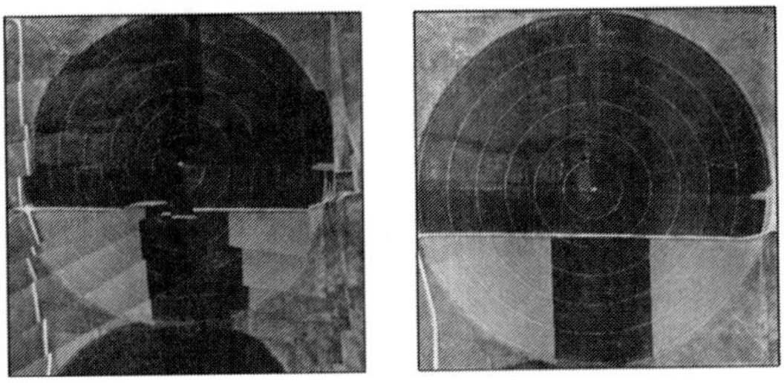

b. c.

Figure 2. (**a**) Some raw natural color images from the unmanned aerial vehicle (UAV) taken from the study area (39°14′N, 112°6′W); (**b**) Orthorectified image using position and orientation of UAV aircraft during image capture; and (**c**) Accurate orthorectified mosaic image from EnsoMOSAIC.

2.3.3. Ground-Based Data Collection

In order to perform ground truthing, at the same time the AggieAir UAV flew over the study area, intensive ground sampling was conducted at precisely determined locations in the field [42]. Soil samples were collected based on a pre-defined spatial distribution map that was developed in light of the crop types and soil characteristics in the field. The data collection included almost 50 samples per AggieAir flight scattered all over the field (minimum of 12 in each quarter) to cover the soil condition properties. Further, the unusable samples were discarded and the data collected from the four days were pooled (making a data set of 184 points) and utilized in the modeling procedure. The research crew collected soil samples from the surface soil and determined gravimetric soil moisture values after the samples were oven dried and weighted. The crew also used a hand-held measuring device to make in-field measurements and double-check the laboratory soil moisture results. The device, manufactured by Decagon Inc. (Pullman, WA, USA), includes a sensor read-out and storage system for real-time readings. Called "Procheck," it was connected to a GS3 soil moisture, temperature, and EC sensor from Decagon Inc as well [43]. Figure 3 illustrates the location of soil moisture samples in the study area.

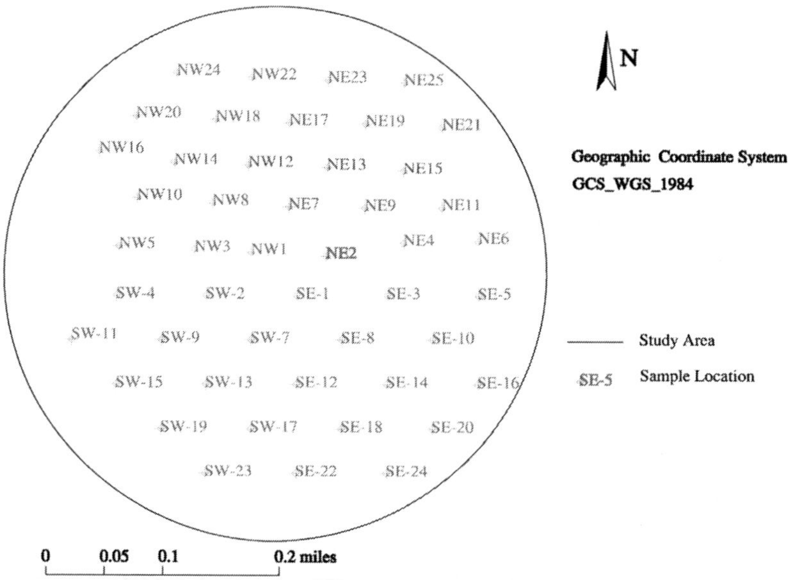

Figure 3. Spatial distribution of soil moisture sample locations in the study area.

2.3.4. Soil Texture Analysis

The upper and lower limits of soil moisture storage in the root zone are a function of soil texture. After the soil has been saturated and drained by gravity, the soil is said to be at "field capacity," and the amount of water that remains in the root zone but which the crop can no longer extract is called the "wilting point" [44]. In order to take these two parameters in to account, 14 different points from around the field were selected for soil texture sampling. After soil type determination, the corresponding field capacity values were acquired from previously published values and considered as model inputs [45].Figure 4 illustrates the soil field capacity map developed by utilizing a Spherical Kriging interpolation method for the information from the 14 available sampling locations.

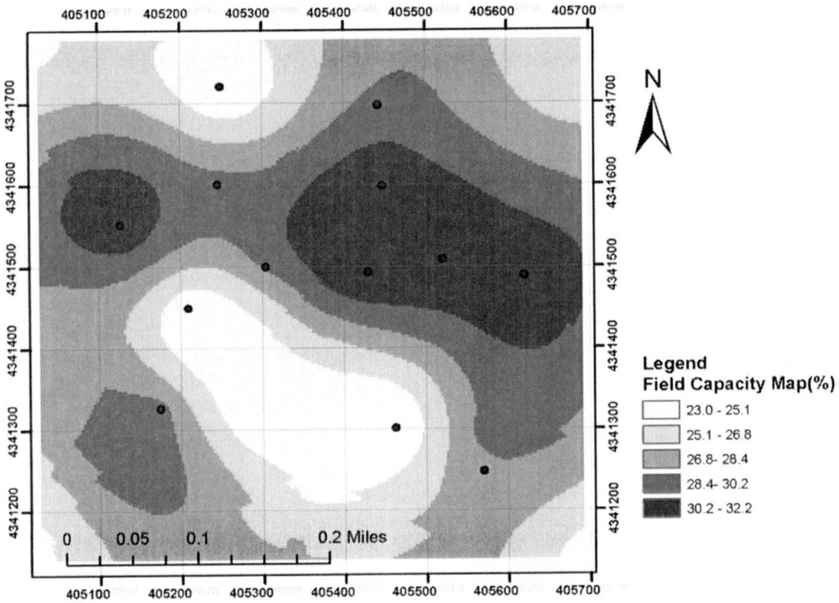

Figure 4. Map of field capacity based on soil texture type and plot of the location of soil samples.

2.3.5. Relevant Vegetation Indices (VIs) from AggieAir Imagery

Visual spectrum (red, green, and blue, or RGB), near-infrared (NIR), and infrared/thermal remotely sensed data and some vegetation indices (VIs) are used as input variables for the soil moisture model. All AggieAir data (RGB, NIR, and thermal imagery), normalized difference vegetation index (NDVI), vegetation condition index (VCI), enhanced vegetation index (EVI), vegetation health index (VHI), and filed capacity were chosen as model inputs with surface soil moisture as the target or output. The VHI was proposed by Kogan (1995), which is an additive combination of VCI and Temperature Condition Index (TCI) [36]. Equations ((1)–(5)) represent the vegetation indices included in this study:

$$\mathbf{NDVI} = \frac{\rho_{\mathrm{NIR}} - \rho_{\mathrm{RED}}}{\rho_{\mathrm{NIR}} + \rho_{\mathrm{RED}}} \quad (1)$$

$$\mathbf{VCI} = 100 \times \frac{\rho_{\mathrm{NDVI}} - \rho_{\mathrm{NDVI_{min}}}}{\rho_{\mathrm{NDVI_{max}}} - \rho_{\mathrm{NDVI_{min}}}} \quad (2)$$

$$EVI = 2.5 \times \frac{\rho_{NIR} - \rho_{RED}}{\rho_{NIR} + C_1 \times \rho_{RED} - C_2 \times \rho_{BLUE} + L} \quad (3)$$

$$TCI = 100 \times \frac{BT_{max} - BT}{BT_{max} - BT_{min}} \quad (4)$$

$$VHI = 0.5 \times VCI + 0.5 \times TCI \quad (5)$$

where ρ_{NIR}, ρ_{RED} and ρ_{BLUE} are NIR, red, and blue reflectance bands; C_1, C_2 and L are the coefficients of the aerosol resistance term, which uses the blue band to correct for aerosol influences in the red band; and BT is the thermal brightness, which is the thermal band reflectance.

2.3.6. Model Validation

A K-fold cross validation was used as the model validation technique in order to generalize an independent dataset. In general, in K-fold cross validation the original dataset (including all samples) is partitioned in to K sub-data sets. Each time, a single sub-data set is retained for evaluation and the remaining (K-1) sub-data sets are used for training. This process repeats K times, and the errors for each time are estimated. Furthermore, the K model errors are averaged to represent the best model [46,47]. Since the authors were not confident about the optimal percentage of data being considered for training, testing and validation to avoid over-fitting, a 5-fold cross validation technique is applied to the original data set and Mean Squared Error (MSE) is the calculated evaluation criterion. The 5-fold cross-validation was done repeatedly, and during the training phase different values for the training technique's parameters were used in concert with different network architectures. Further, the authors ended up with the best values for number of hidden nodes and training parameters. Then with these in hand, finally the network was trained using all the data, with the best umber of hidden nodes and training parameters.

2.3.7. Wrapper Selection

For model construction, it is necessary to identify the best combination of input variables from the available data. A wrapper selection method was used to accomplish this. Guyon (2003) introduced the advantages of applying this method with reference to three main aspects: (1) improving the performance of predictors; (2) obtaining faster and more cost-effective predictors; and (3) providing a better understanding of the underlying process that generated the data [48]. This method is recommended over the backward selection method and is applicable to cases with a small number of inputs. Wrapper selection considers all possible combinations of input variables and develops a separate model for each combination. The models are then scored based on their predictive power, and the best model can be selected based on the corresponding score [48,49]. In order to check the goodness of fit, root mean square error (RMSE), mean absolute error (MAE), coefficient of correlation (r), coefficient of performance (e), and coefficient of determination (R^2) are the statistical parameters that were calculated to evaluate the performance of the many alternative models and score their predictive power [50].

2.3.8. Division Set Up in ANN Model Architecture

The input data division set up can have a significant influence on the performance of an ANN model. Bowden (2002) presented two methodologies for dividing data into representative subsets (training, testing and validation) with similar statistical properties. These methods were proven to develop more robust results compared to conventional approaches in which the dataset was simply divided into arbitrary subsets [51]. The methods were applied by using a 5-fold cross validation method for data generalization. Other water resources related studies have utilized Bowden's approach and concluded that it ensures that the training, testing, and validation sets are representative of the same population [52,53,54,55,56,57].

It is difficult to assess beforehand how large an artificial neural network model should be for a specific application to avoid over-fitting. Model size strongly relates to sample size, and collecting more data and increasing the size of the training set or reducing the size of a network are recommended as solutions [28]. In this study, collecting more data was impossible; therefore, the error of the validation data set was checked as alternative method of investigation [28]. As training initiates, the error for all three data sets (training, testing, and validation) decreases, and in the case of over-fitting, the error for validation set increases while the error in the training set maintains a decreasing trend. If the error in the validation set continues in a reducing trend, there is no danger of over-fitting.

3. RESULTS AND DISCUSSION

3.1. Input Data

3.1.1. Soil Moisture Data Calculation Results

In order to ground truth data and relate soil moisture values to remotely sensed data, gravimetric soil moisture measurements were checked with the corresponding in-field measurements of volumetric soil moisture using soil bulk density values that were extracted from soil texture data. A *t*-test comparing the gravimetric soil moisture measurements against the volumetric soil moisture measurements showed these two data sets are not statistically different at a 95 percent confidence level with *P*-value of 0.3. The results from the *t*-test indicates that either of these data sets can be used for further calculations. Finally, the gravimetric soil moisture values from four flight dates were pooled representing the maximum, minimum and mean values of 30.6, 10.1 and 19.7, respectively, and used as model targets. Also, the spatial distribution of soil moisture from high to low values is in accordance with time after irrigation. The highest values occur immediately behind the center pivot lateral, and the driest spots were concentered in front of the lateral.

3.1.2. Spatial Information of Vegetation Indices

Due to heterogeneity within the field because of the different crop types, an access road, wheel tracks, the center pivot station, and historic locations of fence lines and ditch banks that once occupied the modern field, spatial analysis was required. The significance of spatial information comes from the ability of the human brain to detect spatial patterns in a map or an image. Table 1 represents the temporal and spatial changes of NDVI during the study period. The same information is provided for other three VIs in the supplementary material.

3.2. Wrapper Selection Outcome

Goodness-of-fit statistics were used to test the degree of association between the observed and estimated data. As noted previously, root mean square error (RMSE), mean absolute error (MAE), coefficient of correlation (r), coefficient of performance (e) and coefficient of determination (R^2) were calculated for the models to score their predictive power. In the scoring phase, the authors referred to RMSE and judged the models predictive power based on them, further MAE and R^2were considered and finally e and r came to account.

Table 1. Temporal and spatial changes in normalized difference vegetation index (NDVI) values during the study period.

Crop Type/Date	NDVI(Mean)			
	5/17/2013	6/1/2013	6/9/2013	6/17/2013
Three way. Oat.	0.09	0.34	0.43	0.53
Barley . Wheat	Planting	continued growth	continued growth	full growth
Alfalfa	0.42	0.47	0.53/0.08	0.59/0.13
	continued growth	continued growth	full growth/ after cut	full growth/ after cut
Oat. Alfalfa	0.43	0.48	0.53	0.57
	germination	continued growth	full growth	full growth

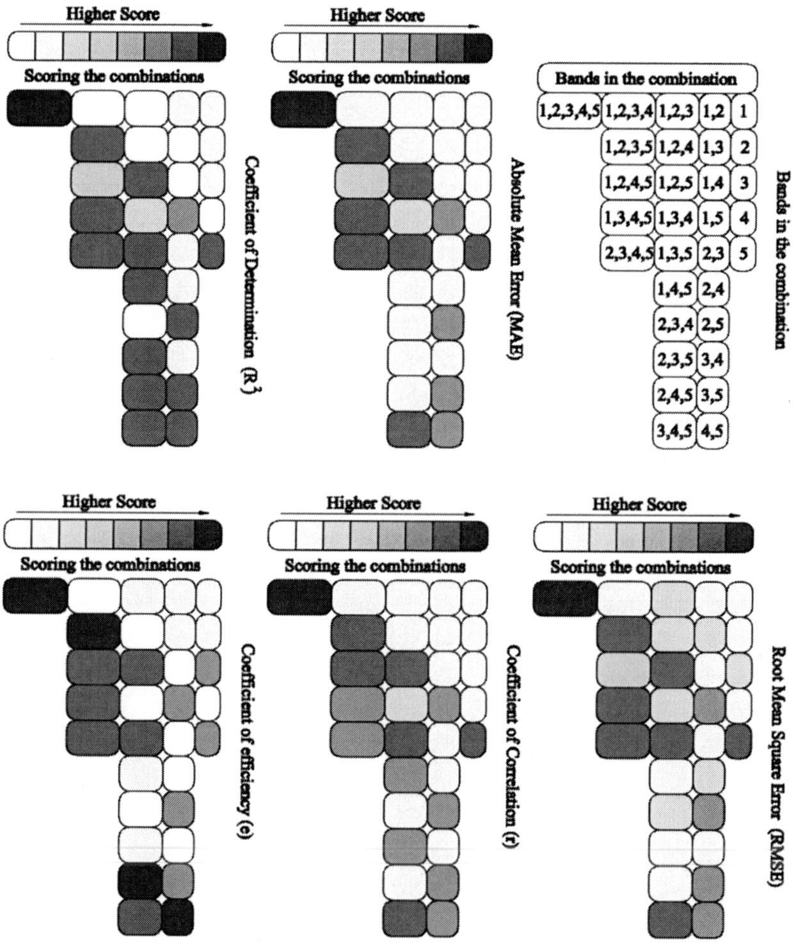

Figure 5. Schematic view of possible performance by 5-input dataset using wrapper selection method and artificial neural network (ANNs).

The models with high but similar predictive power were compared spatially against thermal, NIR and false color images. Also the research crew has collected a set of notes about their observations during the data collection procedure. The notes paid attention to crop types, crops growing stage, location of lateral, irrigation uniformity, wet and dry spots (created due to deficiencies in the irrigation sprinkler system), existence of wind (wind direction if it scatters the water) and weather condition. After the models with high but similar predictive power were developed, the best model was selected visually to accommodate the spatial distribution of above information. Figure 5 illustrates how schematically wrapper selection would evaluate the models for the inputs from AggieAir (RGB, NIR, and Thermal) as an example of wrapper scoring. for this study, 1023 models in 10 sets for all possible combinations of 10 inputs were developed (10 combinations of 1, 45 combinations of 2, 120 combinations of 3, 210 combinations of 4, 252 combinations of 5, 210 combinations of 6, 120 combinations of 7, 45 combinations of 8, 10 combinations of 9 and 1 combination of 10 inputs), and the model results were compared. A trial-and-error approach was utilized to select those models that worked on different numbers of neurons (up to 2 × (number of inputs) + 1 to avoid over-fitting issues), hidden layers, training functions, and division setups [32]. Finally, the model with 8 inputs (red, blue, NIR, thermal, NDVI, VCI, EVI, and field capacity) was selected because it had the best predictive power and best spatial pattern, which was checked visually. Table 2 shows the best model results for all 10 sets of combinations along with their highest predictive power statistics.

Table 2. Goodness-of-fit statistics from Wrapper selection results (1 to 10 inputs) with highest predictive power using ANN.

	ANN Inputs	Division Set up	# of Neurons	RMSE	MAE	r	e	R^2
One Input	Thermal	80/10/10	4	3.0	2.4	0.64	0.4	0.41
Two Inputs	Thermal, Field capacity	75/15/10	5	2.5	1.8	0.78	0.60	0.61
Three Inputs	Red, Blue, Thermal	70/15/15	7	2.7	2.1	0.74	0.54	0.55
Four Inputs	Red, NDVI, VCI, VHI	70/15/15	7	2.5	1.8	0.77	0.59	0.60
Five Inputs	Green, Thermal, VCI, EVI, Field Capacity	80/10/10	9	2.1	1.6	0.84	0.71	0.71
Six Inputs	NIR, Thermal, NDVI, EVI, VHI, Field Capacity	80/10/10	11	2.1	1.5	0.84	0.70	0.71
Seven Inputs	Red, Blue, NIR, Thermal, NDVI, VCI, Field Capacity	80/10/10	12	2.1	1.6	0.86	0.73	0.73
Eight inputs	**Red, Blue, NIR, Thermal, NDVI, EVI, VCI, Field Capacity**	**80/10/10**	**17**	**2.0**	**1.3**	**0.85**	**0.75**	**0.77**
Nine Inputs	Red, Green, Blue, Thermal, NDVI, \EVI, VCI, VHI, Field Capacity	80/10/10	17	2.0	1.4	0.87	0.75	0.75
Ten Inputs	Red, Green, Blue, NIR, Thermal, NDVI, EVI, VCI, VHI, Field Capacity	80/10/10	19	2.0	1.3	0.85	0.73	0.73

3.3. Results Extracted from Artificial Neural Networks (ANNs)

After the intensive trial and error selection procedure using cross validation procedure, a network architecture with one hidden layer and 17 nodes and a division set up of 80:10:10 with trainlm (Levenberg-Marquardt backpropagation) as a training function was selected. Figure 6 illustrates (a) the measured *versus* the estimated soil moisture values of the selected model; (b) the corresponding one-by-one scatter plot (showing that all the points are clustered along the 45° line); (c) residual plot; and (d) residual histogram to demonstrate the validation of the model in the sense of normality, linearity, and equality of variances. Figure 7 represents the performance plot of the selected ANN model with 8 inputs, showing Mean Squared Error (MSE) trend during the learning procedure for the training, testing and validation sets to evaluate possible over-fitting issues. The decreasing trend of the validation set confirms that there is no over-fitting in the model. Figure 8shows estimated surface soil moisture maps (Volumetric Water Content (%) (right column), at four different dates (a) 16 May, (b) 1 June, (c) 9 June and (d) 17 June 2013 (Res. 15 cm) alongside false color composite images (NIR–Green–Blue) (left column). The false color map (NIR-Green-Blue) is related to the relative density of vegetation in the image. Exposed soil (bare) is expected to have lower soil moisture content while areas with high vegetation density the opposite. The concept of using false color composite images was taken from previously published studies [56,58].

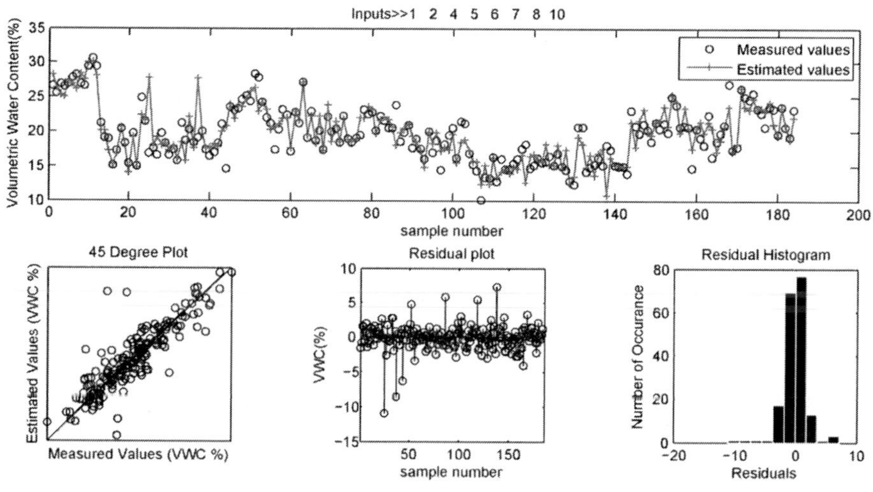

Figure 6. (**a**) The measured soil moisture values *versus* the estimated values of the selected model; (**b**) one-by-one scatter plot; (**c**) residual plot; and (**d**) residual histogram.

As shown in Figure 8, the soil moisture maps have a direct association with the false color composite maps. The field exterior area was not irrigated during the growing cycle and was expected to be less moist. Although the wheel tracks and the access road are located within the irrigation zone, they are expected to be drier since they are covered by bare soil, become more compacted due to traffic over them, and lose moisture rapidly. This assumption also applies to the zones where the crops have been cut.

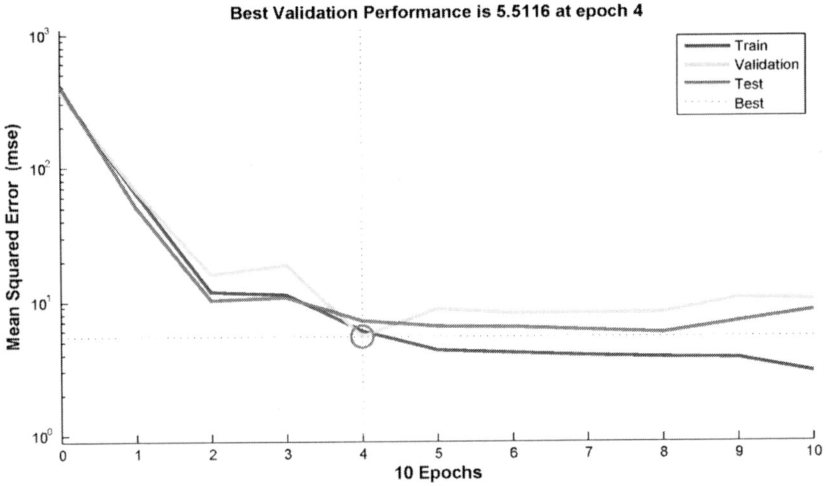

Figure 7. Performance plot of the selected ANN model with 8 inputs showing Mean Squared Error (MSE) trend during the leaning procedure for three data sets (training, testing and validation).

Different crop types have different water demands and water up-take rates that cause surface soil moisture heterogeneity even after a uniform irrigation event. This heterogeneity appeared in the form of cropping patterns in the soil moisture maps. According to the clockwise rotation of the lateral, the spots with the maximum soil moisture values are expected to fall near the lateral and in a counterclockwise direction. This status is clearest in Figure 7a where the field was under a heavy irrigation event at the time the aerial imagery was captured. Table 3 shows the comparison between measured and estimated soil moisture values for different crop type zones. Soil moisture numerical values are presented in the supplementary materials.

The main problem with such modelling procedures is being dependent to site and time. This implies that ground sampling and modeling will be required for every flight to ensure accurate and quality data. So far this is a handicap and should be strengthened with more studies over different types of crops, in different areas at different stages of growth. Having such a model (or a collection of models) makes this practical for routine use (independent of site and time). In addition, the current study was targeted toward showing the detailed information that can be interpreted from high resolution data. Even though such a high resolution might not be required for monitoring agricultural farm conditions that are cropped with inexpensive crops such as alfalfa and oats, this resolution presents its value for other crops that require high resolution data (e.g., vineyards, orchards). These results essentially help to justify future work to look at the value of high resolution data for precision farming activities.

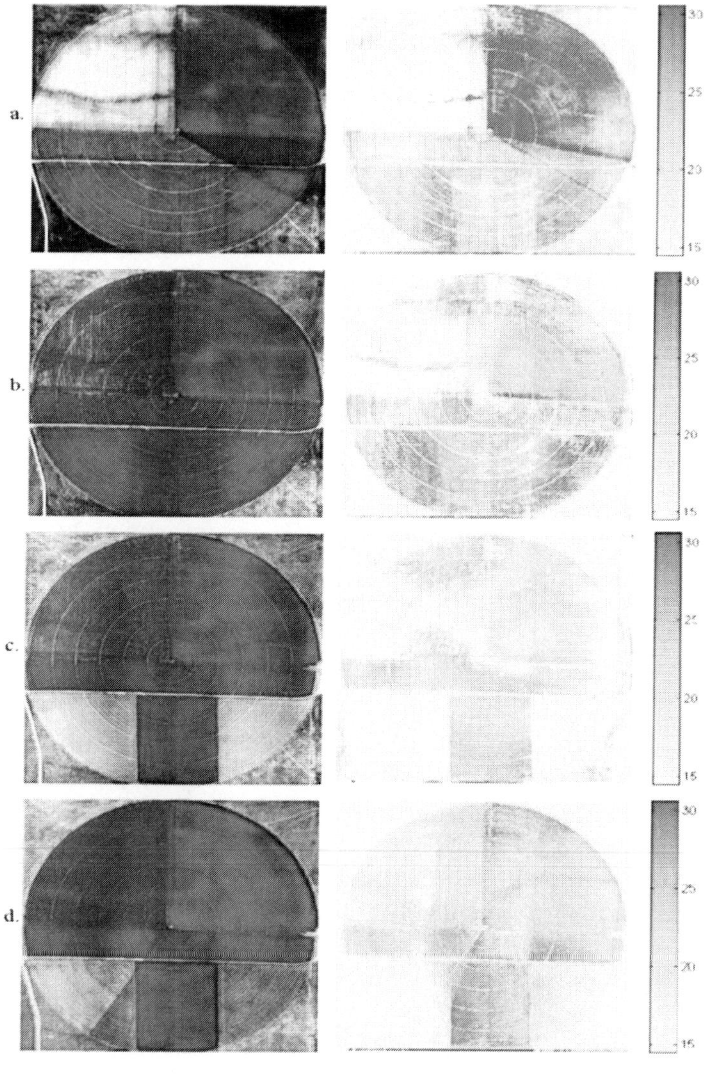

Figure 8. Estimated soil moisture maps (Volumetric Water Content (%)), with ANNs model for four different dates (**a**) 16 May; (**b**) 1 June; (**c**) 9 June and (**d**) 17 June 2013 (Res. 15 cm) (right), false color images (left).

One step forward in generalizing the presented modeling methodology in temporal scale could be the idea of pooling the soil moisture data collected from different dates. In the case of the current study, every single sampling location experiences four different conditions of soil moisture level, which provides a wide range of information about soil moisture status through time. This type of information makes the model more robust in its ability to simulate previously unseen soil moisture conditions through time.

Table 3. Comparison between measured and estimated soil moisture values for different crop type zones.

Crop Type/Date	Soil Moisture (Volumetric Water Content (%)) (Zonal Mean)							
	5/17/2013		6/1/2013		6/9/2013		6/17/2013	
	Measured	Estimated	Measured	Estimated	Measured	Estimated	Measured	Estimated
Three way, Oat, Barley , Wheat	18.9	20.1	21.5	19.0	14.9	15.7	19.3	17.6
Alfalfa	27.6	25.9	25.0	23.5	18.4	18.9	21.2	20.4
Oat, Alfalfa	18.0	18.3	20.5	18.1	15.7	15.8	22.5	18.9

4. CONCLUSIONS

This paper demonstrates the application of a high resolution remote sensing technology (AggieAir) for estimating surface soil moisture as a key piece of information in irrigation water management. High-resolution multi-spectral imagery, in combination with ground sampling, provided enough information for the modelling approaches to accurately estimate spatially distributed surface soil moisture.

This paper presents the results of a modelling approach utilizing ANN in concert with time and site specific information. Parallel to other modeling approaches, such as data mining algorithms or linear regression, the ANN model is calibrated for this study within the conditions of the information collected including soil moisture measurements, soil texture, crop type information, and high resolution multi-spectral imagery. While the site-specific calibrated ANN in this study cannot be used immediately in another location, the modeling procedure (identifying spatial information with the most significant contribution to soil moisture estimation (Table 2)) along with similar field measurements and high resolution multi-spectral imagery and the data mining algorithm, are transferable from this study.

Surface soil moisture estimation was accomplished with an ANN model (RMSE: 2.0, MAE: 1.3, r: 0.87, e: 0.75, R2:0.77) for four dates in 2013 (16 May, 1 June, 9 June, and 17 June). These results show the capability of the model to accurately estimate surface soil moisture. Compared to the traditional soil moisture estimations that are based on a farmer's visual perceptions or a few soil moisture samples averaged across the farm, the modelling approach presented enables greater precision in the application of water and identifies dry/wet spots and water stressed crops.

AggieAir imagery, combined with appropriate analytic tools, allows spatial estimation of surface soil moisture. These estimates were made at much finer resolutions in space and time than those available from conventional remote sensing technologies (e.g., satellite or commercial aerial photography services). Also, the application of data mining algorithms to AggieAir aerial imagery allows for quantification of actionable information for precision agriculture (soil moisture values across the field). The soil moisture maps that are produced can then be related to irrigation water management for scheduling and application rates.

The results from the wrapper selection (Table 2) prove the significance of thermal imagery as the most relevant information in surface soil moisture estimations. In the case of one input, a model with thermal imagery can estimate the soil moisture values with RMSE of approximately 3% (thermal images are provided in supplementary material section).

Soil water holding capacity as a function of soil texture plays an important role in soil moisture values. This parameter was observed by utilizing field capacity as an input to the models. Table 2 shows that field capacity is a component of most of the models. The effect of this parameter is confounded by other important inputs in the spatial distribution of soil moisture in Figure 8. Based on the information presented in Table 2, among the available vegetation indices, NDVI and VCI have a greater explanatory contribution in surface soil moisture estimates. The soil moisture maps have a good association with false color composite maps that allows for distinction of agricultural features in the field.

5. FUTURE WORK

Further studies will involve the estimation of surface soil moisture using other data mining algorithms and its application as a boundary condition to produce remotely sensed estimates of root zone soil moisture. In addition, Pixel-wise estimation of soil moisture could also be applied in a water balance models.

ACKNOWLEDGMENTS

This project was financially supported by the Utah Water Research Laboratory through an MLF Seed Grant. The authors would like to thank the Utah Water Research Laboratory, Utah State University, and the Provo, Office of the US Bureau of Reclamation for their support of this research. The authors acknowledge the extraordinary efforts of the AggieAir flight team and Andres Ticlavilca, Roula Bachour, and Manal Elarab. The authors appreciate the support of Ivan Robins, the farm owner, whose cooperation greatly improved of the data collection procedure.

AUTHOR CONTRIBUTIONS

Austin Jensen guided the AggieAir team in preparing the high-resolution imagery as the required data. Mac McKee and Alfonso Torres-Rua developed the research plan and experiment design and supervised the work required for this paper. Leila Hassan-Esfahani completed the literature review, data acquisition, method selection, modelling, manuscript preparation, and discussions. All authors shared equally in the editing of the manuscript.

CONFLICTS OF INTEREST

The authors declare no conflict of interest.

REFERENCES

1. Grayson, R.B.; Western, A.W. Towards areal estimation of soil water content from point measurements: Time and space stability of mean response. J. Hydrol. 1998, 207, 68–82.

2. Gill, K.; Kemblowski, M.; McKee, M. Soil moisture data assimilation using support vector machines and ensemble Kalman filter. JAWRA J. Am. Water Resour. Assoc. 2007, 43, 1004–1015.

3. Zaman, B.; McKee, M.; Neale, C.M.U. Fusion of remotely sensed data for soil moisture estimation using relevance vector and support vector machines. Int. J. Remote Sens. 2012, 33, 6516–6552.

4. Liu, W.; Baret, F.; Gu, X.; Zhang, B.; Tong, Q.; Zheng, L. Evaluation of methods for soil surface moisture estimation from reflectance data. Int. J. Remote Sens. 2003, 24, 2069–2083.

5. Idso, S.; Jackson, R.; Reginato, R. Estimating evaporation: A technique adaptable to remote sensing. Science 1975, 189, 991–992.

6. Idso, S.; Jackson, R.; Reginato, R. Compensating for environmental variability in the thermal inertia approach to remote sensing of soil moisture. J. Appl. Meteorol. 1976, 15, 811–817.

7. Reginato, R.; Idso, S.; Jackson, R.; Vedder, J.; Blanchard, M.; Goettelman, R. Soil water content and evaporation determined by thermal parameters obtained from ground-based and remote measurements. J. Geophys. Res. 1976, 81, 1617–1620.

8. Reginato, R.; Jackson, R.; Pinter, P. Evapotranspiration calculated from remote multispectral and ground station meteorological data. Remote Sens. Environ. 1985, 18, 75–89.

9. Jackson, T. Soil water modeling and remote sensing. IEEE Trans. Geosci. Remote Sens. 1986, 24, 37–46.

10. Quattrochi, D.; Luvall, J. Thermal infrared remote sensing for analysis of landscape ecological processes: Methods and applications. Landsc. Ecol. 1999, 14, 577–598.

11. Kaleita, A.; Tian, L.; Hirschi, M. Relationship between soil moisture content and soil surface reflectance. Trans. ASAE2005, 48, 1979–1986.

12. Humes, K.; Kustas, W.; Jackson, T.; Schmugge, T.; Moran, M. Combined use of optical and microwave remotely sensed data for the estimation of surface energy balance components over a semi-arid watershed. In Proceedings of the IEEE Topical Symposium on Combined Optical-Microwave Earth and Atmosphere Sensing, Albuquerque, NM, USA, 22–25 March 1993; pp. 86–89.

13. Njoku, E.G.; Jackson, T.J.; Lakshimi, V.; Chan, T.K.; Ngheim, S.V. Soil moisture retrieval from AMSR-E. IEEE Trans. Geosci. Remote Sens. 2003, 41, 215–229.

14. Aritola, J.F.; Pepper, I.L.; Brusseau, M.L. Environmental Monitoring and Characterization; Elsevier Academic Press: San Diego, CA, USA, 2004.

15. Das, N.N.; Mohanty, B.P. Root zone soil moisture assessment using remote sensing and vadose zone modeling. Vadose Zone J. 2006, 5, 296–307.

16. Njoku, E.G.; Ashcroft, P.; Chan, T.K.; Li, L. Global survey and statistics of radio-frequency interference in AMSR-E land observations. IEEE Trans. Geosci. Remote Sens. 2005, 43, 938–947.

17. Crosson, W.L.; Limaye, A.S.; Laymon, C.A. Parameter sensitivity of soil moisture retrievals from airborn C- and X-band radiomter measurements in SMEX02. IEEE Trans. Geosci. Remote Sens. 2005, 43, 2842–2853.

18. Crow, W.T.; Chan, T.K.; Entekhabi, D.; Houser, P.R.; Hsu, A.Y.; Jackson, T.J.; Njoku, E.G.; O'Neill, P.E.; Shi, J.; Zhan, X. An observing system simulation experiment for hydros radiomter-only soil moisture products. IEEE Trans. Geosci. Remote Sens. 2005, 43, 1289–1303.

19. Fernandez-Galvez, J. Errors in soil moisture content estimates induced by uncertainties in the effective soil dielectric constant. Int. J. Remote Sens. 2008, 29, 3317–3323.

20. Cai, G.; Xue, Y.; Hu, Y.; Wang, Y.; Guo, J.; Luo, Y.; Wu, C.; Zhong, S.; Qi, S. Soil moisture retrieval from MODIS data in northern China plain using thermal inertia model. Int. J. Remote Sens. 2007, 28, 3567–3581.

21. Lu, S.; Ju, Z.; Ren, T.; Horton, R. A general approach to estimate soil water content from thermal inertia. Agric. For. Meteorol. 2009, 149, 1693–1698.

22. Verstraeten, W.W.; Veroustraete, F.; van der Sande, C.J.; Grootaers, I.; Feyen, J. Soil moisture retrieval using thermal inertia, determined with visible and thermal spaceborne data, validated for European forests. Remote Sens. Environ.2006, 101, 299–314.

23. Gill, M.; Asefa, T.; Kemblowski, M.; McKee, M. Soil moisture prediction using support vector machines. J. Am. Water Resour. Assoc. 2006, 42, 1033–1046.

24. Jiang, H.; Cotton, W. Soil moisture estimation using an artificial neural network: A feasibility study. Can. J. Remote Sens. 2004, 30, 827–839.

25. Hassan-Esfahani, L; Torres-Rua, A.; Jensen, A.; McKee, M. Topsoil moisture estimation for precision agriculture using unmmaned aerial vehicle multispectral imagery. In Proceedings of the 2014 IEEE International Geoscience and Remote Sensing Symposium (IGARSS), Quebec City, QC, Canada, 13–18 July 2014; pp. 3263–3266.

26. Elshorbagy, A.; Parasuraman, K. On the relevance of using artificial neural networks for estimating soil moisture content. J. Hydrol. 2008, 362, 1–18.

27. Khalil, A.; Gill, M.K.; McKee, M. New applications for information fusion and soil moisture forecasting. In Proceedings of the 8th

International Conference on Information Fusion, Philadelphia, PA, USA, 25–28 July 2005; pp. 1622–1628.

28. Haykin, S. Neural Networks: A Comprehensive Foundation, 2nd ed.; MacMillan: New York, NY, USA, 1999.

29. Khan, M.; Coulibaly, P. Streamflow forecasting with uncertainty estimate using Bayesian learning For ANN. In Proceedings of the 2005 IEEE International Joint Conference on Neural Networks, Montreal, QC, Canada, 31 July–4 August 2005; Volume 5, pp. 2680–2685.

30. Tokar, S.A.; Johnson, P.A. Rainfall-runoff modeling using artificial neural networks. J. Hydrol. Eng. 1999, 4, 232–239.

31. Rumelhart, D.E.; McClelland, J.L.; PDP Research Group. Parallel Distributed Processing. Volume 1: Foundations; The MIT Press: Cambridge, MA, USA, 1986.

32. Hecht-Nielsen, R. Neurocomputing; Addison-Wesley Publishing Company: Reading, MA, USA, 1990.

33. Wasserman, P.D. Neural Computing Theory and Practice; Van Nostrand Reinhold: New York, NY, USA, 1989.

34. Gillies, R.R.; Carlson, T.N. Thermal remote sensing of surface soil water content with partial vegetation cover for incorporation into climate models. J. Appl. Meteorol. 1995, 34, 745–756.

35. Wetzel, P.J.; Woodward, R.H. Soil moisture estimation using GOES-VISSR infrared data: A case study with a simple statistical method. J. Clim. Appl. Meteorol. 1987, 26, 107–117.

36. Kogan, F.N. Droughts of the late 1980s in the United States as derived from NOAA polar-orbiting satellite data. Bull. Am. Meteorol. Soc. 1995, 76, 655–668.

37. Mallick, K.; Bhattacharya, B.K.; Patel, N.K. Estimating volumetric surface moisture content for cropped soils using a soil wetness index based on surface temperature and NDVI. Agric. For. Meteorol. 2009, 149, 1327–1342.

38. Liang, S.; Rui, S.; Xiaowen, Li.; Shunlin,, L.; Renhua, Z. Monitoring surface soil moisture status based on remotely sensed surface temperature and vegetation index information. Agric. For. Meteorol. 2012, 166, 175–187.

39. Bhuiyan, C.; Singha, R.P.; Koganc, F.N. Monitoring drought dynamics in the Aravalli region (India) using different indices based on ground and remote sensing data. Int. J. Appl. Earth Observ. Geoinf. 2006, 8, 289–302.

40. Jensen, A.M. A Geospatial Real-Time Aerial Image Display for a Low-Cost Autonomous Multispectral Remote Sensing. Master's Thesis, Utah State University, Logan, UT, USA, 2009.

41. MosaicMill EnsoMOSAIC. Available online: http://www. ensomosaic. com (accessed on 6 May 2012).

42. Long, D.; Longuevergne, L.; Scanlon, B.R. Uncertainty in evapotranspiration from land surface modeling, remote sensing, and GRACE satellites. Water Resour. Res. 2014, 50, 1131–1151.

43. Decagon Devices, Inc. Available online: http://www.decagon.com/ (accessed on 20 November 2014).

44. Israelson, O.W.; West, F.L. Water holding capacity of irrigated soils. Utah State Agric. Exp. Station Bull 1922, 183, 1–24.

45. Costa, A.; Albuquerque, J.A.; Costa, A.; Pértile, P.; Silva, F.R. Water retention and availability in soils of the State of Santa Catarina-Brazil: Effect of textural classes, soil classes and lithology. Braz. Soil Sci. Soc. 2013, 37, 1535–1548.

46. Kohavi, R. A study of cross-validation and bootstrap for accuracy estimation and model selection. In Proceedings of the Fourteenth International Joint Conference on Artificial Intelligence, Montreal, QC, Canada, 20–25 August 1995; Volume 12, pp. 1137–1143.

47. Geisser, S. Predictive Inference; Chapman and Hall: New York, NY, USA, 1993.

48. Guyon, I. An introduction to variable and feature selection. J. Mach. Learn. Res. 2003, 3, 1157–1182.

49. Kohavi, R.; John, G.H. Wrappers for feature subset selection. Artif. Intell. 1997, 12, 273–324.

50. Glover, D.M.; Jenkins, W.J.; Doney, S.C. Least Squares and Regression Techniques, Goodness of Fit and Tests, Non-Linear Least Squares Techniques; Woods Hole Oceanographic Institute: Woods Hole, MA, USA, 2008.

51. Bowden, G.J.; Maier, H.R.; Dandy, G.C. Optimal division of data for neural network models in water resources applications. Water Resour. Res. 2002.

52. Bowden, G.J.; Maier, H.R.; Dandy, G.C. Input determination for neural network models in water resources applications. Part 2. Case study: Forecasting salinity in a river. J. Hydrol. 2005, 301, 93–107.

53. Anctil, F.; Lauzon, N. Generalisation for neural networks through data sampling and training procedures, with applications to stream flow predictions. Hydrol. Earth Syst. Sci. 2004, 8, 940–958.

54. Jia, Y.; Culver, T.B. Bootstrapped artificial neural networks for synthetic flow generation with a small data sample. J. Hydrol. 2006, 331, 580–590.

55. Lauzon, N.; Anctil, F.; Petrinovic, J. Characterization of soil moisture conditions at temporal scales from a few days to annual. Hydrol. Process. 2004, 18, 3235–3254.

56. Whalley, W.R.; Leeds-Harrison, P.B.; Bowman, G.E. Estimation of soil moisture status using near infrared reflectance.Hydrol. Process. 1991, 5, 321–327.

57. Hassan-Esfahani, L.; Torres-Rua, A.; McKee, M. Assessment of optimal irrigation water allocation for pressurized irrigation system using water balance approach, learning machines, and remotely sensed data. Agric. Water Manage.2015, 153, 42–50.

58. Yin, Z.; Lei, T.; Yan, Q.; Chen, Z.; Dong, Y. A near-infrared reflectance sensor for soil surface moisture measurement.Comput. Electron. Agric. 2013, 99, 101–107.

CHAPTER 10

Advances in Artificial Neural Networks – Methodological Development and Application

Yanbo Huang

United States Department of Agriculture, Agricultural Research Service, Application and Production Technology Research Unit, 141 Experiment Station Road, Stoneville, Mississippi 38776, USA

ABSTRACT

Artificial neural networks as a major soft-computing technology have been extensively studied and applied during the last three decades. Research on backpropagation training algorithms for multilayer perceptron networks has spurred development of other neural network training algorithms for other networks such as radial basis function, recurrent network, feedback network, and unsupervised Kohonen self-organizing network. These networks, especially the multilayer perceptron network with a backpropagation training algorithm, have gained recognition in research and applications in various scientific and engineering areas. In order to accelerate the training process and overcome data over-fitting, research has been conducted to improve the backpropagation algorithm. Further, artificial neural networks have been integrated with other advanced methods such as fuzzy logic and wavelet analysis, to enhance the ability of data interpretation and modeling and to avoid subjectivity in the operation of the training algorithm. In recent years, support vector machines have emerged as a set of high-performance supervised generalized linear classifiers in parallel with artificial neural networks. A review on development history of artificial neural networks is presented and the standard architectures and algorithms of artificial neural networks are described. Furthermore, advanced artificial neural networks will be introduced with support vector machines, and limitations of ANNs will be identified. The future of artificial neural network development in tandem with support vector machines will be discussed in conjunction with further applications to food science and engineering, soil and water relationship for crop management, and decision support for precision agriculture. Along with the network structures and training algorithms, the applications of artificial neural networks will be reviewed as well, especially in the fields of agricultural and biological engineering.

KEYWORDS

artificial neural networks; backpropagation; training algorithm; neuro-fuzzy; wavelet; support vector machines

1. INTRODUCTION

Soft computing is a partnership of computing techniques. The partnership includes fuzzy logic, Artificial Neural Networks (ANNs), and genetic algorithms. Conventionally, a huge set of techniques are referred to as hard computing, such as stochastic and statistical methods bound by the concept called NP (verifiable in Nondeterministic Polynomial time)-complete. Unlike hard computing, soft computing techniques offer "inexact" solutions of very complex problems through modeling and analysis with a tolerance of imprecision, uncertainty, partial truth, and approximation. In fact, soft computing is an integration of biological structures and computing techniques. In the partnership, ANNs provides configurations made up of interconnecting artificial neurons that mimic the properties of biological neurons.

There are a variety of ANN architectures, such as Multilayer Feedforward Network (MFN), Radial Basis Function (RBF) network, recurrent network, feedback network, and Kohonen Self-Organizing Map (SOM) network. Each of these networks has a training algorithm to memorize what it "sees" and project what it does not "see" in the same population. Typical training algorithms include Back-Propagation (BP) for Multi-Layer Perceptron (MLP, which is a MFN model), data clustering and linear weight solution for RBF networks, and unsupervised SOM for Kohonen SOM networks. In

1986, Rumelhart *et al.* [1] reported the BP training algorithm by implementing the Delta rule, a gradient descent learning rule for updating the weights of the artificial neurons in the perceptron-type ANNs, for MFNs in the monograph of the Parallel Distributed Processing (PDP) research group [2]. This work introduced and enhanced recognition of the BP algorithm, gave a strong impulse to the study of the mechanisms and structure of the brain, and provided the catalyst for much of the subsequent research and application of ANNs.

There are many issues and problems in applying ANNs which have inspired studies and research for improving the standard methods to solve problems. Determination of network structure is an issue in identification of ANN models, such as before implementing a MFN, the optimal numbers of hidden layers and nodes need to be decided on [3,4,5], and network pruning for redundancy connection reduction to improve network generalization capacities [6]. The slow convergence speed and local minima sticking with suboptimal solutions of BP training for MLPs have been studied [7,8,9,10,11,12,13,14]. Research has been conducted to improve the standard BP algorithm to overcome data over-fitting [15,16,17,18,19,20,21].

With the development of research and applications, ANNs have been integrated or fused with other methods of soft computing and signal processing such as fuzzy logic [22,23,24,25,26,27] and wavelet analysis [28,29,30,31]. The fusion is to combine or cascade different computing methods with ANNs to improve system performance over an individual technique. In many cases, the problems can be solved more effectively by combining one or two other techniques rather than implementing ANNs exclusively. In this way, the fused methods complement each other to enhance the ability of data interpretation and modeling and to avoid subjectivity in the operation of the training algorithm with ANNs individually.

Support Vector Machines (SVMs) emerged as a method in parallel with artificial neural networks as a set of supervised generalized linear classifiers and often provide higher classification accuracies than MLP neural networks [32,33]. SVMs have a number of advantages over ANNs and have been attracting more and more interest in recent years.

In agricultural and biological engineering, although some early research and applications exist [34,35,36], the interest of ANNs has been growing greatly in the last fifteen years in studies of soil and water regimes related to crop growth, analysis of the operation of food processing, and support of decision-making in precision farming. In summary of papers and reports collected from various sources, particularly through searching in the technical library of ASABE (American Society of Agricultural and Biological Engineers; http://asae.frymulti.com/) and the National Agricultural Library of USDA (United States Department of

Agriculture; http://www.nal.usda.gov/), it was found out that, from early 1990s to the present, there have been 348 related reports and papers (193 peer reviewed) on ANNs. It is interesting that of the 20 reports and papers (13 peer reviewed) on SVMs from 2003 to the present, seven (two peer reviewed) came out in 2008. This may signify more interest of SVMs in the next decade in agricultural and biological engineering.

The purpose of this paper is to overview ANN methodology, to determine the state of the art, to identify the limitations, and to project the future. Specifically, this paper will review the history of the development of ANNs. Then, the standard architectures and algorithms of ANNs will be described. Furthermore, a number of advanced ANN models will be introduced with SVMs, and the limitations of ANNs are identified. The future of research and applications of ANNs will be discussed with the development of SVMs. With the concepts, network structures, and training algorithms, the applications of ANNs will be reviewed with a focus on agricultural and biological engineering as well.

2. HISTORY OF ANN DEVELOPMENT

ANNs provide a method to characterize synthetic neurons to solve complex problems in the same manner as the human brain. For many years, especially since the middle of the last century, an interest in studying the brain's mechanism and structure has been increasing. This growing research interest has led to the development of new computational models, connectionist systems or ANNs, based on the biological background, for solving complex problems like pattern recognition, and fast information processing and adaptation.

In the early 1940s, McCulloch and Pitts [37] studied the potential and capabilities of the interconnection of several basic components, based on the model of a neuron. Later on, others, like Hebb [38], were concerned with the adaptation laws involved in neural systems. In 1958, Rosenblatt [39] coined the name Perceptron and devised an architecture which has subsequently received much attention. In 1960, Widrow and his student, Hoff, [40] presented an important generalization of the perceptron training algorithm as the Least Mean Square (LMS) learning procedure, also known as the Delta rule. The learning rule was applied as the adaptive linear element in the ADALINE (ADAptive LInear Neuron) networks. Then, Minsky and Papert [41] introduced a rigorous analysis of the Perceptron, of which they proved many properties and pointed out limitations of several related models. In the 1970s, the work of Grossberg [42] came to prominence. His work, based on biological and psychological evidence, proposed several novel architectures of nonlinear dynamic systems. In 1982, Hopfield [43] applied a particular nonlinear dynamic structure to solve problems in optimization. All of them conducted pioneer studies on the theoretical aspect of ANNs, particularly starting in the 1950s and 1960s.

In 1986, the PDP research group published a series of algorithms and results [2]. This publication contains an article entitled "Learning Internal Representations by Error Propagation" [1]. This article made the recognition of the BP training algorithm although it was already described in 1974 [44]. The BP algorithm implemented with the general Delta rule [1,7], as the representative of supervised training algorithms, gave a strong impulse to the subsequent research and has resulted in the largest body of research and applications in ANNs although a number of other ANN architectures and training algorithms have been developed and applied at the same time. In 1982, Finnish Professor Teuvo Kohonen [45] first described a different ANN architecture which is trained using an unsupervised SOM learning procedure. This Kohonen SOM algorithm has been populated in many research and practical applications later on. In 1988, RBF neural networks were first used [46], although RBFs were introduced earlier in 1977 [47]. RBF networks have also been widely used with the strong capability of function approximation and, along with MLP perceptron, had the impact on the emergence of SVMs [32,33,48]. In the late 1980s, the standard MFNs were proven as universal approximators on a compact subset of R^n [49,50]. The MFNs have a single hidden layer containing finite number of hidden neurons. The neurons use arbitrary activation function The theory assures that MFNs, including MLP networks, RBF networks and even SVMs under certain conditions [51] can handle problems which are highly complex and nonlinear.

So far, ANNs have been used in many industrial and commercial applications such as process modeling and control [52], character recognition [53], image recognition [54], credit evaluation [55], fraud detection [56,57], insurance [58], and stock forecasting [59]. In a later section, applications of ANNs in agricultural and biological engineering will be reviewed.

3. ANN ARCHITECTURES AND TRAINING ALGORITHMS

3.1. MLP and BP

Because the original perceptrons, single layer feedfoward networks, which were introduced by Rosenblatt [39], are limited to learning linearly separately patterns, nonlinear layers between the input and output are added to separate the data with enough training to model any well-defined function to arbitrary precision. This MFN model is known as a multilayer perceptron. The neural networks formed using the model are universal approximators [49,50]. Figure 1 shows the diagram of a one-hidden-layered MLP network structure. The MLP networks are typically trained with the BP algorithm. The BP algorithm is supervised, which is to map the process inputs to the desired outputs by minimizing the errors between the desired outputs and the calculated outputs driven from the inputs and network learning.

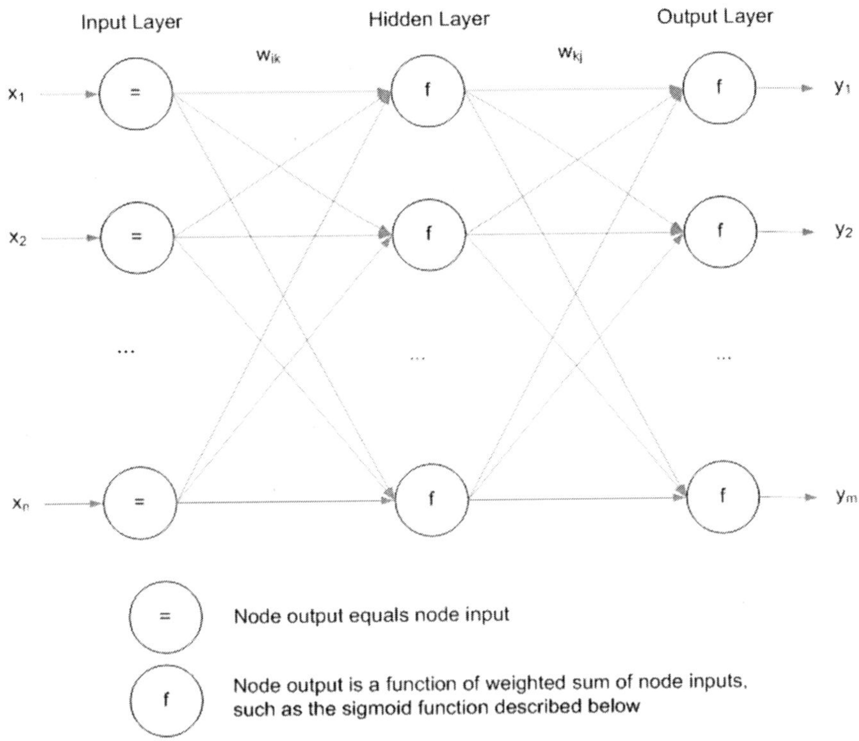

Figure 1. MLP network structure.

The standard BP algorithm [1] is a gradient descent algorithm to minimize the mean square error between the calculated outputs and the desired outputs of the MLP network. Assuming that there are n process inputs, x, and m desired outputs, d, with the network with a sample size of N: $\{x_{it}, d_{jt} \mid i = 1, 2, \dots n; j = 1, 2, \dots, m; t = 1, 2, \dots, N\}$, then the mean square error of the network outputs is:

$$E = \frac{1}{2}\sum_{t=1}^{N}\sum_{j=1}^{m}(d_{jt} - y_{jt})^2 \tag{1}$$

where y_{jt} is the calculated output.

Based on the gradient descent method, each of the network connection weights is updated as follows:

$$\begin{aligned}\Delta w &= -\eta \nabla E(w) \\ &= -\eta \frac{\partial E}{\partial w} \\ &= \sum_{t=1}^{l}\sum_{j=1}^{m}(d_{jt} - y_{jt})\frac{\partial y_{jt}}{\partial w}\end{aligned} \tag{2}$$

where η is a small constant such as 0.1 as the learning rate in the training process of the algorithm. For the MLP in Figure 1, all the weights are the sets of $\{w_{ik} \mid i = 1, 2, \ldots, n;\ k = 1, 2, \ldots, h\}$ and $\{w_{kj} \mid k = 1, 2, \ldots, h;\ j = 1, 2, \ldots, m\}$ where h is the number hidden nodes. Each of the output is as follows:

$$y_{jt} = s(\sum_{k=1}^{h} w_{kj} s(\sum_{i=1}^{n} w_{ik} x_{it})) \tag{3}$$

where s(z) is the activation function of each hidden node and output node. It typically is a sigmoid function as $s(z) = 1/(1 + e^{-z})$. Therefore,

$$\frac{\partial y_{jt}}{\partial w_{kj}} = s'(\sum_{k=1}^{h} w_{kj} s(\sum_{i=1}^{n} w_{ik} x_{it}))[\sum_{k=1}^{h} s(\sum_{i=1}^{n} w_{ik} x_{it})] \tag{4}$$

$$\frac{\partial y_{jt}}{\partial w_{ik}} = s'(\sum_{k=1}^{h} w_{kj} s(\sum_{i=1}^{n} w_{ik} x_{it}))[\sum_{k=1}^{h} s(\sum_{i=1}^{n} w_{ik} x_{it}) + \sum_{k=1}^{h} w_{kj} s'(\sum_{i=1}^{n} w_{ik} x_{it}) x_{it}] \tag{5}$$

and $s'(z) = s(z)(1 - s(z))$ is the first derivative of s(z).

Earlier than Rumelhart *et al.* [1], Werbos [44] gave a different derivation of the BP algorithm. This derivation is more general and mathematically rigorous than the one given by Rumelhart *et al.* In Werbos' derivation, the chain rule is expressed in a convenient form by ordered derivatives. Even so, most research and applications of the BP algorithm still referred to the results of Rumelhart *et al.* due to their milestone publication in 1986 with the PDP research group.

3.2. Radial Basis Function Network

Due to adding hidden nodes and layer(s) and the nonlinearity of the sigmoid activation function of each hidden nodes and/or output nodes in MLPs, the BP algorithm has the possibility to produce complex error surfaces which contain many minima. Since some minima are deeper than others, it is possible that the algorithm will not find a global minima. Instead, the network may fall into local minima, which represent suboptimal solutions. The RBF networks [46] were introduced with the centers of data clustering means, and then the linear solution for the connection weights to the network output directly to a unique local minimum, the global minimum, which shortens network training process greatly.

Figure 2 is a diagram of a one-hidden-layered RBF network structure. The RBF networks also are MFNs. With sufficient larger number of hidden nodes, they are universal approximators as well. The RBF network training also is supervised, involving determination of data clustering centers, c_k ($k = 1, 2, \ldots, h_c$), where h_c also is the number of the centers to be determined, and network connection weights from the hidden layer to the network outputs, w_{kj} ($k = 1, 2, \ldots, h_c$; $j = 1, 2, \ldots, m$). Similar to BP training, the RBF training is to establish the map between process inputs and desired outputs by minimizing the errors between the desired outputs and the calculated outputs driven from the inputs and network learning. The objective function of error minimization is defined as:

$$E = \sum_{t=1}^{N} \sum_{j=1}^{m} (d_{jt} - y_{jt})^2 \tag{6}$$

and:

$$y_{jt} = \sum_{k=1}^{h_c} w_{kj} R\left(\left\| \sum_{i=1}^{n} x_{it} - c_k \right\|\right) \tag{7}$$

where R( ) is a RBF function.

A RBF function is any function R that satisfies $R(x) = R(\|x\|)$. It is a real-valued function that depends on the distance from the origin as $R(x) = R(\|x\|)$ or from other point or center, c, as $R(x - c) = R(\|x - c\|)$. The norm $\|.\|$ is usually Euclidean distance.

The linear solution of output layer weights in equation (7) guarantees the objective function (6) to get the global minima and greatly speed up the training process.

RBF networks integrate nonlinear RBF activation functions in hidden layer and linear combination of weights in output layer. In input data clustering, different techniques can be used, such as k-means clustering [60] and fuzzy c-means [61]. There are choices of RBF functions, such as Gaussian RBF, multiquadric RBF, polyharmonic spline, and thin plate spline. Some of the RBFs will bring more parameters to determine in training process, such as Gaussian RBF, $R(x - c) = \exp(-\beta\|x - c\|^2)$, which brings one more parameter, β, to determine.

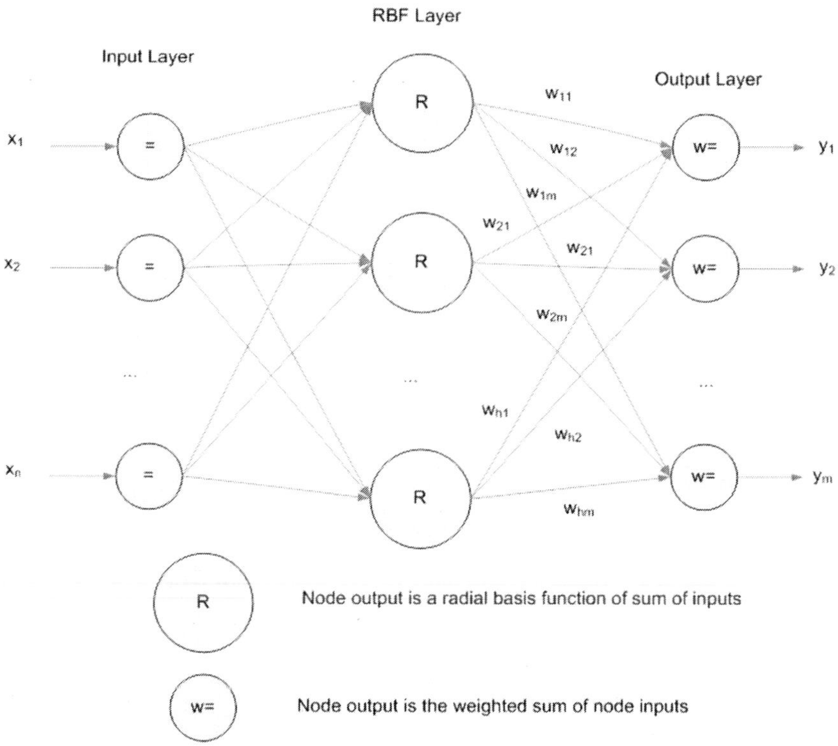

Figure 2. RBF network structure.

3.3. Recurrent and Feedback Networks

A recurrent neural network is a type of ANN where network nodes may have a directed connection. A typical example of this is that either the network's hidden node activation values or network output values are fed back into the network input nodes. These types of networks are good at characterizing process temporal dynamics. Examples of simple recurrent networks are Elman and Jordan networks [62,63].

MLP networks are good for mapping the static relationship of process input and output: x → y. They have been used for identification of dynamic systems [64,65]. However, they require a large number of input nodes which may result in long computing time and being affected by external noise [66]. Recurrent neural networks have been attracted in the field of dynamic system identification since they can avoid the problems MLP networks have [67,68,69].

In time series analysis, ANNs can establish AR (AutoRegressive) function map between process input and output: {y(t-i) | i = 1, 2, …, p} → y(t). In process modeling and control, ANNs can help establish ARX (AutoRegressive with eXogenous input) function map of process input and output: {y(t - i), x(t - j) | i = 1, 2, …, p; j = 1, 2, …, q} → y(t) where p and q are the orders of time lags of y and x, respectively. Figure 3 shows a network structure to map the dynamic process input/output relationship with the ARX function map. Internally, this network is identical to a MLP. Externally, it feeds back the network output back to the network input with a lagged time shift. This network is named External Recurrent Neural Network (ERNN) [70]. Obviously, training of this network should be different from standard BP. Around the network, y(t - i) is a time-delayed output from the network itself. This network model is determined as a specific form of time-lag recurrent networks [4]. For the time-lag recurrent networks with only one recurrent connection, i.e. i = 1, a number of training algorithms have been proposed [71,72,73]. Among the proposed algorithms, the BP Through Time (BPTT) is able to be modified for multiple recurrent connections, i.e. i > 1 using ordered derivatives from Werbos's derivation of the BP algorithm [44]. When the external feedback signal y(t) replaces the input of the network, the change of weights will affect y(t + 1) and thus y(t + 1) all the way to y(N) where T is the length of the data. Thus, the gradient computation has to account for this chaining from t = 0 to t = N. The input layer at time t+1 can be considered as a part of the 3rd layer of the network at time t. When calculating the error of the output layer at time t, it is necessary to calculate the error for the input nodes at time t + 1 up to t + p, all of which are connected to the corresponding output node at time t. In the regular BP algorithm, the error term for the network nodes propagates the required information all the way from t = N back to the current time instant. This is what the BPTT signifies. Based on Werbos's work [44], a modified BPTT algorithm was derived for the ERNN [74] as described above.

This modified algorithm was used for training over the whole trajectory of the training data. The maximum number of prediction steps is equal to the total number of observations. In controller design, the prediction steps would be much smaller. In order to solve this problem, the network can be trained to do shorter prediction with the modified training algorithm [74]. The modified training algorithm is still BPTT, but at each time instant the error term is propagated from the specified prediction step t + L (L << N) back to the current time instant t. This work resulted in an ERNN-based multiple-step-ahead prediction for process controller design [70]:

$$\begin{aligned} & y(t+l \mid t) = f(y(t+l-1 \mid t), \ldots, y(t+l-p \mid t), x(t+l-1), \ldots, x(t+l-q), W) \\ & y(t+l \mid t) = \begin{cases} y(t+l) \text{ when } l > 0 \\ d(t+l) \text{ otherwise} \end{cases} \\ & \quad l = 1, 2, \ldots, L \end{aligned} \tag{8}$$

where $y(t+l \mid t)$ is the l-step-ahead predictor of y(t) at the time instant t, f( ) is the approximation of the function map, and W is the set of weights and bias terms in the network.

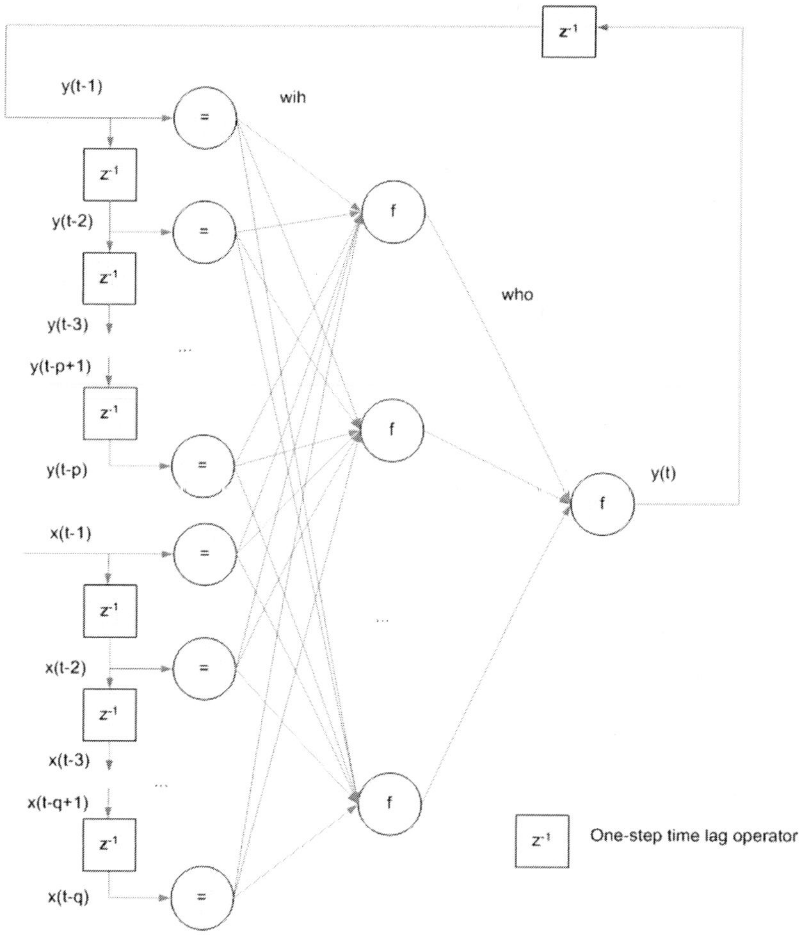

Figure 3. ERNN network structure.

3.4. Kohonen SOM Network and Unsupervised Training

Unlike the trainings of MLP, RBF and ERNN networks, Kohonen SOM networks are trained using an unsupervised algorithm, by which the input data are self-organized into clusters or classes, i.e. similar input data will activate the same network output node. The networks construct one-dimensional, two-dimensional, and even three-dimensional arrays of output nodes to form self-organized feature maps. Figure 4 is the structure of Kohonen SOM network with a two-dimensional array of output nodes used to form feature maps.

This is a two-layer network. The output nodes are orderly arranged in a two-dimensional grid. Each input is connected to all output nodes through a weight. The network training process has two stages. The first is to generate a coarse mapping, i.e. to create some form of topological ordering on the map of randomly oriented nodes. There may be large changes to the orientation of the nodes on the map, so the gain term or adaption rate, η, needs to be kept high to allow large weight modifications and settle into an approximate mapping quickly. Once a stable coarse map is found, the training process goes into the second stage. At this stage, the nodes within the localized regions of the map are fine-tuned to the network inputs. For this fine-tuning much smaller changes of weights must be made at each output node, so the adaptation rate, η, is reduced as training progresses. The weight updating in training is iteratively operated along with winning output node neighborhood:

$$\begin{aligned}&\Delta w_{ijk} = -\eta(w_{ijk} - x_k)\\&ij \in NE_{ij*}\\&(i = 1,2,\ldots,p;\ j = 1,2,\ldots,q;\ k = 1,2,\ldots,n)\end{aligned} \tag{9}$$

where p and q the first and second dimension of the feature map array, respectively, n is the number of network inputs, and ij* is the location of optimal output node. The optimal node is determined by:

$$\begin{aligned}&\underset{ij^*}{MIN}(D_{ij})\\&D_{ij} = \sum_{k=1}^{n}(x_k - w_{ijk})^2\end{aligned} \tag{10}$$

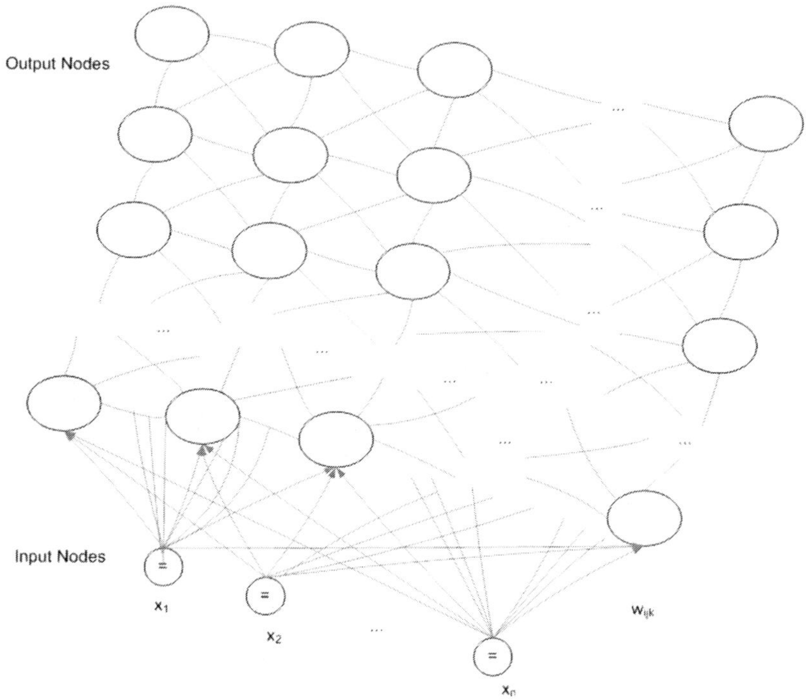

Figure 4. Kohonen SOM network with two-dimensional array of output nodes used to form feature maps.

The neighborhood function NE_{ij} is set large at very beginning of training, and slowly decreases in size with the progress of the training (Figure 5). The output nodes of the network are activated using a winner-take-all method:

$$y_{ij} = \begin{cases} 1 & \text{if } ij = ij^* \\ 0 & \text{otherwise} \end{cases} \tag{11}$$

That is that if (i,j) is the location of the optimal output node, the output node takes the value of 1; otherwise it takes 0. Accordingly, the weights connected to the neighborhood of the optimal node are updated.

4. ADVANCED DEVELOPMENT OF ANNS

With the progress of research and applications, ANN technology has been improved and advanced. Research has been done on modification of the standard BP algorithm to speed up algorithm convergence and avoid local minima in MLP network training. Much work has been devoted to preventing over-fitting in MLPs to improve the generalization ability of the networks. ANNs have been enhanced by fusing with other methods such as fuzzy logic and wavelet analysis. SVMs, emerging as a new soft computing technique, bring up a new wave of research and applications of ANNs.

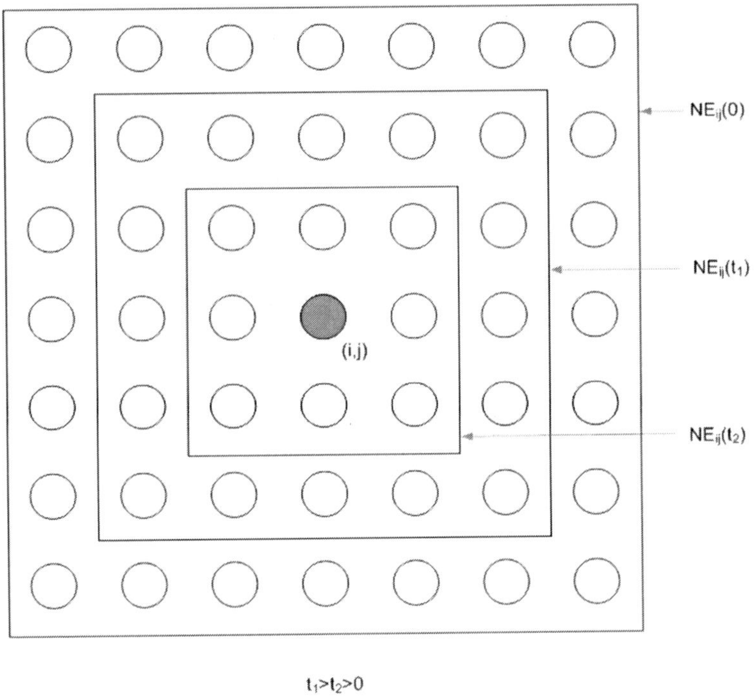

Figure 5. Topological neighborhoods with the progress of training as feature maps are formed.

4.1. Standard BP Enhancement

In standard BP training, it is critical to select a learning rate, η, to let the process converge to the true global minimum of the mean square error of the network outputs with a rather rapid speed. A BP training with a too small learning rate will have a noticeable slow progress. One with a too large learning rate will speed up significantly. However, this may result in oscillations around relatively poor solutions. Evidence shows that the use of a momentum term in the BP algorithm can be helpful in speeding the convergence and avoiding local minima (http://www. cs. bham. ac.uk/~pxt/NC/ ASSIGNMENT/MICHAEL/ momentum.html). The momentum term is defined as a fraction of the previous weight change $\alpha\Delta w^{-}$, and added to equation (2) from the standard BP algorithm [7]:

$$\Delta w = -\eta\nabla E(w) + \alpha\Delta w^{-} \tag{12}$$

where α is taken $0 \leq \alpha \leq 0.9$.

In this way, in BP training, the momentum is used to stabilize the weight change using a combination of the gradient decreasing term with a fraction of the previous weight change.

With the concept of momentum in BP training, the method has been further developed. Yu *et al.* [9] developed an adaptive momentum algorithm which can update the momentum coefficient automatically in every iteration step. The result of this adaptive process is equivalent to adding a momentum term to the standard BP algorithm. The momentum coefficient is updated automatically in every iteration. Numerical simulations show that the adaptive momentum algorithm can eliminate possible divergent oscillations during the initial training, and can also accelerate the learning process and result in a lower error when the final convergence is reached. Gerke and Hoyer [10] presented an analysis of fuzzy adaption of training parameters (learning rate and momentum) to accelerate BP learning in MFNs.

There are other studies on BP training acceleration and local minima. Huang *et al.* [75] investigated the efficiency of the training processes of MFNs using the gradient descent method in the standard BP algorithm and Levenberg-Marquardt method in backpropagation. It was found that in the case of low epoch training (below several thousand epochs) using the gradient descent algorithm, the Levenberg-Marquardt algorithm was less efficient, and in the case of high epoch training (above several thousand epochs) using the gradient descent algorithm, the Levenberg-Marquardt algorithm was more efficient. Jeenbekov and Sarybaeva [11] described properties of various parameters of sigmoid activation function mathematically with their influence on the speed of convergence of the BP training algorithm for MFNs. Wang *et al.* [12] proposed an improved BP algorithm intended to avoid the local minima problem caused by neuron saturation in the hidden layer. Each training pattern has its own activation functions of neurons in the hidden layer. When the network outputs have not received their desired signals, the activation functions are adapted so as to prevent neurons in the hidden layer from saturating. Bi *et al.* [13] proposed a modified error function with two terms. By adding one term to the conventional error function, the modified error function can harmonize the update of weights connected to the hidden layer and those connected to the output layer. Thus, it can avoid thelocal minima problem caused by update disharmony between weights connected to the hidden layer and the output layer. Simulations on some benchmark problems and a real classification task have been performed to test the validity of the modified error function.

Otair and Salameh [14] proposed an algorithm used for training that depends on a multilayer neural network with a very small learning rate, especially when using a large training set size. It can be applied in a generic manner for any network size that uses a BP algorithm through an optical time (seen time).

4.2. Network Generalization

Although MFNs can be universal approximators with a sufficiently large number of hidden nodes, excessive number of nodes in the hidden layer may endanger the network to become memorized, which may lead to overfit the input variables [76,77,78]. The overfitted networks may fit training data points well but may not be able to well interpolate and extrapolate testing data points. The testing data points can be between the training points or out of the range of the training points. Overfitting is a very important problem in MLPs, and much work has been devoted to preventing overfitting to improve network generalization with techniques such as early stopping, weight decay, and pruning. Figure 6 shows typical error curves of MLP training and testing. With the increase of the hidden nodes, the training error is large at the beginning and then keeps decreasing with a gradual decline of the curve slope. Similarly, the testing error is also larger at the beginning and then keeps decreasing until it reaches a point, h*. Starting from this point, with further increase of the hidden nodes, the testing error increases gradually while the training error continues to decrease. During training, if the minimal point on the testing curve can be found, the overfitting problem may be solved. The training and testing curves with training steps have a similar behavior as shown in Figure 6, which can be used to investigate overfitting, as can happen with too many training steps.

Based on the explanation of Figure 6, during MLP training, the training error and testing error are checked at every step, and the training process terminates as long as the minimal point of testing error is reached. This is so called early stopping method [20]. Further, a three-set approach was proposed [15]. This approach is to divide a data set into three nonoverlapping subsets: one for training to update network weights, one for testing to terminate training to prevent overfitting, and one for validation to evaluate the trained network. This approach gives a reasonable estimate of the network's generalization ability.

Adding a penalty term, E_p, to the mean square error of the network outputs in the standard BP algorithm, a new objective function is formulated for MLP training:

$$E_w = E + \lambda E_p \tag{13}$$

where λ is a pre-set constant for the penalty term. The training algorithm using this objective function causes network weights to converge to smaller absolute values than they would in the standard case. This method is called weight decay [16,75].The generalization ability of a network depends on the adjustment of the decay constant, λ. With weight decay training, the network can avoid oscillation in the outputs caused by large weights.

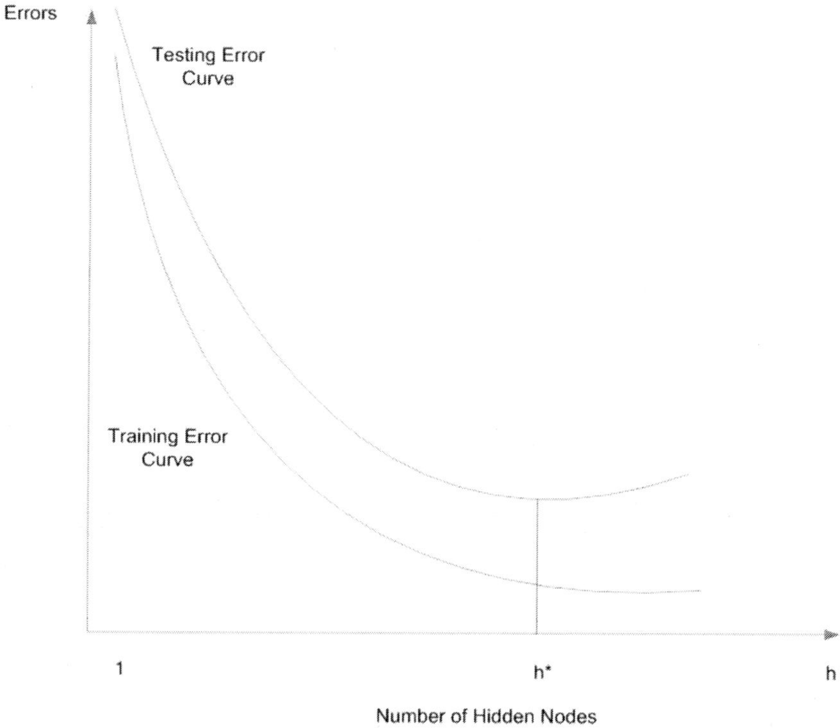

Figure 6. MLP error curves vs. number of hidden nodes.

Network pruning is another method used to prevent overfitting. This method begins with a fully connected network and the network is made smaller by iteratively eliminating least effective nodes in the hidden layer(s) or interconnections between nodes. The pruned network provides a structure that has a greater capacity for generalization [6,18,19].

4.3. Neuro-Fuzzy Systems

Fuzzy logic is a form of multi-valued logic derived from fuzzy set theory [79] to deal with reasoning that is approximate, rather than precise. In contrast to yes/no or 0/1 binary logic, Fuzzy logic provides a set of membership values inclusively between 0 and 1 to indicate the degree of truth. Fuzzy logic can be combined with ANNs to form a learning machine that is able to determine the parameters for a fuzzy system by ANN computing. A fuzzy system comprises of linguistic rules such as IF THEN from prior knowledge. Input and output variables of the system are described linguistically. If the knowledge is incomplete, wrong, or contradictory, the fuzzy system needs to be tuned. Fuzzy system is incapable of learning; it is a mechanism of straight interpretation and implementation. ANNs are capable of learning with a sufficient amount of observed examples for a problem. These observations are used to train the network model. The result can provide data for parameterization of fuzzy rules.

The fuzzy-neural systems have been studied and applied. Takagi and Hayashi [22] proposed a neural-fuzzy reasoning system that is capable of automatic determination of inference rules and adjustment according to the time-variant reasoning environment with the use of NN in fuzzy reasoning. Horikawa *et al.* [23] presented a fuzzy modeling method using fuzzy neural networks with the BP algorithm. The method can identify the fuzzy model of a nonlinear system automatically. Nie and Linkens [24] approximated reasoning through a BP Neural Network with the aid of fuzzy set theory. An example of multivariable fuzzy control of blood pressure was examined for the underlying principles. Simpson and Jahns [25] introduced the fuzzy min-max function approximation neural network. The function approximation network is realized by modifying the developed fuzzy min-max clustering network [80] to include an output layer that sums and thresholds the hidden layer membership functions. Jang [81] introduced the framework of ANFIS (Adaptive-Network-based Fuzzy Inference System). Jang and Sun [27] reviewed fundamental and advanced developments in neuro-fuzzy systems for modeling and control. They introduced design methods for ANFIS in modeling and control applications.

A fuzzy-neural system can be represented as a MFN. The first layer is for input variables. The middle or hidden layer(s) are for encoding fuzzy rules and fuzzy inference. The last layer is for output variables. The fuzzy inferences are converted as network connection weights. Figure 7 shows the structure of a fuzzy-neural system based on the idea of ANFIS [81].

For n input variables $\{x_i \mid i = 1, 2, \ldots, n\}$, suppose that there are m fuzzy if-then rules:

Rule 1: If x_1 is M_{11}, x_2 is M_{21}, …, and x_n is M_{n1}, then $f_1=a_{11}x_1+a_{12}x_2+\ldots+a_{1n}x_n+r_1$
Rule 2: If x_1 is M_{12}, x_2 is M_{22}, …, and x_n is M_{n2}, then $f_2=a_{21}x_1+a_{22}x_2+\ldots+a_{2n}x_n+r_2$ (14)
… … …
Rule m: If x_1 is M_{1m}, x_2 is M_{2m}, …, and x_n is M_{nm}, then $f_m=a_{m1}x_1+a_{m2}x_2+\ldots+a_{mn}x_n+r_m$

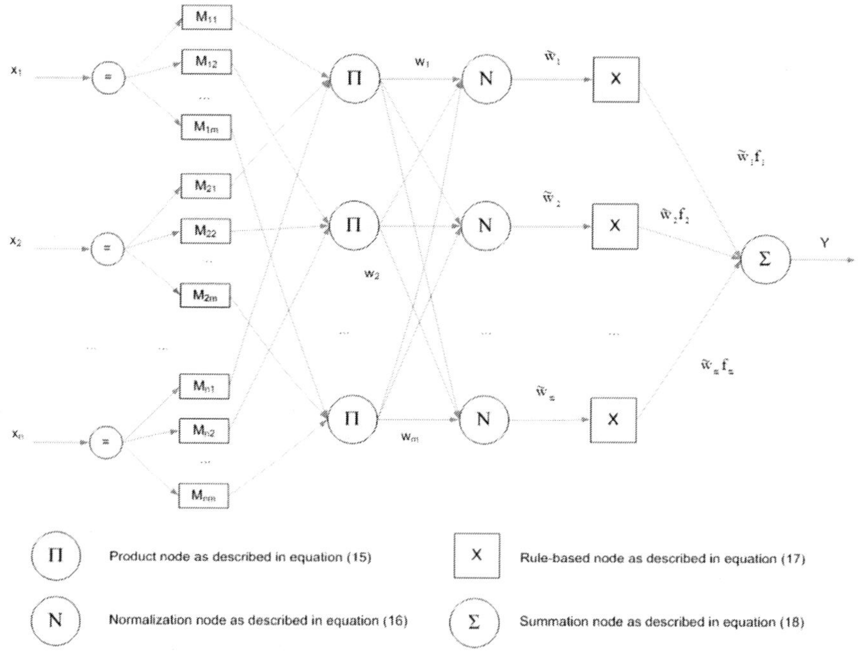

Figure 7. Structure of a fuzzy-neural system.

where M_{ij} is a linguistic label (small, medium, and large) of a membership function, $\mu_{Mij}(x)$, of x_i with the associated node j. The membership function can be in different forms with different parameter sets, such as the generalized bell-shaped function,

$$\mu_{M_{ij}}(x) = \frac{1}{1+[(\frac{x-c_{ij}}{a_{ij}})^2]^{b_{ij}}}$$

or the Gaussian function, $\mu_{M_{ij}}(x) = \exp[-(\frac{x-b_{ij}}{a_{ij}})^2]$. In the layer, each node labeled Π multiplies the incoming membership values and sends the product out to the next layer as connection weights:

$$w_j = \mu_{M_{1j}}(x) \times \mu_{M_{2j}}(x) \times \ldots \times \mu_{M_{nj}}(x) \quad (j = 1, 2, \ldots, m) \tag{15}$$

In the next layer labeled N, the connection weights are normalized:

$$\tilde{w}_j = \frac{w_j}{\sum_{i=1}^{m} w_i} \quad (j = 1, 2, \ldots, m) \tag{16}$$

In the next layer the outputs are:

$$\begin{aligned} y_j &= \tilde{w}_j f_j \\ &= \tilde{w}_j (a_{j1} x_1 + a_{j2} x_2 + \ldots + a_{jn} x_n + r_j) \end{aligned} \quad (j = 1, 2, \ldots, m) \tag{17}$$

In the last node, the node generates the overall output of the network by summing up from all incoming connections:

$$\begin{aligned} Y &= \sum_{i=1}^{m} y_i \\ &= \sum_{i=1}^{m} \tilde{w}_i f_i \\ &= \frac{\sum_{i=1}^{m} w_i f_i}{\sum_{i=1}^{m} w_i} \end{aligned} \tag{18}$$

The parameters included in the fuzzy and neural network equations above can be updated using the method of gradient descent in the standard BP algorithm. Other methods were suggested also to update the parameters [81].

4.4. Wavelet-Based Neural Networks

As discussed above, the standard BP algorithm may produce problems in training, for example, converge to local optima. Besides, the determination of the network structure in BP training, such as the number of hidden nodes, is still subjective. With the intelligent integration of wavelet transform, ANNs can be able to approximate data with non-linear and non-stationary properties and to conduct self-learning adaptive structure identification and parameter estimation in modeling process.

Wavelets are "small" waves that should integrate to zero around the x-axis. They localize functions well and a "mother" wavelet can be translated and dilated into a series of wavelet basis functions. The basic idea of wavelets can be traced back to very early in this century. However, the development of the construction of compactly supported orthonormal wavelets [82] and the wavelet-based multiresolution analysis [83] has resulted in extensive research and applications of wavelets in recent years. In comparison to the more widely known Fourier transform, wavelet basis functions are local both in frequency and time while Fourier basis functions are local only in frequency. Wavelet analysis has been widely used in a variety of fields, especially in signal and image processing.

Wavelet analysis is becoming a common tool for analyzing localized variations of power within a time series [84]. By decomposing a time series into time-frequency space, one is able to determine both the dominant modes of variability and how those modes vary in time. Wavelet-based ANNs (WANNs) have been developed and used for function approximation [28,29,30]. WANNs have been further developed for nonlinear time series forecasting [31]. WANNs have a strong ability of function approximation for optimal time series pattern recognition and their modeling algorithms are different from conventional BP neural network training algorithms. Therefore, WANNs have the potential to overcome the problems that conventional ANNs had.

WANN is a MFN model with wavelet analysis. It replaces the sigmoid function in the MLP with nonlinear wavelet basis function, $\varphi(\frac{t-b}{a})$, where a and b are the scale factor and translation factor of the wavelet function, respectively. In this model, nonlinear time series is expressed as a linear combination of selected nonlinear wavelet basis so that a finite number of wavelet series items are used to approximate time series function, i.e.

$$y(t) = \sum_{i=1}^{h} w_i \varphi(\frac{t-b_i}{a_i}) \tag{19}$$

Figure 8 shows the structure of the WANN. This is a single hidden-layered feedforward network. It only has one input node and one output node. The objective of the modeling task is to determine the network parameters, w_i, a_i, b_i and h, to achieve an optimal fitting between the desired output series d(t) and the calculated output series y(t):

$$J_h = \frac{1}{2}\sum_{t=1}^{N}[d(t) - y(t)]^2 \tag{20}$$

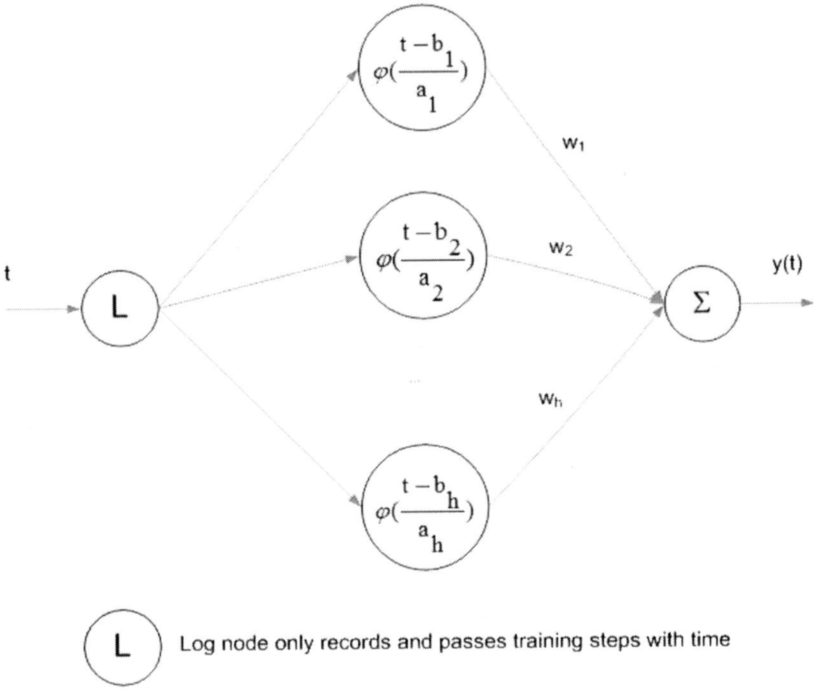

Figure 8. Structure of WANN.

Unlike in the standard BP algorithm, in WANN training determination of h is a step-wise process. Preset a tolerance ε for nonlinear time series fitting error. Starting h from 1 and calculate J_1. If $J_1 < \varepsilon$, the optimal h, h* = 1; otherwise, set h = 2 and calculate J_2. If $J_2 < \varepsilon$, h* = 2; otherwise, keep on going until h* is found with $J_{h^*} < \varepsilon$.

Using Newton's method it can be derived that the parameter updating equations can be as follows in training:

$$\begin{aligned}
\Delta w_i &= \sum_{t=1}^{N}[d(t) - y(t)]\varphi(\frac{t - b_i}{a_i}) \\
\Delta a_i &= \sum_{t=1}^{N}[d(t) - y(t)]\frac{w_i}{a_i^2}(t - b_i)\varphi'(\frac{t - b_i}{a_i}) \\
\Delta b_i &= -\sum_{t=1}^{N}[d(t) - y(t)]\frac{w_i}{a_i}\varphi'(\frac{t - b_i}{a_i})
\end{aligned} \tag{21}$$

The forms of $\varphi(t)$ and $\varphi'(t)$ depend on the selected wavelet function. For example, for Morlet mother wavelet, $\varphi(t) = \cos(1.75t)\exp(-t^2/2)$. Then its first derivative is $\varphi'(t) = -1.75\sin(1.75t)\exp(-t^2/2)$ $-tcons(1.75t)\exp(-t^2/2)$.

4.5. SVMs

SVMs were proposed by Vapnik [85]. Based on the description of Burges [32] and Cristianini and Taylor [33], SVMs are a set of supervised generalized linear classifiers that have often been found to provide higher classification accuracies than widely used MLP classifiers. The growing interest in SVMs is shown by their successful application in many areas such as automobile [86], time series analysis [87,88], medicine [89], pattern recognition [90], geophysics [91,92], computational biology [93], and control engineering [94]. The interest resulted from SVMs' intrinsic effectiveness with respect to traditional classifiers. This effectiveness results in high classification accuracies and very good generalization capabilities. In architecture design, SVMs requires limited effort to configure few control parameters compared with traditional classifiers such as MLPs. Also, based on the universal approximation capability of their standard MFN counterparts, the approximation capability of SVMs was investigated [51]. It was shown that an SVM with polynomial kernel of degree *n*-1, which is trained on a training set of size *n* can approximate the n training points up to any accuracy.

A SVM performs classification through mapping input vectors into a higher-dimensional space and constructing a hyperplane that optimally separates the data in the higher-dimensional space. A SVM model using sigmoid kernel function is equivalent to a two-layer perceptron neural network. However, the operation of the SVM is different. In training, a SVM implements a simple linear mapping or linear classifier together with a prior fixed nonlinear mapping in order to make data separable. This implementation allows training to not suffer from the problem of local minima, and to focus on function optimization with respect to its generalization ability. These advantages often lead to better results for SVMs compared to MFNs.

The goal of SVM modeling is to find the optimal hyperplane that separates clusters of vectors, a set of features, such that data points with one class of the variable are on one side of the plane and ones with the other class are on the other side of the plane. The vectors near the hyperplane are the support vectors. Let's look at the 2-dimensional case as an example. Assume that for classification, the data are from a categorical variable with two classes, and there are two prediction variables, X and Y, with continuous values. The data points using the value of X on the X axis and Y on the Y axis may be plotted as Figure 9.

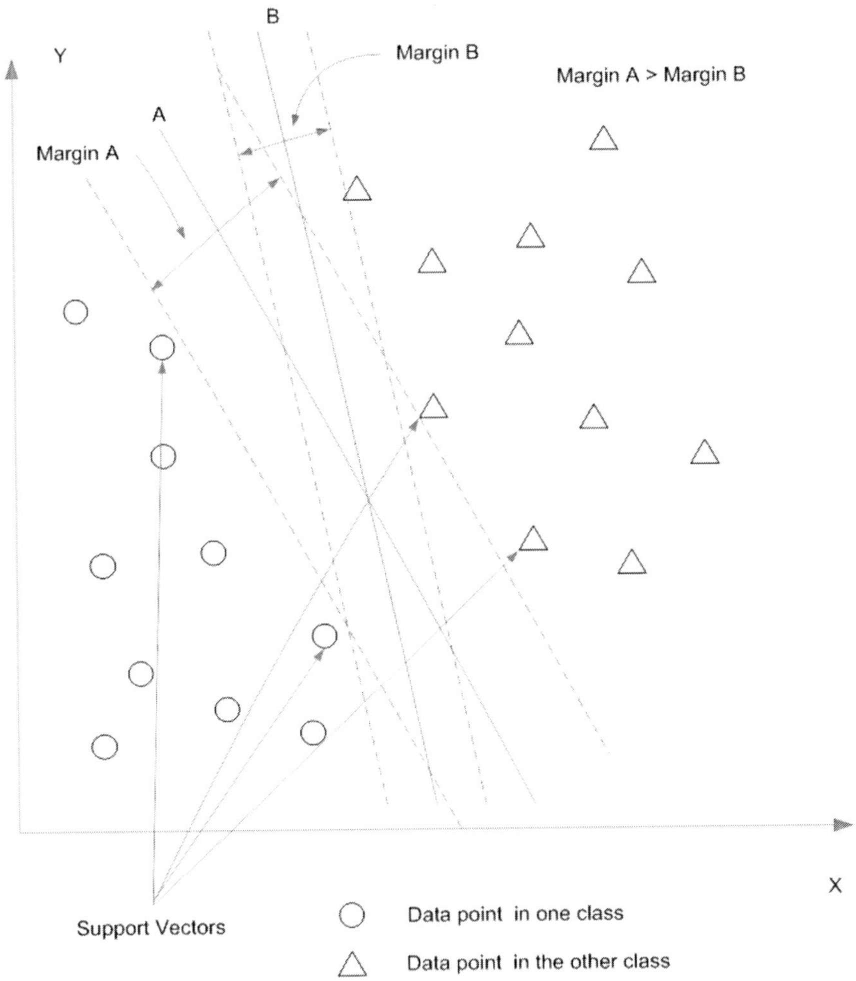

Figure 9. Margin between support vectors in two-dimension.

In the plot it can be seen what classes each (X,Y) point belongs to. In this example, the data points with one class are in the lower left corner of the plot and the data points with the other classes are in the upper right corner. The SVM analysis is find a one-dimensional hyperplane, i.e. a line, to separate the data points based on their target classes. Obviously, there are an infinite number of possible such lines. Two of them are shown in the plot. The task of SVM now is to determine which line is optimal. For this optimization two parallel (dashed) lines are constructed, one on each side of the separating line. The two lines are pushed up against the two sets of data points belonged to two classes. They label the distance between the separating line and the closest vectors to the line. The two lines are used to calculate the margin, which is defined as the distance between them. The SVM analysis is to find the line (hyperplane in general) that makes the margin between the support vectors maximized. In Figure 9, the line A is superior to the line B. In this sense, SVM classifiers are also known as maximum margin classifiers.

Mathematically, suppose having a training data pairs $\{(x_i, c_i) \mid i= 1, 2, \ldots, N\}$ with $x_i \in R^n$ and $c_i \in \{1, -1\}$, indicate what class the data point x_i belongs to. In a simple case, a hyperplane can be defined as $F(x) = w^t x + w_0$ where w is the adaptable weight vector and w_0 is the bias term, and T is the vector transpose operator. In training, the data points x_i are projected into the higher-dimensional space, and the classification is conducted through sign[f(x)] where f(x) is the estimate of F(x). The SVM algorithm searches for a hyperplane f(x) that maximizes the margin between the two sets of data points in class 1 and -1.

Further, the margin maximum problem can be solved in any high-dimensional space by introducing a kernel function [95,96]. With a nonlinear kernel function, the low-dimensional input space is nonlinearly transformed into a high-dimensional feature space such that the probability that the feature space is linear separable becomes higher. Theoretically, the kernel function is able to implicitly map the input space into an arbitrary high-dimensional feature space that can be linearly separable even if the input space may not be linearly separable. Some commonly used kernel functions are polynomial, Gaussian, Sigmoid, and RBF.

In general, SVMs's training is to solve the following optimization problem:

$$\underset{w, w_0, \xi}{\mathrm{MIN}}\left(\frac{1}{2} w^T w + \lambda \sum_{i=1}^{N} \xi_i\right)$$
$$\text{s.t. } c_i[w^T \phi(x_i) + w_0] \geq 1 - \xi_i \qquad (22)$$

where λ is a user defined positive constant, a penalty parameter of the error term, ξ_i (≥ 0) is the slack variable to measure the degree of misclassification of x_i, and $\varphi(x)$ is the function to map data points x_i from the input space to a higher-dimensional feature space.

With equation (22), data points xi are mapped into a higher-dimensional space by the function of $\varphi(x)$. Then a linearly separable hyperplane ($w^T \phi(x_i) + w_0$) is found with the maximal margin in the higher-dimensional space. Further, the mapping function $\varphi(x)$ is determined with a specified kernel function $K(x_i,x_j)$:

$$K(x_i,x_j) = \phi(x_i)\,\phi(x_j) \tag{23}$$

If RBF is chosen as the kernel function, then $K(x_i, x_j) = \exp(-\gamma\|x_i - x_j\|^2)$ $(\gamma > 0)$. The RBF kernel is good at handling nonlinear classification for SVMs.

5. LIMITATIONS OF ANNS

As described above, ANNs are powerful computing techniques, which are designed to mimic human learning processes by establishing linkages between process input and output data. These techniques have been widely applied with advanced development with their unique advantages, such as no underlying assumption about the distribution of data, arbitrary decision boundary capabilities, universal approximation capabilities, easy adaptation to different types and structures of data, ability to fuzzy output values to enhance classification, and good generalization capabilities. However, ANNs have some disadvantages in common, which need to be considered in practical application:

- Black box

ANNs are black box in nature. Therefore, if the problem is to find the output response to the input such as system identification [96], ANNs can be a good fit. However, if the problem is to specifically identify causal relationship between input and output, ANNs have only limited ability to do it compared with conventional statistical methods.

- Long computing time

ANN training needs to iteratively determine network structure and update connection weights. This is a time-consuming process. With a typical personal computer or work station, the BP algorithm will take a lot of memory and may take hours, days and even longer before the network converges to the optimal point with minimum mean square error. Conventional statistical regression with the same set of data, on the contrary, may generate results in seconds using the same computer.

- Overfitting

With too much training time, too many hidden nodes, or too large training data set, the network will overfit the data and have a poor generalization, i.e. high accuracy for training data set but poor interpolation of testing data. This is an important issue being investigated in ANN research and applications as described above.

Also, SVMs were originally developed to solve binary classification problems. How to effectively extend it for multiclass classification is still an ongoing research issue. Typically, a multiple class classifier can be constructed by combining several binary classifiers. Further, all classes can be considered at once. Hsu and Lin [98] gave decomposition implementations for two such "all-together" methods. They compared the performance with three methods based on binary classifications: one-against-all, one-against-one, and Directed Acyclic Graph SVM (DAGSVM). The experiments indicated that the one-against-one and DAG methods are more suitable for practical use than the other methods.

6. ANN APPLICATIONS IN AGRICULTURAL AND BIOLOGICAL ENGINEERING

ANNs have the largest body of applications in agricultural and biological engineering compared to other soft computing techniques such as fuzzy logic and genetic algorithms. In summary of related 348 papers and reports, ANNs have been applied in solving problems in food quality and safety (35.3%), crop (22.7%), soil and water (14.4%), precision agriculture (6.6%), animal management (5.2%), post harvest (2.6%), food processing (2.3%), greenhouse control (2%), agricultural vehicle control (1.2%), agricultural machinery (1.2%), agricultural pollution (1.2%), agricultural biology (1.2%), ecology and natural resources (1.4%), agricultural robotics (0.3%), chemical application (0.3%), and others (2.3%) such as bioenergy and agricultural facilities (Figure 10). These ANN applications have been created mainly through classification (45.1%), modeling and prediction (44%), control (4%), and simulation (2.6%), parameter estimation (2%), detection (1.2%), data clustering (0.6%), optimization (0.3%) and data fusion (0.3%) as well (Figure 11).

Early applications include modeling sensory color quality of tomato and peach [99], investigation of combined effects of CO_2 and sucrose on the growth of alfalfa cuttings using a Kalman filter neural network [100], classification of apple surface features [101], corn kernel breakage classification [102], and machine vision inspection of potatoes [103].

Recently, the BP ANN method was used in regression for modeling the correlation between crop yield and 10 yield components of chickpea [104]. A BP ANN was employed and trained with extracted features from the data collected by electronic nose to identify the wheat age [105]. BP neural network and generalized regression neural network techniques were developed and compared to model source gas and PM10 concentration and emission rate (GPCER) generated and emitted from swine deep-pit finishing buildings as affected by time of day, season, ventilation rates, animal growth cycles, in-house manure storage levels, and weather conditions [106]. MLP models were developed and used to predict or classify water stress in samples of Sunagoke moss using the texture features extracted from color co-occurrence matrix (CCM) and gray-level co-occurrence matrix (GLCM) methods [107]. A BP ANN was developed for predicting fungal infections using input from principal components extracted from sample visible and near-infrared (NIR) reflectance spectroscopy for early detection of Botrytis cinerea on eggplant leaves [108].

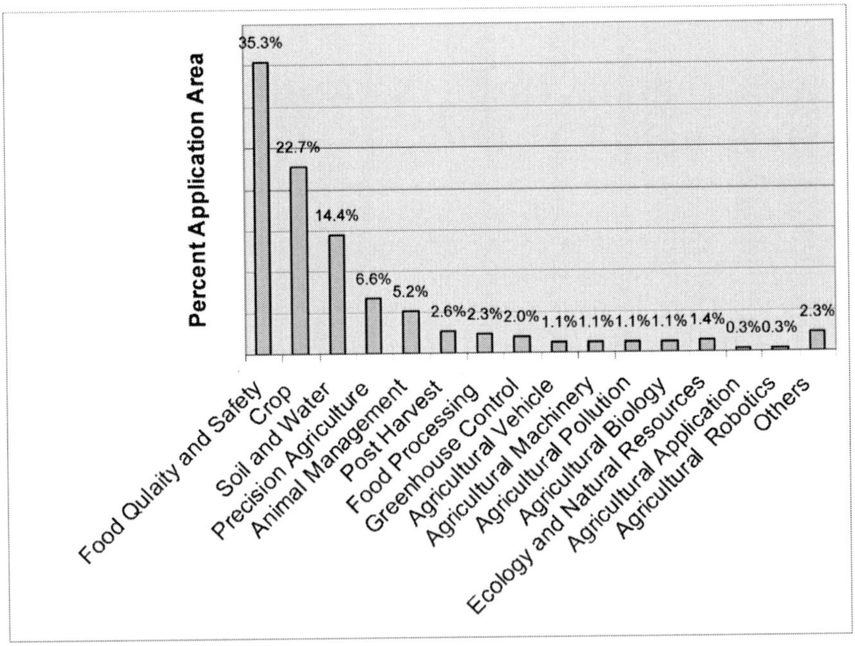

Figure 10. Application area distribution of ANNs in agricultural and biological engineering.

Similar to applications in general engineering fields, most of the work in agricultural and biological engineering has been accomplished by multilayer feedforward ANN trained by the famous BP algorithm, which was inspired by the work of Rumelhart *et al.* in 1986. Among the 348 collected papers and reports, besides 83 of them in which the ANNs' structure and training algorithm were not explicitly stated, 210 of them (60%) were based on MFNs trained by the BP algorithm. Twelve of them (3.5%) were based on PNN (Probabilistic Neural Network). Eleven of them (3%) were based on MLP using different training algorithms such as Levenberg-Marquardt optimization procedure, Broyden-Fletcher-Goldfarb-Shanno (BFGS) optimization procedure, and genetic algorithm. Seven of them (2%) were based on Kohonen SOM. Six of them (2%) were based upon other unsupervised training algorithms such as fuzzy art, ART2 and Auto-Associative network. Four of them (1%) were based on RBF networks. Three of them (0.8%) were based on LVQ (Learning Vector Quantization). Two of them (0.6%) were based on GRNN (Generalized Regression Neural Network). Ten of them (3%) were based on other network structures such as Counter-Propagation (CP) and Adaptive Logic Network (ALN). Figure 12 shows the ANN method distribution of applications in agricultural and biological engineering.

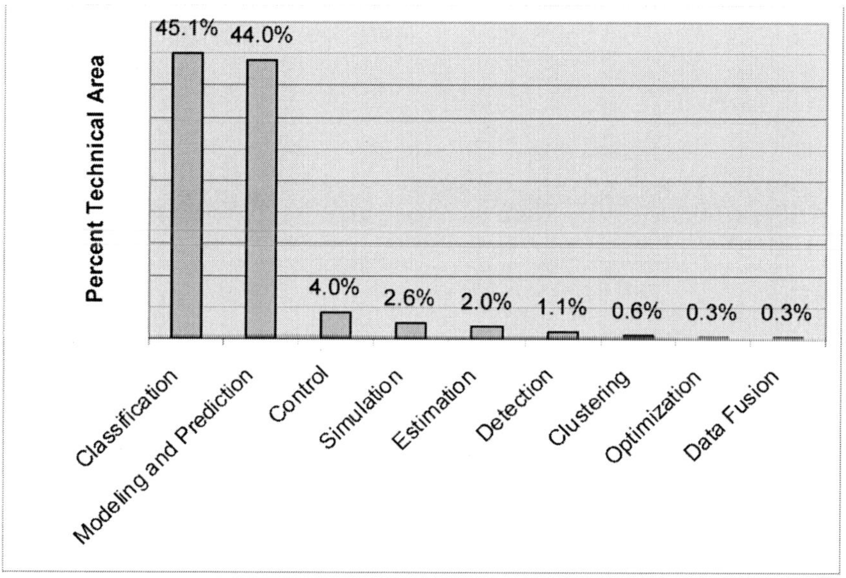

Figure 11. Technical area distribution of ANNs in agricultural and biological engineering.

Fuzzy-neural systems have been developed and used in agricultural and biological engineering. Table 1 listed papers and reports on fuzzy-neural methods and systems in applications in agricultural and biological engineering. Linko *et al.* [109] applied neural network modeling for fuzzy food extrusion control. Kim and Cho [110] used the BP algorithm in training ANN models for prediction of volume, browning and bread temperatures. The outputs of the ANN models were used for fuzzy control simulation of the oven used in the baking process. Morimoto *et al.* [111] proposed and then applied a new fuzzy control technique, which efficiently selects optimal membership functions and control rules by using neural networks and genetic algorithms to the control of relative humidity in a fruit-storage house. The response of relative humidity, as affected by ventilation, was first identified using neural networks, and then optimal membership functions and control rules were sought through simulation of the identified model using GAs. Odhiambo *et al.* [112] developed a strategy consisting of fusing the fuzzy logic and ANN on a conceptual and structural basis for an easy and efficient means of tuning fuzzy ET (Evapotranspiration) models. The neural component provided supervised learning capabilities for optimizing the membership functions and extracting fuzzy rules from a set of input–output examples selected to cover the data hyperspace of the sites evaluated.

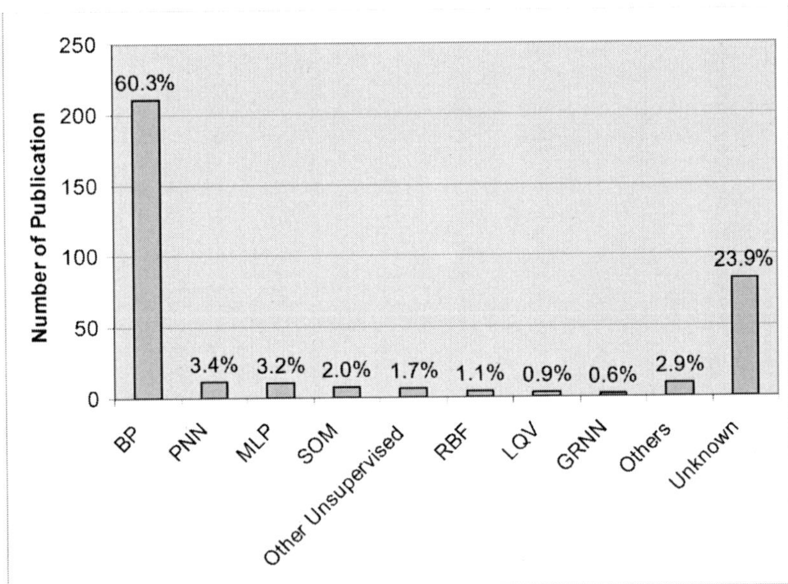

Figure 12. ANN method distribution of applications in agricultural and biological engineering.

Table 1. Papers and reports on fusion of fuzzy logic and ANNs in agricultural and biological engineering.

Year	Author	Fusion Type	Application Area
1992	Linko *et al.* [109]	ANN modeling for fuzzy control	Extrusion control
1997	Kim and Cho [110]	ANN modeling plus fuzzy control simulation	Bread baking process control
1997	Morimoto *et al.* [111]	ANN modeling plus GA parameter optimization for fuzzy control	Fruit storage control
2001	Odhiambo *et al.* [112]	Conceptual and structural fusion of fuzzy logic and ANN	ET model optimization
2003	Andriyas *et al.* [118]	FCM clustering for RBF training	Prediction of the performance of vegetative filter strips
2003	Chtioui *et al.* [113]	SOM with FCM clustering	Color image segmentation of edible beans
2003	Lee *et al.* [119]	ANFIS modeling	Prediction of multiple soil properties
2003	Neto *et al.* [120]	ANFIS classification	Adaptive image segmentation for weed detection
2004	Odhiambo *et al.* [115]	Fuzzy-Neural Netwok unsupervised classification	Classification of soils
2004	Meyer *et al.* [114]	ANFIS classification	Classification of uniform plant, soil, and residue color images
2004	Goel *et al.* [121]	Fuzzy c-means clustering for RBF training	Prediction of sediment and phosphorous movement through vegetative filter strips
2006	Hancock and Zhang [115]	ANFIS classification	Hydraulic vane pump health classification
2007	Xiang and Tian [117]	ANN modeling plus ANFIS training of fuzzy logic controller	Outdoor automatic camera parameter control

Chtioui *et al.* [113] developed a novel segmentation approach that partitions color images into two uniform regions is described. This unsupervised procedure is based on a SOM neural network and fuzzy c–means clustering (FCM). The SOM allows the mapping of a color image related to edible beans into a consistent two–dimensional table through a non–linear projection. Fuzzy clustering is then applied to the Kohonen map to determine the two cluster centers. Meyer *et al.* [114] conducted a digital camera operation study for classifying uniform images of grass, bare soil, corn stalks residue, wheat straw residue, and a barium sulfate reference panel, based on color. The classifications were conducted with fuzzy inference systems, built with subtractive clustering, an ANFIS. Odhiambo *et al.* [115] applied a fuzzy-neural network (F-NN) classifier for unsupervised clustering and classification of soil profiles using ground-penetrating radar (GPR) imagery. Hancock and Zhang [116] developed an on-line hydraulic vane pump fault detection system. This fault detection system decomposed vertical pump vibration signals using wavelet packet analysis. Packets containing signal features distinguishing normal and failed pump operation were entered into an ANFIS for pump health classification. Xiang and Tian [117] developed an artificially intelligent controller based on an ANN and an

ANFIS for implementing controller to automatically adjust multispectral camera parameter to compensate for changing natural lighting conditions and to acquire white-balanced images.

The interest of SVMs in agricultural and biological engineering is steadily growing (Figure 13). The figure indicates that although the number of peer reviewed publications on this topic did not increase steadily over the past few years, a trend exists for total number of papers and report to increase in the past few years and to be able to project this trend in the next few years. The successful implementations of SVMs in agricultural and biological engineering are listed in Table 2, which includes classification of intact and cracked eggs [122], classification of forest data cover types [123], classification of meat with small data set [124], black walnut shell and meat classification using hyperspectral fluorescence imaging [125], classification to differentiate individual fungal infected and healthy wheat kernels [126], classification for weed and nitrogen stress detection in corn [127], discrimination for screening of compound feeds using NIR hyperspectral data [128], identification of tea varieties by computer vision [129], discrimination of wheat classes with NIR spectroscopy [130], detection of underdeveloped hazenuts from fully developed nuts by impact acoustics [131], classification of electronic nose data [132], classification of modified starches [133], classification of milk with an electronic nose [134], and recognition of plant disease [135]. In these applications, Zhang *et al.* [126] fused SVM with kernel of RBF neural network. Pardo and Sberveglieri [132] implemented SVM with RBF kernel. Brudzewski *et al.* [134] applied SVM neural network with one hidden layer of non-linear neurons, one-output linear neuron and specialized learning procedure leading to the global minimum of the error function and excellent generalization ability of the trained network.

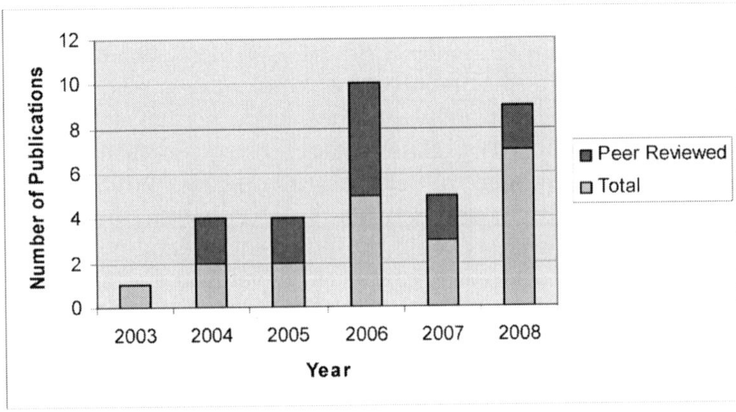

Figure 13. Publications on applications of SVMs in agricultural and biological engineering.

Table 2. Papers and reports on applications of SVMs in agricultural and biological engineering.

Year	Author	Application Method	Application Area
2003	Fletcher and Kong [136]	SVM classification	Classifying feature vectors and decide whether each pixel in hyperspectral fluorescence images of poultry carcasses falls in normal or skin tumor categories
2004	Brudzewski *et al.* [134]	SVM neural network classification	Classification of milk by an electronic nose
2004	Tian *et al.* [135]	SVM classification	Classification for recognition of plant disease
2005	Pardo and Sberveglieri [132]	SVM with RBF kernel of RBF	Classification of electronic nose data
2005	Pierna *et al.* [133]	SVM classification	Classification of modified starches by Fourier transform infrared spectroscopy
2006	Chen *et al.* [129]	SVM classification	Identification of tea varieties by computer vision
2006	Karimi *et al.* [127]	SVM classification	Classification for weed and nitrogen stress detection in corn
2006	Onaran *et al.* [131]	SVM classification	Detection of underdeveloped hazenuts from fully developed nuts by impact acoustics
2006	Pierna *et al.* [128]	SVM classification	Discrimination of screening of compound feeds using NIR hyperspectral data
2006	Wang and Paliwal [130]	Least-Squares SVM classification	Discrimination of wheat classes with NIR spectroscopy
2007	Jiang *et al.* [125]	Gaussian kernel based SVM classification	Black walnut shell and meat classification using hyperspectral fluorescence imaging
2007	Oommen *et al.* [137]	SVM modeling and prediction	Simulation of daily, weekly, and monthly runoff and sediment yield fron a watershed
2007	Oommen *et al.* [137]	SVM modeling and prediction	Simulation of daily, weekly, and monthly runoff and sediment yield fron a watershed
2007	Zhang *et al.* [126]	Multi-class SVM with kernel of RBF neural network	Classification to differentiate individual fungal infected and healthy wheat kernels.
2008	Fu *et al.* [138]	Least Squares SVM modeling and prediction	Quantification of vitamin C content in kiwifruit using NIR spectroscopy
2008	Khot *et al.* [124]	SVM classification	Classification of meat with small data set
2008	Kovacs *et al.* [139]	SVM modeling and prediction	Prediction of different concentration classes of instant coffee with electronic tongue measurements
2008	Peng and Wang [140]	Least-Squares SVM modeling and prediction	Prediction of pork meat total viable bacteria count with hyperspectral imaging
2008	Sun *et al.* [106]	SVM modeling and prediction	On-line assessing internal quality of pears using visible/NIR transmission
2008	Trebar and Steele [123]	SVM classification	Classification of forest data cover types
2008	Yu *et al.* [9]	Least-Squares SVM modeling and prediction	Rice wine composition prediction by visible/NIR spectroscopy
2009	Deng *et al.* [122]	SVM classification	classification of intact and cracked eggs

7. CONCLUSION AND FUTURE DIRECTIONS

As described above, ANNs have their advantages and have been used as a powerful tool in solving problems in scientific research and engineering applications. ANNs have their own limitations to restrict them as a substitute of traditional methods such as statistical regression, pattern recognition, and time series analysis. In the next decade with advanced development of computer power, ANNs will continue to develop new applications in various fields, including agricultural and biological engineering. As a powerful alternative to conventional methods, ANNs will be studied more to develop approaches to overcoming problems of ANNs in general or in a specific research area. For effective research and development, the guidelines may be generated for predetermining best ANN structure and training algorithms for a given problem. For process statistical modeling, additional research topics may be to establish generic procedures to include significant variables to and exclude non-significant ones from ANN models and to add confidence limits on the output predictions and parameter estimations. These works will let ANNs inherit advantages from conventional statistics.

The fusion of ANNs with other advanced computing methods will continue. These methods will include but are not limited to fuzzy logic, genetic algorithms, decision tree, and wavelet analysis. The fusion will allow the system to inherit the advantages of both of the paradigms and avoid the drawbacks.

SVMs are an alternative learning method for polynomial, RBF and MLP classifiers in which the weights of the network are found by solving a quadratic programming problem with linear constraints, rather than by solving a non-convex, unconstrained minimization problem as in standard neural network training. With the advantages of SVMs over ANNs and the growing interests of SVMs, it can be expected that in the next decade SVMs will be more actively used in various fields, including agricultural and biological engineering, although the results were not already as expected [130].

In food science and engineering, soil and water relationship for crop management, and decision support for precision agriculture, more applications of ANNs and SVMs will be expected. These techniques will be applied standalone or fusion with other soft and hard techniques. Areas of study can involve classification and prediction of food quality and safety, classification for agricultural soil spatial distribution, water resource optimization for irrigation planning, detection and classification of crop stress and pests (weeds, insects and diseases) detection, analysis of remote sensing imagery, study of crop and yield, and field prescriptions for variable rate chemical application.

ACKNOWLEDGEMENTS

Special thanks to Steven J. Thomson, Lead Scientist of Application and Production Technology Research Unit in United States Department of Agriculture, Agricultural Research Service, for his careful review and support.

REFERENCES

1. Rumelhart, D.E.; Hinton, G.E.; Williams, R.J. Learning internal representations by error propagation. In Parallel Distributed Processing: Explorations in the Microstructures of Cognition; Rumelhart, D.E., McClelland, J.L., Eds.; MIT Press: Cambridge, MA, USA, 1986; Vol. I, pp. 318–362.

2. Rumelhart, D.E.; McClelland, J.L. Parallel Distributed Processing: Explorations in the Microstructures of Cognition; MIT Press: Cambridge, MA, USA, 1986.

3. Kavuri, S.N.; Venkatasubramanian, V. Solving the hidden node problem in networks with ellipsoidal unitsand related issues. In Proceedings of International Joint Conference on Neural Networks, Baltimore, MA, USA, June 7-11, 1992; Vol. I, pp. 775–780.

4. Su, H.; McAvoy, T.J.; Werbos, P.J. Long-term predictions of chemical processes using recurrent neural networks: a parallel training approach. Ind. Eng. Chem. Res. 1992, 31, 1338–1352.

5. Jou, I.C.; You, S.S.; Chang, L.W. Analysis of hidden nodes for multi-layer perceptron neural networks. Patt. Recog.1994, 27, 859–864.

6. Sietsma, J.; Dow, R.J.F. Neural net pruning: why and how. In Proceedings of IEEE Int. Conf. Neural Networks, San Diego, CA, USA, July 24-27, 1988; Vol. I, pp. 325–333.

7. Rumelhart, D.E.; Hinton, G.E.; Williams, R.J. Learning representations by back-propagating errors. Nature 1986, 323, 533–536.

8. McClelland, J.L.; Rumelhart, D.E. Explorations in parallel distributed processing: A handbook of models, programs, and exercises; MIT Press: Cambridge, MA, USA, 1988.

9. Yu, X.; Loh, N.K.; Miller, W.C. A new acceleration technique for the backpropagation algorithm. In Proceedings of IEEE International Conference on Neural Networks, San Diego, CA, USA, March 28 – April 1, 1993; Vol. III, pp. 1157–1161.

10. Gerke, M.; Hoyer, H. 1997. Fuzzy backpropagation training of neural networks. In Computational Intelligence Theory and Applications; Reusch, B., Ed.; Springer: Berlin, Germany, 1997; pp. 416–427.

11. Jeenbekov, A.A.; Sarybaeva, A.A. Conditions of convergence of back-propagation learning algorithm. In Proceedings of SPIE on Optoelectronic and Hybrid Optical/Digital Systems for Image and Signal Processing, Bellingham, WA, USA, 2000; Vol. 4148, pp. 12–18.

12. Wang, X.G.; Tang, Z.; Tamura, H.; Ishii, M.; Sun, W.D. An improved backpropagation algorithm to avoid the local minima problem. Neurocomputing 2004, 56, 455–460.

13. Bi, W.; Wang, X.; Tang, Z.; Tamura, H. Avoiding the local minima problem in backpropagation algorithm with modified error function. IEICE Trans. Fundam. Electron. Commun. Comput. Sci. 2005, E88-A, 3645–3653.

14. Otair, M.A.; Salameh, W.A. Speeding up back-propagation neural networks. In Proceedings of the 2005 Informing Science and IT Education Joint Conference, Flagstaff, AZ, USA, June 16-19, 2005; pp. 167–173.

15. Burke, L. Assessing a neural net. PC AI. 1993, 7, 20–24.

16. Finnoff, W.; Hergert, F.; Zimmermann, H.G. Improving model selection by nonconvergent methods. Neural Netw.1993, 6, 771–783.

17. Wang, J.H.; Jiang, J.H.; Yu, R.Q. Robust back propagation algorithm as a chemometric tool to prevent the overfitting to outliers. Chemom. Intell. Lab. Syst. 1996, 34, 109–115.

18. Kavzoglu, T.; Mather, P.M. Assessing artificial neural network pruning algorithms. In Proceedings of the 24th Annual Conference and Exhibition of the Remote Sensing Society, Cardiff, South Glamorgan, UK, September 9-11, 1998; pp. 603–609.

19. Kavzoglu, T.; Vieira, C.A.O. An analysis of artificial neural network pruning algorithms in relation to land cover classification accuracy. In Proceedings of the Remote Sensing Society Student Conference, Oxford, UK, April 23, 1998; pp. 53–58.

20. Caruana, R.; Lawrence, S.; Giles, C.L. Overfitting in neural networks: backpropagation, conjugate gradient, and early stopping. In Proceedings of Neural Information Processing Systems Conference, Denver, CO, USA, November 28-30, 2000; pp. 402–408.

21. Lawrence, S.; Giles, C.L. Overfitting and neural networks: conjugate gradient and backpropagation. In Proceedings of International Joint Conference on Neural Networks, Como, Italy, July 24-27, 2000; pp. 114–119.

22. Takagi, T.; Hayashi, I. NN–driven fuzzy reasoning. IJAR 1991, 5, 191–212.

23. Horikawa, S.; Furuhashi, T.; Uchikawa, Y. On fuzzy modelling using fuzzy neural networks with back propagation algorithm. IEEE Trans. Neural Netw. 1992, 3, 801–806.

24. Nie, J.; Linkens, D. Neural network–based approximate reasoning: Principles and implementation. Int. J. Contr. 1992,56, 399–413.

25. Simpson, P.K.; Jahns, G. Fuzzy min–max neural networks for function approximation. In Proceedings of IEEE International Conference on Neural Networks, San Francisco, CA, USA, March 28 – April 21, 1993; Vol. 3, pp. 1967–1972.

26. Mitra, S.; Pal, S.K. Logical operation based fuzzy MLP for classification and rule generation. Neural Netw. 1994, 7, 353–373.

27. Jang, R.J.S.; Sun, C.T. Neuro–fuzzy modelling and control. Proc. IEEE 1995, 83, 378–406.

28. Cristea, P.; Tuduce, R.; Cristea, A. Time series prediction with wavelet neural networks. In Proceedings of IEEE Neural Network Applications in Electrical Engineering, Belgrade, Yugoslavia, September 25-27, 2000; pp. 5–10.

29. Shashidhara, H.L.; Lohani, S.; Gadre, V.M. Function learning wavelet neural networks. In Proceedings of IEEE International Conference on Industrial Technology, Goa, India, January 19-22, 2000; Vol. II, pp. 335–340.

30. Ho, D.W.C.; Zhang, P.A.; Xu, J. Fuzzy wavelet networks for function learning. IEEE Trans. on Fuzzy Syst. 2001, 9, 200–211.

31. Zhou, B.; Shi, A.; Cai, F.; Zhang, Y. Wavelet neural networks for nonlinear time series analysis. Lecture Notes Comput. Sci. 2004, 3174, 430–435.

32. Burges, C.J.C. A tutorial on support vector machines for pattern recognition. Data Min. Knowl. Disc. 1998, 2, 121–167.

33. Cristianini, N.; Taylor, J.S. An Introduction to Support Vector Machines and Other Kernel-Based Learning Methods; Cambridge University Press: New York, NY, USA, 2000.

34. Whittaker, A. D.; Park, B.P.; McCauley, J.D.; Huang, Y. Ultrasonic signal classification for beef quality grading through neural networks. In Proceedings of Automated Agriculture for the 21st Century, Chicago, IL, USA, December 10-14, 1991; pp. 116–125.

35. Zhang, Q.; Litchfield, J.B. Advanced process controls: Applications of adaptive, fuzzy and neural control to the food industry. In Food Processing Automation; ASAE: St. Joseph, MI, USA, 1992; Vol. II, pp. 169–176.

36. Eerikäinen, T.; Linko, P.; Linko, S.; Siimes, T.; Zhu, Y.H. Fuzzy logic and neural networks applications in food science and technology. Trends Food Sci. Technol. 1993, 4, 237–242.

37. McCulloch, W.S.; Pitts, W.H. A logical calculus of the ideas immanent in nervous activity. Bull. Math. Biophys. 1943, 5, 115–133.

38. Hebb, D.O. The organization of behavior; Wiley: New York, NY, USA, 1949.

39. Rosenblatt, F. The perceptron: A probabilistic model for information storage and organization in the brain. Psycho. Rev. 1958, 65, 386–408.

40. Widrow, B.; Hoff, M.E. Adaptive switching circuits. In WESCON Convention Record; Institute of Radio Engineers: New York, NY, USA, 1960; Vol. VI, pp. 96–104.

41. Minsky, M.; Papert, S.A. Perceptrons: An Introduction to Computational Geometry; MIT Press: Cambridge, MA, USA, 1969.

42. Grossberg, S. Adaptive pattern classification and universal recoding, 1: Parallel development and coding of neural feature detectors. Biol. Cybernetics 1976, 23, 187–202.

43. Hopfield, J.J. Neural networks and physical systems with emergent collective computational abilities. In Proceedings of the National Academy of Sciences of the USA; National Academy of Sciences: Washington, DC, USA, 1982; Vol. 79, 8, pp. 2554–2558.

44. Werbos, P.J. Beyond Regression: New Tools for Prediction and Analysis in the Behavioral Sciences. In Doctoral Dissertation; Applied Mathematics, Harvard University: Boston, MA, USA, 1974.

45. Kohonen, T. Self-organized formation of topologically correct feature maps. Biol. Cybernetics 1982, 43, 59–69.

46. Broomhead, D.S.; Lowe, D. Multivariable functional interpolation and adaptive networks. Comp. Syst. 1988, 2, 321–355.

47. Powell, M.J.D. Restart procedures for the conjugate gradient method. Math. Program. 1977, 12, 241–254.

48. Dong, C.X.; Yang, S.Q.; Rao, X.; Tang, J.L. An algorithm of estimating the generalization performance of RBF-SVM. In Proceedings of 5th International Conference on Computational Intelligence and Multimedia Applications, Xian, Shanxi, China, September 27-30, 2003; pp. 61–66.

49. Hornik, K.; Stinchcombe, M.; White, H. Multilayer feedforward networks are universal approximators. Neural Netw.1989, 2, 359–366.

50. Cybenko, G.V. Approximation by superpositions of a sigmoidal function, Math. Contr. Signal. Syst. 1989, 2, 303–314.

51. Hammer, H.; Gersmann, K. A note on the universal approximation capability of support vector machines. Neural Process. Lett. 2003, 17, 43–53.

52. Lennox, B.; Montague, G.A.; Frith, A.M.; Gent, C.; Bevan, V. Industrial application of neural networks – an investigation. J. Process Contr. 2001, 11, 497–507.

53. Hussain, B.; Kabuka, M.R. A novel feature recognition neural network and its application to character recognition.IEEE Trans. Patt. Anal. Mach. Intell. 1998, 6, 98–106.

54. Ma, L.; Khorasani, K. Facial expression recognition using constructive feedforward neural networks. IEEE Trans. Syst. Man Cybernetics B 2004, 34, 1588–95.

55. Piramuthu, S. Financial credit-risk evaluation with neural and neurofuzzy systems. Eur. J. Operat. Res. 1999, 112, 310–321.

56. Barson, P.; Field, S.; Davey, N.; McAskie, G.; Frank, G. The detection of fraud in mobile phone networks. Neural Netw.World 1996, 6, 477–484.

57. Ghosh, S.; Reilly, D.L. Credit card fraud detection with a neural-network. In Proceedings of the 27th Annual Hawaii International Conference on System Science, Maui, HI , USA, January 4-7, 1994; Vol. III, pp. 621–630.

58. Braun, H.; Lai, L.L. A neural network linking process for insurance claims. In Proceedings of 2005 International Conference on Machine Learning and Cybernetics, Guangzhou, Guangdong, China, August 18-21, 2005; Vol. I, pp. 399–404.

59. Fu, T.C.; Cheung, T.L.; Chung, F.L.; Ng, C.M. An innovative use of historical data for neural network based stock prediction. In Proceedings of Joint Conference on Information Sciences, Kaohsiung, Taiwan, October 8-11, 2006.

60. Hartigan, J.A.; Wong, M.A. A k-means clustering algorithm. Appl. Statistics 1979, 28, 100–108.

61. Zhu, Y.; He, Y. Short-term load forecasting model using fuzzy c means based radial basis function network. In Proceedings of 6th International Conference on Intelligence Systems Design and Applications, Jinan, China, October 16-18, 2006; Vol. I, pp. 579–582.

62. Jordan, M.I. Attractor dynamics and parallelism in a connectionist sequential machine. In Proceedings of 8th Annual Conference of Cognitive Science Society, Amherst, MA, USA, August 15-17, 1986; pp. 531–546.

63. Elman, J.L. Finding structure in time. Cogn. Sci. 1990, 14, 179–211.

64. Pham, D.T.; Liu, X. Neural Networks for Identification, Prediction and Control; Springer-Verlag: London, UK, 1995.

65. Yun, L.; Haubler, A. Artificial evolution of neural networks and its application to feedback control. Artif. Intell. Eng.1996, 10, 143–152.

66. Pham, D.T.; Karaboga, D. Training Elman and Jordan networks for system identification using genetic algorithms.Artif. Intell. Eng. 1999, 13, 107–117.

67. Pham, D.T.; Liu, X. Dynamic system identification using partially recurrent neural networks. J. Syst. Eng. 1992, 2, 90–97.

68. Pham, D.T.; Liu, X. Training of Elman networks and dynamic system modeling. Int. J. Syst. Sci. 1996, 27, 221–226.

69. Ku, C.C.; Lee, K.Y. Diagonal recurrent neural networks for dynamic systems control. IEEE Trans. Neural Netw. 1995, 6, 144–156.

70. Huang, Y.; Whittaker, A.D.; Lacey, R.E. Neural network prediction modeling for a continuous snack food frying process. Trans. ASAE 1998, 41, 1511–1517.

71. Pineda, F.J. Recurrent backpropagation and the dynamical approach to adaptive neural computation. Neural Comput.1989, 1, 167–172.

72. William, R.J.; Zipser, D.A. A learning algorithm for continually running fully recurrent neural networks. Neural Comput. 1989, 1, 270–280.

73. Pearlmutter, A.B. Dynamic recurrent neural networks; Technical Report CMU-CS-90-196; Carnegie Mellon University: Pittsburgh, PA, USA, 1990.

74. Huang, Y. Sanck food frying process input-output modeling and control through artificial neural networks. Ph.D. Dissertation, Texas A&M University, College Station, TX, USA, 1995.

75. Huang, Y.; Lacey, R.E.; Whittaker, A.D. Neural network prediction modeling based on ultrasonic elastograms for meat quality evaluation. Trans. ASAE 1998, 41, 1173–1179.

76. Haykin, S. Neural Networks A Comprehensive Foundation, 2nd Edition ed; Prentice Hall Inc.: Upper Saddle River, NJ, USA, 1999.

77. Bishop, C.M. Neural Networks for Pattern Recognition; Oxford University Press: Oxford, UK, 1995.

78. Bishop, C.M. Pattern Recognition and Machine Learning; Springer-Verlag New York, Inc.: Secaucus, NJ, USA, 2006.

79. Zadeh, L.A. Fuzzy sets. Inf. Contr. 1965, 8, 338–353.

80. Simpson, P.K. Fuzzy min-max neural networks – part 2: clustering. IEEE Trans. Fuzzy Syst. 1992, 1, 32–45.

81. Jang, J.S.R. ANFIS: adaptive-network-based fuzzy inference system. IEEE Trans. Syst. Man Cybernetics 1993, 23, 665–685.

82. Daubechies, I. Orthonormal bases of compactly supported wavelets. Comm. Pure Appl. Math. 1988, 41, 909–996.

83. Mallat, S. A theory of multiresolution signal decomposition: The wavelet representation. IEEE Trans. Patt. Anal. Mach. Intell. 1989, 11, 674–693.

84. Percival, D.B.; Walden, A.T. Wavelet Methods for Time Series Analysis; Cambridge University Press: Cambridge, UK, 2000.

85. Vapnik, V. The Nature of Statistical Learning Theory; Springer-Verlag: London, UK, 1995.

86. Rychetsky, M.; Ortmann, S.; Glesner, M. Support vector approaches for engine knock detection. In Proceedings of International Joint Conference on Neural Networks (IJCNN 99), Washington, DC, USA, July 10-16, 1999; Vol. II, pp. 969–974.

87. Müller, K.R.; Smola, A.; Rätsch, G.; Schölkopf, B.; Kohlmorgen, J.; Vapnik, V. Using support vector machines for time series prediction. In Advances in Kernel Methods; Schölkopf, B., Burges, C.J.C., Smola, A.J., Eds.; MIT Press: Cambridge, MA, USA, 1999; pp. 242–253.

88. Iplikci, S. Dynamic reconstruction of chaotic systems from inter-spike intervals using least squares support vector machines. Phys. D. 2006, 216, 282–293.

89. Lee, Y.J.; Mangasarian, O.L.; Wolberg, W.H. Breast cancer survival and chemotherapy: a support vector machine analysis. In DIMACS Series in Discrete Mathematics and Theoretical Computer Science; American Mathematical Society: Providence, RI, USA, 2000; Vol. 55, pp. 1–10.

90. Kim, K.I.; Jung, K; Park, S.H.; Kim, H.J. Support vector machines for texture classification. IEEE Trans. Patt. Anal. Mach. Intell. 2002, 24, 1542–1550.

91. Hidalgo, H.; Sosa, S.; Gómez-Treviño, E. Application of the kernel method to the inverse geosounding problem. Neural Netw. 2003, 16, 349–353.

92. Pal, M. Support vector machines-based modelling of seismic liquefaction potential. Int. J. Num. Anal. Meth. Geomech.2006, 30, 983–996.

93. Schölkopf, B.; Smola, A.J. Learning with Kernels; MIT Press: Cambridge, MA, USA, 2002.

94. Rangwala, H.; Karypis, G. Profile-based direct kernels for remote homology detection and fold recognition.Bioinformatics 2005, 21, 4239–4247.

95. Iplikci, S. Support vector machines-based generalized predictive control. Int. J. Rob. Nonl. Contr. 2006, 16, 843–862.

96. Cortes, C.; Vapnik, V. Support vector networks. Mach. Learn. 1995, 20, 273–297.

97. Ljung, L. Perspectives on system identification; Division of Automatic Control, Linköpings Universitet: Linköping, Sweden, 2008.

98. Hsu, C.W.; Lin, C.J. A comparison of methods for multi-class support vector machines. IEEE Trans. Neural Netw. 2002,13, 415–425.

99. Thai, C.N.; Shewfelt, R.L. Modeling sensory color quality of tomato and peach: neural networks and statistical regression. Trans. ASAE 1991, 34, 950–955.

100. Tani, A.; Murase, H.; Kiyota, M.; Honami, N. Growth simulation of alfalfa cuttings in vitro by Kalman filter neural network. Acta Horticul. 1992, 319, 671–676.

101. Yang, Q. Classification of apple surface features using machine vision and neural networks. Comput. Electron. Agric.1993, 9, 1–12.

102. Liao, K.; Paulsen, M.R.; Reid, J.F. Corn kernel breakage classification by machine vision using a neural network classifier. Trans. ASAE 1993, 36, 1949–1953.

103. Deck, S.H.; Morrow, C.T.; Heinemann, P.H.; Sommer III, H.J. Comparison of a neural network and traditional classifier for machine vision inspection of potatoes. Appl. Eng. Agric. 1995, 11, 319–326.

104. Khazaei, J.; Naghavi, M.R.; Jahansouz, M.R.; Salimi-Khorshidi, G. Yield estimation and clustering of chickpea genotypes using soft computing techniques. Agron. J. 2008, 100, 1077–1087.

105. Zhang, H.; Wang, J. Identification of stored-grain age using electronic nose by ANN. Appl. Eng. Agric. 2008, 24, 227–231.

106. Sun, G.; Hoff, S.J.; Zelle, B.C.; Nelson, M.A. Development and comparison of backpropagation and generalized regression neural network models to predict diurnal and seasonal gas and PM10 concentrations and emissions from swine buildings. Trans. ASABE 2008, 51, 685–694.

107. Ondimu, S.N.; Murase, H. Comparison of plant water stress detection ability of color and gray-level texture in sunagoke moss. Trans. ASABE 2008, 51, 1111–1120.

108. Wu, D.; Feng, L.; Zhang, C.; He, Y. Early detection of botrytis cinerea on eggplant leaves based on visible and near-infrared spectroscopy. Trans. ASABE 2008, 51, 1133–1139.

109. Linko, P.; Zhu, Y.H.; Linko, S. Application of neural network modeling in fuzzy extrusion control. Food Bioprod. process. 1992, 70, 131–137.

110. Kim, S.; Cho, I. Neural network modeling and fuzzy control simulation for bread-baking process. Trans. ASAE 1997,40, 671–676.

111. Morimoto, T.; Suzuki, J.; Hashimoto, Y. Optimization of a fuzzy controller for fruit storage using neural networks and genetic algorithms. Eng. Appl. Artif. Intell. 1997, 10, 453–461.

112. Odhiambo, L.O.; Yoder, R.E.; Yoder, D.C.; Hines, J.W. Optimization of fuzzy evapotranspiration model through neural training with input-output examples. Trans. ASAE 2001, 44, 1625–1633.

113. Chtioui, Y.; Panigrahi, S.; Backer, L.F. Self-organizing map combined with a fuzzy clustering for color image segmentation of edible beans. Trans. ASAE 2003, 46, 831–838.

114. Meyer, G.E.; Hindman, T.W.; Jones, D.D.; Mortensen, D.A. Digital camera operation and fuzzy logic classification of uniform plant, soil, and residue color images. Appl. Eng. Agric. 2004, 20, 519–529.

115. Odhiambo, L.O.; Freeland, R.S.; Yoder, R.E.; Hines, J.W. Investigation of a fuzzy-neural network application in classification of soils using ground-penetrating radar imagery. Appl. Eng. Agric. 2004, 20, 109–117.

116. Hancock, K.M.; Zhang, Q. A hybrid approach to hydraulic vane pump condition monitoring and fault detection.Trans. ASABE 2006, 49, 1203–1211.

117. Xiang, H.; Tian, L.F. Artificial intelligence controller for automatic multispectral camera parameter adjustment. Trans. ASABE 2007, 50, 1873–1881.

118. Andriyas, S.; Negi, S.C.; Rudra, R.P.; Yang, S.X. Modelling total suspended solids in vegetative filter strips using artificial neural networks; ASAE: St. Joseph, MI., USA, 2003; ASAE number: 032079.

119. Lee, K.H; Zhang, N.; Das, S. Comparing adaptive neuro-fuzzy inference system (ANFIS) to partial least-squares (PLS) method for

simultaneous prediction of multiple soil properties; ASAE: St. Joseph, MI., USA, 2003; ASAE paper number: 033144.

120. Neto, J.C.; Meyer, G.E.; Jones, D.D.; Surkan, A.J. Adaptive image segmentation using a fuzzy neural network and genetic algorithm for weed detection; ASAE: St. Joseph, MI., USA, 2003; ASAE paper number: 033088.

121. Goel, P.K.; Andriyas, S.; Rudra, R.P.; Negi, S.C. Modeling sediment and phosphorous movement through vegetative filter strips using artificial neural networks and GRAPH; ASAE: St. Joseph, MI., USA, 2004; ASAE paper number: 042263.

122. Deng, X.; Wang, Q.; Wu, L.; Gao, H.; Wen, Y.; Wang, S. Eggshell crack detection by acoustic impulse response and support vector machine. African J. Agric. Res. 2009, 4, 40–48.

123. Trebar, M.; Steele, N. Application of distributed SVM architectures in classifying forest data cover types. Comput. Electron. Agric. 2008, 63, 119–130.

124. Khot, L.R.; Panigrahi, S.; Woznica, S. Neural-network-based classification of meat: evaluation of techniques to overcome small dataset problems. Biol. Eng. 2008, 1, 127–143.

125. Jiang, L.; Zhu, B.; Rao, X.; Berney, G.; Tao, Y. Discrimination of black walnut shell and pulp in hyperspectral fluorescence Imagery using Gaussian kernel function approach. J. Food Eng. 2007, 81, 108–117.

126. Zhang, H.; Paliwal, J.; Jayas, D.S.; White, N.D.G. Classification of fungal infected wheat kernels using near-infrared reflectance hyperspectral imaging and support vector machine. Trans. ASABE 2007, 50, 1779–1785.

127. Karimi, Y.; Prasher, S.O.; Patel, R.M.; Kim, S.H. Application of support vector machine technology for weed and nitrogen stress detection in corn. Comput. Electron. Agric. 2006, 51, 99–109.

128. Pierna, J.A.F.; Baeten, V.; Dardenne, P. Screening of compound feeds using NIR hyperspectral data. Chemom. Intell. Lab. Syst. 2006, 84, 114–118.

129. Chen, Q.; Zhao, J.; Cai, J.; Wang, X. Study on identification of tea using computer vision based on support vector machine. Chinese J. Sci. Instrum. 2006, 27, 1704–1706.

130. Wang, W.; Paliwal, J. Spectral data compression and analyses techniques to discriminate wheat classes. Trans. ASABE2006, 49, 1607–1612.

131. Onaran, I.; Pearson, T.C.; Yardimci, Y.; Cetin, A.E. Detection of underdeveloped hazelnuts from fully developed nuts by impact acoustics. Trans. ASABE 2006, 49, 1971–1976.

132. Pardo, M.; Sberveglieri, G. Classification of electronic nose data with support vector machines. Sens. Actuat. B 2005,107, 730–737.

133. Pierna, J.A.F.; Volery, P.; Besson, R.; Baeten, V.; Dardenne, P. Classification of modified starches by Fourier transform infrared spectroscopy using support vector machines. J. Agric. Food Chem. 2005, 53, 6581–6585.

134. Brudzewski, K.; Osowski, S.; Markiewicz, T. Classification of milk by means of an electronic nose and SVM neural network. Sens. Actuat. B 2004, 98, 291–298.

135. Tian, Y.; Zhang, C.; Li, C. Study on plant disease recognition using support vector machine and chromaticity moments. Trans. Chinese Soc. Agric. Mach. 2004, 35, 95–98.

136. Fletcher, J.T.; Kong, S.G. Principal component analysis for poultry tumor inspection using hyperspectral fluorescence imaging. In Proceedings of the International Joint Conference on Neural Networks, Portland, Oregon, USA, July 20-24, 2003; Vol. I, pp. 149–153.

137. Oommen, T.; Misra, D.; Agarwal, A.; Mishra, S.K. Analysis and application of support vector machine based simulation for runoff and sediment yield; ASABE: St. Joseph, MI., USA, 2007; ASABE paper number: 073019.

138. Fu, X.; Ying, Y.; Xu, H.; Yu, H. Support vector machines and near infrared spectroscopy for quantification of vitamin C content in kiwifruit; ASABE: St. Joseph, MI., USA, 2008; ASABE number: 085204.

139. Kovacs, Z.; Kantor, D.B.; Fekete, A. Comparison of quantitative determination techniques with electronic tongue measurements; ASABE: St. Joseph, MI., USA, 2008; ASABE paper number: 084879.

140. Peng, Y.; Wang, W. Prediction of pork meat total viable bacteria count using hyperspectral imaging system and support vector machines. In Proceedings of the Food Processing Automation Conference, Providence, RI, USA, June 28-29, 2008. CD-ROM.

Index